COMPLEX VARIABLE THEORY
AND TRANSFORM CALCULUS

W0042223

COMPLEX VARIABLE THEORY

AND

TRANSFORM CALCULUS

WITH TECHNICAL APPLICATIONS

BY

N. W. M^CLACHLAN

D.SC. (ENGINEERING), LONDON

Hon. Member, British Institution of Radio Engineers;
Professor of Electrical Engineering, Emeritus,
University of Illinois;
Walker-Ames Professor of Electrical Engineering,
University of Washington (in 1954)

SECOND EDITION

CAMBRIDGE
AT THE UNIVERSITY PRESS
1963

CAMBRIDGE UNIVERSITY PRESS
Cambridge, New York, Melbourne, Madrid, Cape Town, Singapore,
São Paulo, Delhi, Dubai, Tokyo, Mexico City

Cambridge University Press
The Edinburgh Building, Cambridge CB2 8RU, UK

Published in the United States of America by Cambridge University Press, New York

www.cambridge.org
Information on this title: www.cambridge.org/9780521154154

First edition 1939
Second edition 1953
Reprinted 1955
Reprinted 1963
First published under the title *Complex Variable and Operational Calculus
with Technical Applications*
First paperback edition 2010

A catalogue record for this publication is available from the British Library

ISBN 978-0-521-05651-9 Hardback
ISBN 978-0-521-15415-4 Paperback

CONTENTS

PART IV. APPENDICES AND LIST OF
REFERENCES

PREFACE TO THE SECOND EDITION

In 1936–7 when the MS. of the first edition was prepared, the degree of rigour seemed to be adequate, but certain pure mathematicians (and physicists!) who reviewed the book disagreed. In the interim, the standard of technical mathematics has improved, and it is now possible to be more rigorous than before. Accordingly the chapters on Complex Integration Theory have been rewritten, amplified, and made rigorous enough for all but the pure mathematician, *to whom the book is not addressed*. Sections of the text set in small type deal with more recondite topics, and may be omitted in the first reading, reference being made to them when required. The rest of the book has been revised completely and brought up to date. Certain of the old sections have been removed to make way for more important subject matter, e.g. repeated impulses, Fourier transforms, and frequency spectra have been added to Chapter x. In solving ordinary differential equations, use is made of either a list of p-multiplied Laplace transforms or the Mellin inversion theorem, according to the problem under consideration. The sections involving partial differential equations have been recast. The approach is via Laplace transform, thereby permitting the initial conditions to be incorporated easily. The Mellin theorem is used for inversion.* Practical data for loaded and unloaded submarine telegraph cables have been given. The calculated and actual shapes of received signals, together with diagrams of the circuits used at both ends of the cable, are reproduced in Chapter XIII. The data for loaded cables are due to Mr A. L. Meyers, and for unloaded cables to the Author. There are additional sections on electrical filters. By aid of a new theorem,† the solution for a dissipative filter can be expressed concisely in the form of a definite integral. The number and variety of the problems to be

---

* Iterated use of the Laplace transform for solving partial differential equations is exemplified in [266].

† N. W. McLachlan, *Math. Gaz.* **30**, 85, 1946.

worked out by the reader has been increased, while the reference list has been extended. An Appendix on convergence of many of the series which occur in the text is given, and should be useful. The list of p-multiplied Laplace transforms covers merely what is needed for the text, since an extended list is available elsewhere [235 a, b].

The present work and that entitled *Modern Operational Calculus* [236] are complementary, and may be used together. The latter proceeds via real variable and Laplace transform method, which carries the subject to a stage where it may profitably be taken over by the complex variable method, as exemplified herein. Complex integration is needed in solving many of the technical problems involving partial differential equations, and in deriving asymptotic formulae.

N. W. M.

MAY 1952

PREFATORY NOTE

Drs A. J. Macintyre and C. Strachan read parts of the MS. of the second edition, while Profs. T. J. Higgins and E. J. Scott read proofs. I wish to thank these gentlemen for their very helpful criticisms and suggestions. I welcome the opportunity to thank Prof. P. Humbert for his kind gesture in obtaining publication of the lists of transforms in references [235 a, b].

In this new impression, I am much indebted to Prof. A. Erdélyi for his valuable criticism and suggestions. I have made some minor alterations in the text.

APRIL 1962   N.W.M.

PREFACE TO THE FIRST EDITION

The purpose of this book is to provide a modern treatment of the so-called operational method, and to illustrate its application to problems in various branches of technology. Although it is written primarily for the mathematical technologist,* certain parts of the text may be useful to others....

The reader may wonder why p is used outside the Laplace integral (1) § 8·11. The reasons for this are as follows: (i) By retaining p, the operational forms† of various functions are identical with those obtained by the Heaviside method. Such forms are of long standing and widespread use, so that an alteration now would be inexpedient;‡ (ii) The operational form of t^n is $n!/p^n$. Thus if t and p are considered to have dimensions d and d^{-1}, respectively, and if $f(t)$ and its operational form can be expanded in absolutely convergent series, the corresponding terms are identical dimensionally. This is useful for checking purposes.

The book is divided into four parts, (I) Complex Variable, (II) Operational Calculus, (III) Technical Applications and examples to be worked out by the reader, (IV) Appendices and list of references. Each of the first three parts is preceded by a foreword,§ which the reader should peruse carefully before commencing to study the part in question. *Part I must be understood thoroughly.* After reading each chapter, the corresponding problems at the end of Part III ought to be worked out....When Part I has been assimilated, a knowledge of the early parts of chaps. VIII–X will enable the reader to pass on to Part III. To avoid interpolated explanations of cognate points in the text, a number of Appendices is given in Part IV. Frequent reference is made to these throughout the book....

* A person who uses mathematics to solve technical problems of various kinds, e.g. acoustical, aeronautical, chemical, electrical, mechanical, thermal, etc. The term also applies to the mathematician engaged in industrial and applied research work.

† Designated p-multiplied Laplace transforms in the second edition.

‡ This reason is now inapplicable (1952).

§ Omitted from the second edition.

The technologist is not fitted by training, nor has he the time, to delve into rigour to the last epsilon. Just as the mathematician does not need to be versed in thermodynamics and internal combustion engine design to drive a motor car, the technologist need not know how to prove all the theorems he uses. But like the mathematical motorist, he must be acquainted with the highway code. In other words some rigour is needed, and I hope that in this volume a happy mean has been struck between the demands of the mathematician on the one hand, and the requirements of the practical man on the other.

It may be argued by some, that, on the whole, the text is difficult, because complex integration plays such an important part therein. Looking back half a century, we find that engineers regarded the differential and integral calculus as a mystery beyond the reach of the majority. Nowadays the engineering student takes the calculus (at least the small and inadequate dose administered to him) in his stride. Consequently if this book is the means of introducing complex integration to the mathematical technologist who reads the English language, it will justify itself in this respect alone. Moreover, after the customary lapse of valuable time, those who teach the 'young technical idea' will no longer be panic-stricken by a subject which has graced the curricula of continental technical institutions for many years.

Symbols. A new symbol, namely ⟹, has been introduced to signify 'Laplace transform of', for reasons stated in [131]*. The round end points to the transform, and the open end to the corresponding function t, e.g. $f(t) \Rightarrow \phi(p)$. m, n, r are used for integers, while μ, ν denote unrestricted numbers. $R(\nu)$ means 'the real part of ν'. Symbols in heavy type indicate per unit length or area, as the case may be. $\sim$ signifies that the right-hand side is an asymptotic formula for the left-hand side when the variable is large enough. $\simeq$ signifies 'is approximately equal to', $\neq$ signifies 'is not equal to', $O(1/z^2)$ signifies 'is of order $1/z^2$', $\exp\{f(z)\}$ signifies $e^{f(z)}$, $\to \pm\, 0$ means 'tends to zero' from the positive or the negative side, $\to \pm\, \infty$ means 'tends to infinity' from the positive or the negative side, $\sqrt{i^{\frac{1}{2}}} = i = e^{\frac{1}{2}\pi i} = (1+i)/\sqrt{2}$ and kindred symbols are illustrated in Fig. 21 (c) on p. 63.

* The numbers in [   ] are references on pp. 366–377.

Acknowledgments. I have been exceedingly fortunate in enlisting the help of a number of friends, namely, Professors W. N. Bailey, W. G. Bickley, T. A. A. Broadbent, E. T. Copson, Drs J. M. Jackson, A. T. McKay and Mr A. L. Meyers. They have read various parts of the MS. critically and/or corrected proofs. I owe a great deal to their valuable comments, and accord them my warmest thanks. I am also much indebted to Professor A. R. Collar and Dr R. A. Fraser for suggestions regarding § 11·41* et seq. Best thanks for the loan of blocks are due to the Delegates of the Oxford Press (Figs. 23, 49a, 56, 57), the Editors of the *Philosophical Magazine* (Figs. 44, 45), and the Editor of the *Wireless Engineer* (Figs. 42, 43, 46–48, 58, 60). Permission has kindly been granted by the Council of the Physical Society to reproduce the table on p. 272 from a paper by Dr A. T. McKay, and by the Bell Telephone Laboratories to reproduce Fig. 56 from a paper by L. J. Sivian in the *Bell System Technical Journal.*

N. W. M.

LONDON
MARCH 1939

* The numbers in this preface refer to the first edition.

PART I
THEORY OF COMPLEX INTEGRATION

I

FUNCTIONS OF A COMPLEX VARIABLE

1·1. Definition and pictorial representation. If a complex number $w = u + iv$ is connected with another complex number $z = x + iy$, so that the value of w can be determined uniquely by that of z, then w is said to be a function of z. The variables u, v, x, y are real, u and v being real functions of x, y.

Fig. 1 illustrates the well-known Argand or vector diagram, where P is the representative point of the complex number $x + iy$, in what we regard as the z-plane. Although in Fig. 1, $\theta < \frac{1}{2}\pi$, the general value of PON is $(\theta + 2n\pi)$, n being any non-zero integer. Let the vector OP rotate counter-clockwise—this being taken as the positive direction—about centre O. Each time OP passes through the position shown, the value of z corresponding thereto is obtained.

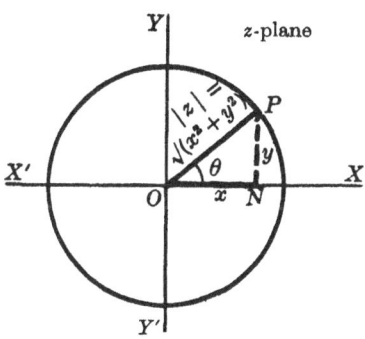

Fig. 1. $X'OX$ is the axis for real and $Y'OY$ that for imaginary numbers. $\theta = P\hat{O}N = $ phase z.

Since $x = r\cos\theta$, $y = r\sin\theta$, the general value of z is

$$z = r\{\cos(\theta + 2n\pi) + i\sin(\theta + 2n\pi)\} = r\,e^{i(\theta + 2n\pi)} = r\,e^{i\theta}, \quad (1)$$

where $r = |z| = \sqrt{(x^2 + y^2)}$ and $e^{2n\pi i} = 1$, n being taken positive.

Consider the function
$$w = z^2. \quad (2)$$

If for z we substitute its general value from (1), then

$$w = r^2\,e^{i(2\theta + 4n\pi)}. \quad (3)$$

Now $e^{i(2\theta + 4n\pi)} = e^{2i\theta}$, so if both r and θ are fixed, w is identical for all values of n. Hence to every value of z there corresponds only one value of w, so the latter is said to be a *single-valued* function of z. Moreover, for all finite values of z, w is defined uniquely.

The relationship $y^2 = x$ for the real variable entails that $y = \pm \sqrt{x}$, so two equal but opposite values of y correspond to each value of $x > 0$. Thus, y is not determined *uniquely* by any positive value of x except zero. Herein we shall avoid this type of ambiguity where functions of a complex variable are concerned. To do so, the variable z will be limited in such a way that the function under consideration is defined *uniquely* for all values of z in a prescribed angle range. This is treated later on.

If w and z were real variables, their relationship, could be represented geometrically by a plane curve, in the well-known way. With complex numbers this is impossible, since there are four variables. Thus, if $z = x + iy$, we obtain from (2)

$$w(z) = u(x, y) + i\, v(x, y) = (x + iy)^2 = (x^2 - y^2) + 2ixy, \qquad (4)$$

so
$$u(x, y) = (x^2 - y^2) \quad \text{and} \quad v(x, y) = 2xy. \qquad (5)$$

Moreover, u and v are functions of both x and y. Each relationship in (5) is the symbolical representation of a surface. By adopting some form of pictorial projection, the two surfaces can be represented. A more satisfactory procedure, however, is to calculate the moduli of w corresponding to various values of x and y; then using isometric graph paper, the modular surface can be plotted. The idea of plotting the modular surface is due to Heffter.* The case of $w = z^2$ is illustrated in Fig. 2, where

$$|w| = \sqrt{(u^2 + v^2)} = x^2 + y^2. \qquad (6)$$

The complex variable z may be represented by any point in the z-plane containing the axes $X'OX$, $Y'OY$. Thus in Fig. 2 the range of z is unlimited. The modular surface and the function $w = z^2$ are continuous, and $|w|$ is finite provided z is finite. The slope of the surface at every point, except infinity, is finite and continuous.

1·2. Continuity. A function $w = f(z)$ is said to be continuous at the point $z = a$, if corresponding to a given positive quantity ϵ, however small, a finite positive quantity η can be found such that

$$|f(z) - f(a)| < \epsilon, \quad \text{when} \quad |z - a| < \eta. \qquad (1)$$

* Z. Math. Phys. **44**, 235, 1899.

We express this by saying that $f(z)$ is continuous at $z = a$ if $f(a)$ has a definite value and if $\lim_{z \to a} f(z) = f(a)$. In Fig. 3 (a) P represents the point a, while z is any point on the small circle described

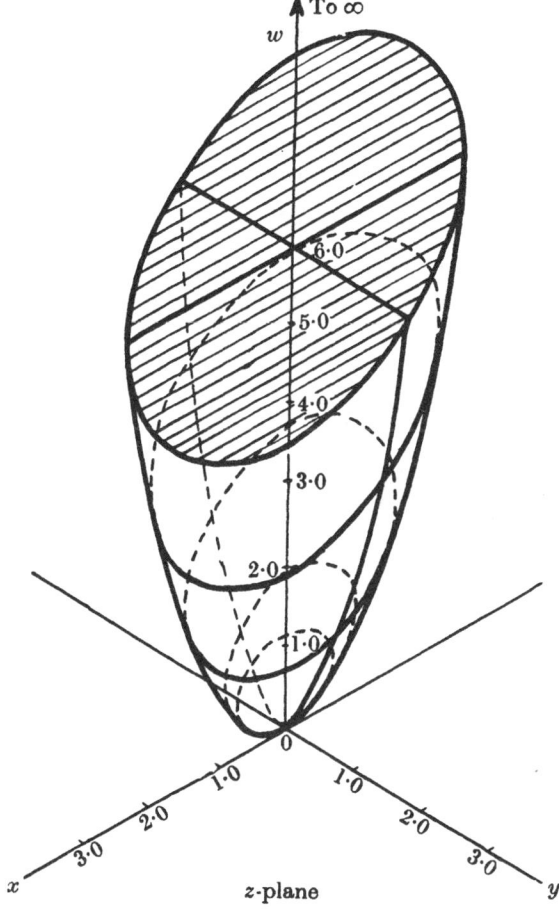

Fig. 2. Modular surface (paraboloid of revolution) of the function $w = z^2$; isometric plotting.

about P. The preceding condition for continuity of $f(z)$ at a is independent of the direction from which z approaches a. For instance, the approach might be along NP, MP, or any curve, e.g. RP. As an example consider the function $w = (z-a)^2$, which is the same as (2) §1·1, except that the origin is moved to the point $z = a$. Taking P in Fig. 3 (*b*) as a subsidiary origin,

$$PN = r = |z - a| \quad \text{and} \quad (z - a) = r\,e^{i\theta} = r(\cos\theta + i\sin\theta).$$

Thus as N approaches P from *any* point on the small circle and along *any* path, $r \to 0$ and $z \to a$, so in the limit $r = 0$ and $z = a$. Now

$$|f(z) - f(a)| = |(z - a)^2| \leqslant |(z - a)(z + a)| < 2R\,|z - a|, \quad (2)$$

provided that z and a both lie within a given circle centre O and radius R, e.g. the broken-line circle in Fig. 3 (*b*). Hence if $|z - a| < \epsilon/2R$,

$$|f(z) - f(a)| < \epsilon, \quad (3)$$

and the function $w = (z - a)^2$ is continuous at the point $z = a$. In this instance, the appropriate value of η is $\epsilon/2R$.

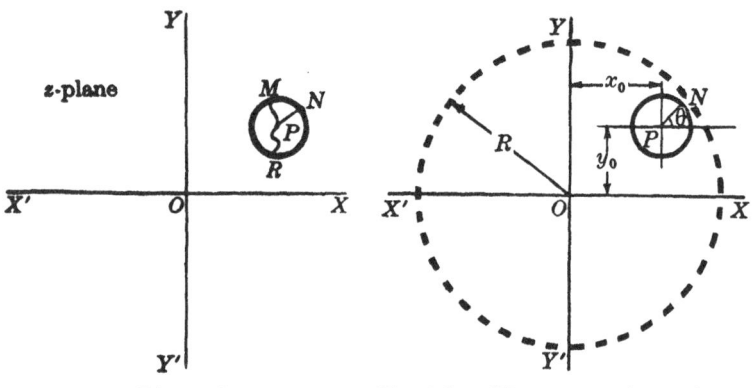

Fig. 3 (*a*).

Fig. 3 (*b*). $OP = a = x_0 + iy_0$. z is any point on the small circle, e.g. $R > |z|$ and $|a|$, so $2R > |z + a|$.

To illustrate a discontinuity, consider the function

$$w = 1/(z - a).$$

As z approaches the point a (see Fig. 3 (*b*)) from *any* direction, $1/(z - a)$ increases without limit, i.e $\underset{z \to a}{f(z)} \to \infty$. Thus the function has no definite value at $z = a$, and is discontinuous, as also is its first derivative.

$f(z)$ is said to be continuous in a given region if it is continuous at all points of the region. For example, $w = z^2$ is continuous for all values of z except infinity, while $w = 1/z$ is continuous at all points except the origin. Any polynomial of the nth degree,

$$w = a_0 + a_1 z + a_2 z^2 + \ldots + a_n z^n,$$

is single-valued and continuous in the finite part of the z-plane.

A continuous function of z is also a continuous function of the two real variables x, y. In the case of $w = z^2$, taking $a = x_0 + iy_0$, we have (see Fig. 3 (b))

$$z^2 = x^2 - y^2 + 2ixy = u(x,y) + i\,v(x,y), \tag{4}$$

$$a^2 = x_0^2 - y_0^2 + 2ix_0 y_0 = u_0(x, y) + iv_0(x, y), \tag{5}$$

so $$z^2 - a^2 = (u - u_0) + i(v - v_0), \tag{6}$$

and $$|u - u_0| \leqslant |z^2 - a^2| < \epsilon, \tag{7}$$

provided that $|z - a| < \eta$, the equality sign holding for $v = v_0$. Now

$$|z - a| = \sqrt{\{(x - x_0)^2 + (y - y_0)^2\}}, \tag{8}$$

and this is less than η, if both $(x - x_0)$ and $(y - y_0) < \eta/\sqrt{2}$, a condition which can be satisfied. Thus

$$|u - u_0| < \epsilon, \tag{9}$$

so u is a continuous function of both x and y, and represents a continuous surface. The function v may be treated in like manner. Since u, v are continuous functions of x and y, it follows that $w = z^2$ is also a continuous function of these variables.

1·31. Definition of derivative. Let z be any point on the small circle in Fig. 3 (b) whose centre is at $z = a$. Suppose that *after* choosing a real positive number ϵ, however small, a real number η can be found such that

$$\left| \frac{f(z) - f(a)}{z - a} - f'(a) \right| < \epsilon \tag{1}$$

when the radius of the small circle, i.e. $|z - a|. < \eta$. Then (1) may be written

$$f(z) = f(a) + (z - a)f'(a) + \lambda(z - a), \tag{2}$$

where $|\lambda| < \epsilon$. In the limit when $z \to a$, and $|\lambda| \to 0$, (2) may be expressed in the form

$$f'(a) = \lim_{z \to a} \frac{f(z) - f(a)}{z - a}. \tag{3}$$

If we take z as the centre of the small circle and $|h|$ its radius, $(z+h)$ is any point on the circumference. Hence writing z for a, and $(z+h)$ for z in (3), we obtain

$$f'(z) = \lim_{|h|\to 0} \frac{f(z+h)-f(z)}{h}. \tag{4}$$

Formula (4) appears to be identical with that for the real variable, but there is a difference in the interpretation which is explained in § 1·32.

1·32. Example. Consider the exponential function e^z. It is finite and single-valued in any finite region of the z-plane. Let z be the point P at the centre of the small circle of radius $|h|$ in Fig. 3 (a), and $(z+h)$ any point on the circumference. Then by (4) § 1·31

$$f'(z) = \lim_{|h|\to 0} \frac{e^{z+h}-e^z}{(z+h)-z} = \lim_{|h|\to 0} \frac{e^z(e^h-1)}{h} \tag{1}$$

$$= \lim_{|h|\to 0} e^z\left(1 + \frac{h}{2!} + \frac{h^2}{3!} + \ldots\right), \tag{2}$$

so

$$\frac{d(e^z)}{dz} = e^z. \tag{3}$$

Taking P in Fig. 3 (a) as a subsidiary origin,

$$h = r(\cos\theta + i\sin\theta) = r\,e^{i\theta},$$

represents any point on the small circle MNR of radius

$$r = PN = |h|.$$

Substituting for h into (2) leads to

$$\frac{e^{z+h}-e^z}{(z+h)-z} = e^z\left(1 + \frac{r\,e^{i\theta}}{2!} + \frac{r^2\,e^{2i\theta}}{3!} + \ldots\right), \tag{4}$$

and this $\to e^z$ *independently* of θ as $r = |h| \to 0$. That is to say, the limit is independent of the direction from which h on the small circle approaches the point P. With the real variable, h would tend to zero in a direction parallel to the x-axis, and herein lies the difference between the two types of limit. The derivative of any analytic function of z is identical for real and complex variables.

1·33. Definition of analytic, regular, or holomorphic function. A function $f(z)$ of the complex variable $x + iy$ is said to be analytic, regular, or holomorphic at a point z_0 in the z-plane if it is single-valued and has a derivative $f'(z)$ in some neighbourhood of $z = z_0$. If a function is analytic, and, therefore, continuous at all points in a bounded region of the z-plane, it is said to be analytic throughout the region.

Example. Since e^z and its first derivative are unique and finite in any bounded region of the z-plane, the exponential function is analytic therein. As shown in §1·53, it is not analytic at infinity.

1·34. Cauchy-Riemann conditions for analytic functions. A continuous function of the complex variable may be expressed in the form

$$w(z) = u(x,y) + iv(x,y), \tag{1}$$

where $u(x,y)$ and $v(x,y)$ are real functions of x, y, e.g. (4), (5), §1·1. Then

$$\frac{\delta w}{\delta z} = \frac{\delta u + i\,\delta v}{\delta x + i\,\delta y}, \tag{2}$$

and we have to obtain the conditions under which this tends to a definite limit as $\delta z \to 0$. By §1·32 the derivative is independent of the way in which $\delta z \to 0$, so we may let δx and $\delta y \to 0$ in succession.

In (1) with y constant, let x increase by a small amount δx, and we have

$$\frac{\delta w}{\delta z} = \frac{u(x + \delta x, y) - u(x,y)}{\delta x} + i\,\frac{v(x + \delta x, y) - v(x,y)}{\delta x}. \tag{3}$$

For (3) to tend to a limit as $\delta x \to 0$, u, v must have partial derivatives with respect to x, so

$$\frac{dw}{dz} = \lim_{\delta x \to 0} \frac{\delta w}{\delta z} = \frac{\partial u}{\partial x} + i\,\frac{\partial v}{\partial x}. \tag{4}$$

Taking x constant and y variable, we get

$$\frac{dw}{dz} = \lim_{\delta y \to 0} \frac{\delta w}{\delta z} = \frac{\partial v}{\partial y} - i\,\frac{\partial u}{\partial y}. \tag{5}$$

For $\lim \delta w/\delta z$ to be identical in (3), (4), we must have

$$\frac{\partial u}{\partial x} = \frac{\partial v}{\partial y} \quad \text{and} \quad \frac{\partial u}{\partial y} = -\frac{\partial v}{\partial x}. \tag{6}$$

Provided all members of (6) are continuous, these Cauchy-Riemann conditions are *necessary* for $\delta w/\delta z$ to have a unique limit. Assuming such conditions to be satisfied at a point z, we shall show that they are *sufficient*. Then u, v and their first partial derivatives are continuous in the neighbourhood of z, so [250]

$$\delta u = \left(\frac{\partial u}{\partial x}\right)\delta x + \left(\frac{\partial u}{\partial y}\right)\delta y + \lambda_1\,\delta x + \lambda_2\,\delta y, \tag{7}$$

and

$$\delta v = \left(\frac{\partial v}{\partial x}\right)\delta x + \left(\frac{\partial v}{\partial y}\right)\delta y + \lambda_3\,\delta x + \lambda_4\,\delta y, \tag{8}$$

where $\lambda_1, ..., \lambda_4 \to 0$ as δx and $\delta y \to 0$. Substituting from the second part of (6) into (7) and (8), we get

$$\delta w = \delta u + i\,\delta v$$
$$= \frac{\partial v}{\partial y}(\delta x + i\,\delta y) - i\frac{\partial u}{\partial y}(\delta x + i\,\delta y) + (\lambda_1 + i\lambda_3)\,\delta x + (\lambda_2 + i\lambda_4)\,\delta y. \tag{9}$$

Hence by (2), (9)

$$\frac{\delta w}{\delta z} = \frac{\partial v}{\partial y} - i\frac{\partial u}{\partial y} + (\lambda_1 + i\lambda_3)\frac{\delta x}{\delta z} + (\lambda_2 + i\lambda_4)\frac{\delta y}{\delta z}. \tag{10}$$

Now $\left|\dfrac{\delta x}{\delta z}\right|$ and $\left|\dfrac{\delta y}{\delta z}\right| \leqslant 1$, so that as $\delta z \to 0$ the last two terms in (10) do likewise, and we obtain (5). The result at (4) may be derived in a similar way. We have proved, therefore, that if the Cauchy-Riemann conditions are satisfied, $w'(z)$ exists, so $w(z)$ is analytic at the point z. Hence these conditions are *necessary and sufficient* for this purpose.

It may be shown also that

$$\frac{dw}{dz} = \frac{\partial u}{\partial x} - i\frac{\partial u}{\partial y} = \frac{\partial v}{\partial y} + i\frac{\partial v}{\partial x}, \tag{11}$$

so dw/dz can be written in four different ways. In virtue of the relationships (4), (5), u and v are sometimes called conjugate* functions. Since $w(z)$ is analytic at the point z, it follows by § 2·41 that u and v have continuous second derivatives with respect to x and y. Thus from (6)

$$\frac{\partial^2 u}{\partial x^2} = \frac{\partial^2 v}{\partial y\,\partial x} \quad \text{and} \quad \frac{\partial^2 u}{\partial y^2} = -\frac{\partial^2 v}{\partial x\,\partial y}, \tag{12}$$

so by addition

$$\frac{\partial^2 u}{\partial x^2} + \frac{\partial^2 u}{\partial y^2} = 0. \tag{13}$$

Similarly, we find that

$$\frac{\partial^2 v}{\partial x^2} + \frac{\partial^2 v}{\partial y^2} = 0, \tag{14}$$

and, therefore, both u, v satisfy Laplace's equation in two independent variables x, y.

---

* This word is *not* used here in the sense that $u \pm iv$ are conjugate.

Polar form of Cauchy-Riemann conditions. Corresponding to (6), we have

$$\frac{\partial u}{\partial r} = \frac{1}{r}\frac{\partial v}{\partial \theta}, \quad \text{and} \quad \frac{\partial v}{\partial r} = -\frac{1}{r}\frac{\partial u}{\partial \theta}. \tag{15}$$

The derivation of these formulae is left as an exercise for the reader.

1·35. Examples.

1°. (4) § 1·1 is an analytic function; for

$$\partial u/\partial x = \partial v/\partial y = 2x, \quad \text{and} \quad \partial u/\partial y = -\partial v/\partial x = -2y, \tag{1}$$

so (6) § 1·34 is satisfied, while u, v and their derivatives are continuous. The reader should confirm that (13), (14) § 1·34 are satisfied also.

2°. Take $w(z) = 1/z = 1/(x+iy) = (x-iy)/(x^2+y^2)$.

Then $u = x/(x^2+y^2) \quad \text{and} \quad v = -y/(x^2+y^2)$,   (2)

so $\partial u/\partial x = \partial v/\partial y = (y^2-x^2)/(x^2+y^2)^2$,   (3)

and $\partial u/\partial y = -\partial v/\partial x = -2xy/(x^2+y^2)^2$.   (4)

These derivatives are unique and determinate except when $z = 0$ ($x = y = 0$), so $1/z$ is analytic except at the origin. The reader should confirm that (1) satisfies (13), (14) § 1·34.

3°. $w(z) = x - iy$ is *not* an analytic function; for $u = x$, $v = -y$, $\partial u/\partial x = 1$, $\partial v/\partial y = -1$, which are equal but *opposite*.

4°. The result given at (3) § 1·32 may be derived also in the following way:

$$e^z = e^{x+iy} = e^x(\cos y + i\sin y), \tag{1}$$

so $u = e^x\cos y$ and $v = e^x\sin y$, both of which are continuous and have continuous partial derivatives with respect to x, y. Applying (4) § 1·34 to (1) leads to

$$\frac{d(e^z)}{dz} = \frac{\partial u}{\partial x} + i\frac{\partial v}{\partial x} = e^x(\cos y + i\sin y) \tag{2}$$

$$= e^z. \tag{3}$$

In general one of (4), (5), (11) § 1·34 may be used to find the derivative of a function $w(z)$ in a region of the z-plane where it is analytic.

1·4. Singularities.

For every function except a constant, there are one or more points in the z-plane at which it ceases to be analytic. These exceptional points are called singularities of the function. At all other points the function $f(z)$ is said to be analytic. For example, if $w = 1/z$, $dw/dz = -1/z^2$, and both the function and its first derivative are infinite at $z = 0$, so w is not analytic there. The origin is, therefore, a singular point, and it is the only singularity of the function. A second example is that of $w = e^{1/z}$, giving $dw/dz = -z^{-2}e^{1/z}$. As $z \to 0$, $w \to \infty$, $dw/dz \to \infty$, so the function has a singularity at the origin. With a polynomial of the nth degree,

$$w = a_0 + a_1 z + a_2 z^2 + \ldots + a_n z^n, \qquad (1)$$

there is no singularity in the finite part of the z-plane, the only singularity being at infinity. For let $z = 1/\zeta$, then

$$w = a_0 + a_1/\zeta + a_2/\zeta^2 + \ldots + a_n/\zeta^n, \qquad (2)$$

and $w \to \infty$ as $\zeta \to 0$, i.e. as $z \to \infty$. This substitution is equivalent to treating 'infinity' as a point (the so-called point at infinity). For the behaviour of $w(z)$ at infinity is made to depend upon that of $w(1/\zeta)$ at $\zeta = 0$. The singularity of $w(z)$ at infinity is the same kind as that of $w(1/\zeta)$ at the origin.

1·51. Types of singularity.

Mathematical functions of the complex variable are distinguished by their singularities, of which there are three kinds: (1) poles or unessential singularities, (2) essential singularities, (3) branch points, which occur because for certain z the function has more than one value, thereby introducing ambiguity.

1·521. Poles.

Formal definitions of poles and essential singularities are based upon Laurent's theorem which is given in § 2·62. The subject is introduced here by means of examples of functions having these singularities. A pole is regarded as an *isolated* singularity, because the function is analytic everywhere in the neighbourhood,* but *not at* the pole itself. If $w = 1/z$, the function has a simple pole (unit order) at $z = 0$, and as $z \to 0$, $w \to \infty$, irrespective of the direction in which the origin is approached. The terminology 'pole' will be understood if we consider the modulus of w as $z \to 0$, when plotted pictorially, using polar coordinates. Writing $z = r e^{i\theta}$, $w = (1/r) e^{-i\theta}$, the modulus being $1/r$, which is plotted against r and θ in Fig. 4. As r decreases, the modulus increases, and as $r \to 0$, $1/r \to \infty$. Consequently the modular surface in the neighbourhood of the

---

* We shall frequently use 'near' to mean 'in the neighbourhood of'.

origin takes the form of a tapered *pole* whose height $\to \infty$ as
$r \to 0$. The function $w = 1/(z-a)$ has a simple pole at $z = a$, so
the modular surface near this point is identical with that of

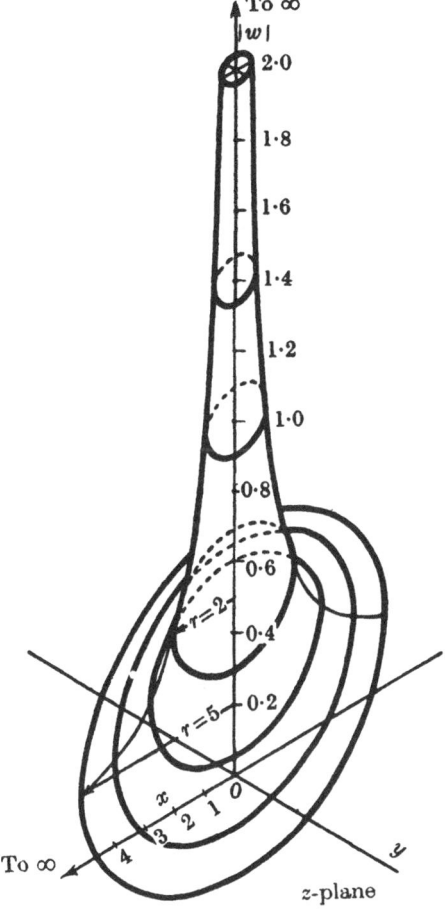

Fig. 4. Modular surface (hyperboloid of revolution)
of $w = z^{-1} = r^{-1}e^{-i\theta}$; isometric plotting.

$w = 1/z$ near the origin. $w = 1/(z-a)^n$, n an integer $\geqslant 1$, has a
pole of order n at $z = a$. The function (2) § 1·4 has a pole of order
n at $\zeta = 0$, so (1) § 1·4 has this type of singularity at infinity. The
above functions are *single-valued* for all z except singularities,
where they are undefined.

A pole is said to be an unessential singularity, because it may be removed, in effect, if the function is multiplied by a suitable factor whose index is *finite*. Thus if $1/(z-a)^n$ is multiplied by $(z-a)^n$, the desired result is obtained.

A function whose only singularities are poles (finite in number) is a *rational* function.* Conversely, the only singularities of a rational function are poles. Such a function is single-valued, except at its singularities where it is undefined. The rational function $w = 1/z\,(z-a)$ has two poles, one at $z = 0$ the other at $z = a$. Each pole can be removed—in effect—if w is multiplied by the appropriate factor. Thus to remove the pole at $z = a$, multiply by $(z-a)$. In analytical work each pole is removed, in effect, individually, but temporarily, in a certain evaluation at the pole in question.

The poles of either $1/\sin z$ or $1/\cos(1/z)$ may be considered with reference to the infinite products which represent the denominators of these functions. Thus

$$\sin z = z\left(1-\frac{z}{\pi}\right)\left(1+\frac{z}{\pi}\right)\left(1-\frac{z}{2\pi}\right)\left(1+\frac{z}{2\pi}\right)\cdots$$

$$= z\prod_{n=1}^{\infty}\left\{1-\left(\frac{z}{n\pi}\right)^2\right\}, \tag{1}$$

and

$$\cos\frac{1}{z} = \left(1-\frac{2}{\pi z}\right)\left(1+\frac{2}{\pi z}\right)\left(1-\frac{2}{3\pi z}\right)\left(1+\frac{2}{3\pi z}\right)\cdots$$

$$= \prod_{n=0}^{\infty}\left\{1-\left[\frac{2}{\pi z(2n+1)}\right]^2\right\}. \tag{2}$$

Since $\sinh z = -i\sin iz$, it follows that the poles of $1/\sinh z$ occur when $iz = n\pi$, or $z = in\pi$, n being an integer. Hence the distribution of poles on the imaginary axis is identical with those of $1/\sin z$ on the real axis. From (1)

$$\sinh z = z\prod_{n=1}^{\infty}\left\{1+\left(\frac{z}{n\pi}\right)^2\right\}, \tag{3}$$

---

* A rational function is the quotient of one polynomial by another, which may be supposed prime to each other. (1) § 1·4 is a rational function; a constant may be regarded as a polynomial.

and since $\cosh z = \cos iz$, (2) yields

$$\cosh z = \prod_{n=0}^{\infty} \left\{ 1 + \left[\frac{2z}{(2n+1)\pi} \right]^{2} \right\}. \tag{4}$$

1·522. Behaviour of function near a pole. We shall consider $w = 1/z$ as $z \to 0$. Then

$$w = e^{-i\theta}/r = (1/r)(\cos\theta - i\sin\theta) = u + iv, \tag{1}$$

with $\qquad\qquad u = (\cos\theta)/r \quad \text{and} \quad v = -(\sin\theta)/r. \tag{2}$

Select any value of θ, say θ_0. Then the vector for $w = e^{-i\theta_0}/r$ has variable length $1/r$, but fixed angle $-\theta_0$. Thus when $z = r\,e^{i\theta_0} \to 0$, in *any* fixed direction θ_0, $r \to 0$, $1/r \to \infty$, so $1/z \to \infty$, and the function is analytic throughout the neighbourhood of the origin but *not at* it.

1·523. Limit point of poles. If the only singularities of a function in the entire z-plane (including infinity) are poles, there is a finite number of them. Consider the function $w = 1/\sin z$, which *appears* to have an infinite number of poles at $z = n\pi$, $n = -\infty$ to $+\infty$. If we put $z = 1/\zeta$, then $w = 1/\sin(1/\zeta)$, and as $\zeta \to 0$, $z \to \infty$, so the number of poles increases without limit. Within any *finite* circle of radius R, however large, there is only a *finite* number of poles of $w = 1/\sin z$; but outside the circle there is an infinite number of them. Then we regard infinity as a limit point of the poles of $1/\sin z$. Similarly, the origin is a limit point of the poles of $w = 1/\cos(1/z)$. Since $\cos(1/z) = 0$ when $1/z = \frac{1}{2}(2n+1)\pi$, n being any integer, the poles occur on the real axis at $z = 2/(2n+1)\pi$. The outermost poles occur when $n = 0, -1$, i.e. $z = \pm 2/\pi$. As $n \to \pm\infty$, $z \to 0$ and the line density of the poles $\to \infty$, so $z = 0$ is a limit point of the poles of $1/\cos(1/z)$. Thus inside a circle of radius, say $1/\pi$, about the origin, there is a limit point, but only a finite number of poles outside it. In virtue of the behaviour of the function near $z = 0$, the singularity is not a pole. The singular points of a *single-valued* function which are not poles are defined to be essential singularities. If the function is analytic in the neighbourhood of the singularity, it is said to be an isolated singularity. Since $1/\cos(1/z)$ is not analytic as $z \to 0$ along the real axis, the point $z = 0$ is a non-isolated essential singularity.

1·53. Essential singularities.

A function of the type

$$e^{1/z} = 1 + \frac{1}{z} + \frac{1}{2! \, z^2} + \frac{1}{3! \, z^3} + \cdots + \frac{1}{n! \, z^n} + \cdots, \qquad (1)$$

which is expressible in an infinite series of inverse (negative) *integral* powers of z, has an essential singularity at the origin. As $z \to 0$, each term (except the first) in the expansion $\to \infty$, and one might be tempted to regard the singularity as a pole of infinite order. There is, however, a marked difference between the behaviour of a function near a pole and that near such an essential singularity, as we shall now demonstrate.

Behaviour of $e^{1/z}$ as $z \to 0$. By (1) § 1·522, $1/z = (1/r)(\cos\theta - i\sin\theta)$, so

$$w = e^{1/z} = e^{\cos\theta/r}[\cos(\sin\theta/r) - i\sin(\sin\theta/r)]. \qquad (2)$$

Thus in the w-plane we have

$$w = u + iv = r_1 e^{-i\theta_1}, \qquad (3)$$

where $r_1 = e^{\cos\theta/r}$, $\theta_1 = \sin\theta/r$, $u = r_1\cos\theta_1$, $v = -r_1\sin\theta_1$. Choose any value θ_0 in $0 < \theta < \tfrac{1}{2}\pi$ in the z-plane. Then in the w-plane, $r_1 = e^{\cos\theta_0/r}$ and $\theta_1 = \sin\theta_0/r$ increase rapidly as $r \to 0$. Accordingly, the vector $r_1(\cos\theta_1 - i\sin\theta_1)$, corresponding to w, rotates with ever-increasing length and angular velocity,[*] whereas near a pole, the angle in the w-plane remains fixed (see § 1·522). Thus the representative point of w (end of vector) traces out a spiral of ever-increasing radius. Hence $e^{1/z}$ assumes an infinity of values as $z \to 0$, for any θ_0 in $0 < \theta < \tfrac{1}{2}\pi$ (z-plane). When $\theta_0 = 0$, $z \to 0$ along the positive real axis and $e^{1/z} = e^{1/r} \to \infty$. If $\theta = \tfrac{1}{2}\pi$, $e^{1/z} = \cos(1/r) - i\sin(1/r)$, so both circular functions oscillate rapidly as $r \to 0$. The vector in the w-plane is now of unit length, and while it rotates, $e^{1/z}$ takes successively all values having unit modulus. For $\theta = \pi$, $e^{1/z} = e^{-1/r} \to 0$ as z and $r \to 0$, and so on. Consequently we say that $e^{1/z}$ is indeterminate and has a singularity at $z = 0$. Although the behaviour of the function throughout the neighbourhood of the origin is peculiar compared with that near a pole, the function is analytic except at $z = 0$, so the singularity is said to be isolated. The first derivative of $e^{1/z}$, namely, $-e^{1/z}/z^2$, is also indeterminate and has a singularity at the origin. There is no pole due to $1/z^2$, the singularity being an essential one, as may be deduced from the behaviour of the function (see § 2·68).

Writing $z = (\zeta + a)$, thereby moving the origin to $z = a$, the function $e^{1/(z-a)}$ becomes $e^{1/\zeta}$, and by what precedes it is seen to have an isolated essential singularity at $\zeta = 0$, i.e. at $z = a$.

[*] For the sake of illustration we may suppose that $r \to 0$ linearly with time.

The function e^z has an isolated essential singularity at infinity. For writing $z = 1/\zeta$, we get

$$e^z = e^{1/\zeta} = 1 + \frac{1}{\zeta} + \frac{1}{2!\,\zeta^2} + \frac{1}{3!\,\zeta^3} + \ldots + \frac{1}{n!\,\zeta^n} + \ldots, \qquad (4)$$

which has the same form as (1). Since $z \to \infty$ as $\zeta \to 0$, the essential singularity of e^z occurs when z is infinite. It follows that if t is real and > 0, e^{zt} has an isolated essential singularity at infinity. The reader should now be able to show that $e^{-1/z}$, $e^{i/z}$, $e^{-i/z}$, $\sin (1/z)$, and $\cos (1/z)$ have isolated essential singularities at the origin, while e^{-z}, e^{iz}, e^{-iz}, $\sinh z$, and $\cosh z$ have them at infinity. An example of a non-isolated essential singularity where the function is not analytic in the neighbourhood thereof was cited in § 1·523. The function $1/\sin (1/z)$ has a non-isolated essential singularity at the origin.

1·541. Branch points. The important subject of branch points is approached most easily perhaps by reference to the real variable. The function $y = x^n$ is single-valued for all positive integral values of n. Three cases are illustrated in Fig. 5 (a), these corresponding to $n = 1$, 2 and 3. If a line $P'NP$ is drawn parallel to $Y'OY$, it cuts each curve in one point only, which exemplifies the single-valuedness of the function. When $n = \frac{1}{2}$, $y = \sqrt{x}$ or $x = y^2$, which is illustrated in Fig. 5 (b). The line $P'NP$ cuts the curve in two points, so that to each value of x, except $x = 0$, there correspond two different values of y. Here the values NP and NP' are equal but of opposite sign. We can consider the function $y = \sqrt{x}$ to have two branches which meet at the origin, which is termed a branch point.

1·542. We have now to enter the realm of the complex variable by passing from the case where z is on the real axis and phase $z = 0$, to the z-plane where the variable $z = x + iy$ can have any absolute value and phase, see § 1·1. Consider the function $w = \sqrt{z}$ with $z = r\,e^{i\theta}$, as at (1) § 1·1. Then we have

$$w = \sqrt{z} = \sqrt{r}(\cos \tfrac{1}{2}\theta + i \sin \tfrac{1}{2}\theta).$$

Let A in Fig. 6 (a) represent the complex number z, with $OB = r\cos\theta$ and $BA = r\sin\theta$. In Fig. 6 (b) P represents w,

$$ON = \sqrt{r}\cos\tfrac{1}{2}\theta, \quad NP = \sqrt{r}\sin\tfrac{1}{2}\theta,$$

$$OP = \sqrt{(OA)} \quad \text{and} \quad P\hat{O}N = \tfrac{1}{2}A\hat{O}B.$$

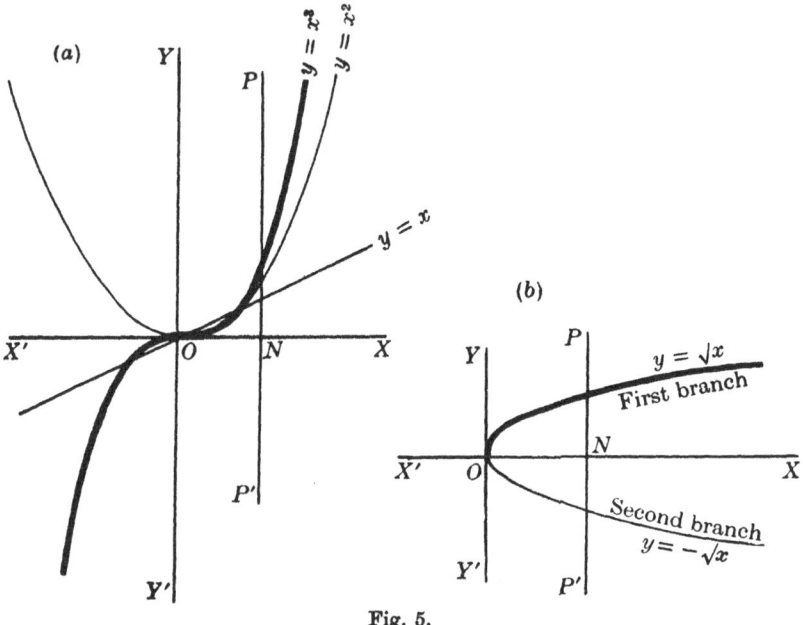

Fig. 5.

If OA rotates counter-clockwise (this being regarded as the positive direction) with constant angular velocity, OP will rotate positively with half this velocity. When OA has turned through 2π radians, thereby regaining its initial position, OP has described π radians and occupies the position OP_1. From (1) and Fig. 6 (b) we see that for each angle in the range θ to $(\theta + 2\pi)$ there corresponds a different value of w. When OA rotates from $(\theta + 2\pi)$ to $(\theta + 4\pi)$, P moves round the semicircle P_1Q_1P, thereby regaining its initial position. All the values of w whose representative points lie on the semicircle P_1Q_1P differ from those on the semicircle PQP_1. When $A\hat{O}X = \theta + 2\pi$, P is at P_1 and $\hat{UOP_1} = \frac{1}{2}\theta + \pi$, so from (1) the values of w at P and P_1 are equal but opposite. For at P_1,

$$w = \sqrt{r}\left[\cos\left(\tfrac{1}{2}\theta + \pi\right) + i\sin\left(\tfrac{1}{2}\theta + \pi\right)\right] = -\sqrt{r}\left[\cos\tfrac{1}{2}\theta + i\sin\tfrac{1}{2}\theta\right], \quad (2)$$

which is evident from Fig. 6(b), since $NP = -N_1P_1$ and $ON = -ON_1$. If OA is rotated beyond $(\theta + 4\pi)$, the preceding values of w are merely repeated. Consequently w is a two-valued function of z, and each of the parts (1), (2) is called a branch. One circuit of OA brings the function to the second branch, while after two circuits the function returns to its original value.

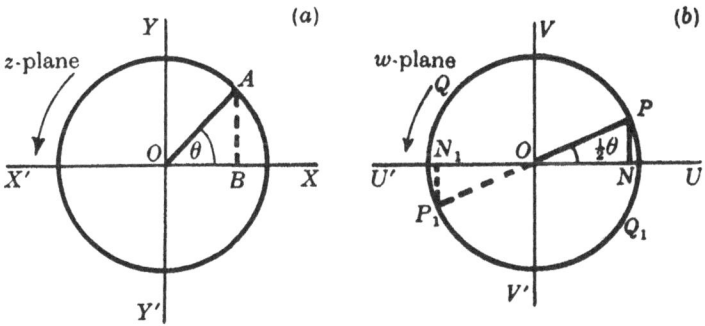

Fig. 6. The point A in the z-plane is defined to be $z = OB + iBA = x + iy$.
$\theta = $ phase $z = \tan^{-1}y/x$.

1·543. In Fig. 6(a) suppose that OA starts from OX where $\theta = 0$. After one turn round the origin, $\theta = 2\pi$, so by (1) § 1·542 w has *two* values $\pm\sqrt{r}$ at OX. Consequently $w = \sqrt{z}$ is not analytic on the positive real axis when the angle range $0 \leqslant \theta \leqslant 2\pi$ is used. The same argument applies on the radial line at the terminal values of the range $\alpha \leqslant \theta \leqslant (2\pi + \alpha)$, α being arbitrary. Now the functions used in complex theory must be analytic and single-valued, so we adopt a convention to ensure this property. Imagine a barrier or cut to exist along OX, so that θ cannot take the values 0, $2n\pi$, $n = 1, 2, \ldots$. Then the first angle range is $0 < \theta < 2\pi$, and w has the desired properties at all points in the finite part of the z-plane, except on the barrier. This angle range may be regarded as the first or principal branch of the function. For the second branch, we may imagine OA in Fig. 6(a) to pass beneath the barrier, so the angle range will be $2\pi < \theta < 4\pi$, w having the desired properties as before. For any value of z, those of w are equal but opposite on the two branches.

O is called a branch point because, in absence of the barrier, there is no neighbourhood of the origin throughout which w is analytic, owing to its being two-valued. To make $w = \sqrt{z}$ unique on each branch, the barrier must start from the branch point. It may be a curve, but a straight line is more convenient, for then the angle range is identical for all values of $|z| = r$. The angular position of the barrier is arbitrary. For a line making an angle α with the positive real axis, the principal and second branches of $w = \sqrt{z}$ are $\alpha < \theta < (2\pi + \alpha)$, and $(2\pi + \alpha) < \theta < (4\pi + \alpha)$, respectively. Taking $\alpha = -\pi$, the two ranges are $-\pi < \theta < \pi$ and $\pi < \theta < 3\pi$, and these ranges are often used in analysis.

The function $w = z^\nu$, ν real and non-integral, has a branch point at the origin. In technical applications herein, the indices associated with branch points are usually half-odd integers, e.g. $\nu = \frac{3}{2}, \frac{1}{2}, -\frac{1}{2}$.

1·611. Phase change in relation to branch points. In Fig. 6 (a) the representative point of z, namely A, moves round a circle with the origin as centre. This is a particular case, for the point z may traverse *any* open or closed curve in the z-plane. Consider the function $w = \sqrt{z}$, where z has successively all values on the closed curve C illustrated in Fig. 7 (a).

Starting at OA_1 which is tangential to C, A the representative point of z travels round C to A_2, OA_2 being tangential to C. Meanwhile phase z varies from θ_1 to θ_2. Translated into symbolic form, the two positions OA_1, OA_2 are, respectively,

$$w = \sqrt{r_1}(\cos \tfrac{1}{2}\theta_1 + i \sin \tfrac{1}{2}\theta_1) \tag{1}$$

and $$w = \sqrt{r_2}(\cos \tfrac{1}{2}\theta_2 + i \sin \tfrac{1}{2}\theta_2), \tag{2}$$

where $$OA_1 = r_1 \quad \text{and} \quad OA_2 = r_2.$$

When the rotation of A is continued positively from A_2, phase z decreases, e.g. it is θ_3 at A_3, until finally after one circuit of the contour OA regains its initial position OA_1. Now the angle swept out by OA is zero, since it merely oscillates to A_2 and back.

Hence the final value of the function is that given at (1), so w returns to its original value. It should now be evident that unless C encloses the branch point at O, the angle θ_1 cannot alter, however many revolutions the point A may make.

The function $w = \sqrt{z}$ has two branches. If we choose $-\pi < \theta < \pi$ as the angle range of the principal branch, that of the second branch is $\pi < \theta < 3\pi$. If the principal branch is used when z traverses a closed curve excluding the origin, the function is single-valued and analytic everywhere in the finite part of the z-plane except when $\theta = -\pi$.

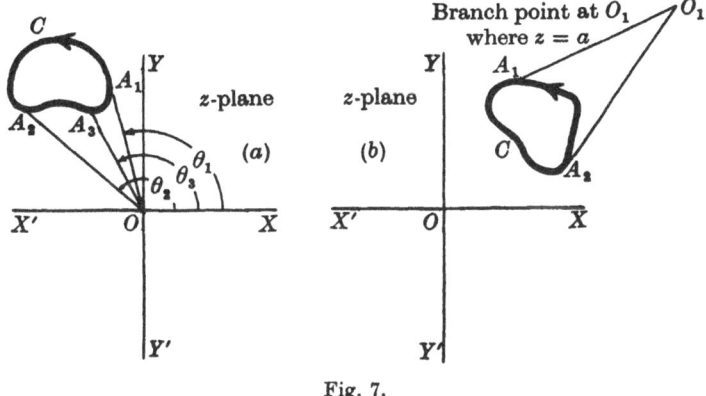

Fig. 7.

1·612. If $w = z^{\frac{1}{3}} = r^{\frac{1}{3}} e^{\frac{1}{3}i\theta}$, there is a branch point at the origin. After one circuit of a curve enclosing the origin, the second branch of the function is reached, where $w = r^{\frac{1}{3}} e^{i(\theta+2\pi)/3}$; after two circuits the third branch is reached with $w = r^{\frac{1}{3}} e^{i(\theta+4\pi)/3}$. Finally, a third circuit restores the function to its original value, for then $w = r^{\frac{1}{3}} e^{i(\theta+6\pi)/3} = r^{\frac{1}{3}} e^{\frac{1}{3}i\theta}$, since $\cos\frac{1}{3}(\theta+6\pi) + i\sin\frac{1}{3}(\theta+6\pi) = \cos\frac{1}{3}\theta + i\sin\frac{1}{3}\theta$. Similarly, if n is a positive integer > 1, on the n branches of $w = z^{1/n}$, we have $w = r^{1/n} e^{i(\theta+2m\pi)/n}$ ($0 \leqslant m \leqslant n$). The phase change in w after m circuits about the origin is $2m\pi/n$, so if $m = 2$, $n = 3$, on reaching the third branch of $w = z^{\frac{1}{3}}$, the phase change is $\frac{4}{3}\pi$. If the angle range of the principal branch is $-\pi < \theta < \pi$, that of the second and third branches is $\pi < \theta < 3\pi$, and $3\pi < \theta < 5\pi$, respectively.

1·62. Suppose the branch point occurs at $z = a$, as shown in Fig. 7 (b), and consider z to traverse any curve C which does not enclose a. By moving the origin to $z = a$, it is evident that the line $O_1 A_1$ regains its original position after each circuit of the contour, i.e. it merely oscillates to and fro, so the function

$$w = \sqrt{(z-a)} \tag{1}$$

resumes its original value.

1·63. The function $w = \sqrt{\{z(z-a)\}}$ has two branch points, one at $z = 0$, the other at $z = a$. If a closed curve encircles either $z = 0$ or $z = a$, but not both of them, after z makes one circuit of the contour, w commences a new branch, while after two circuits w resumes its initial value. When both branch points lie within the contour, one circuit restores w to its original value. To illustrate these matters more vividly, we take the simple case shown in Fig. 8 (a), where a is real and positive on the x-axis.

In Fig. 8 (a) only the branch point at the origin is enclosed by the contour. Put $z = r_1 e^{i\theta_1}$ at 0, and $(z-a) = r_2 e^{i\theta_2}$ at a, giving

$$w = \sqrt{(r_1 r_2)}\, e^{\frac{1}{2}(\theta_1+\theta_2)i} = \sqrt{(r_1 r_2)}\, [\cos \tfrac{1}{2}(\theta_1 + \theta_2) + i \sin \tfrac{1}{2}(\theta_1+\theta_2)]. \tag{1}$$

Then for case (a) after one circuit of P round O, the angle θ_1 increases to $\theta_1 + 2\pi$, while θ_2 merely oscillates without change. Thus the final value of w is

$$w = \sqrt{(r_1 r_2)}\, [\cos \tfrac{1}{2}(\theta_1 + \theta_2 + 2\pi) + i \sin \tfrac{1}{2}(\theta_1 + \theta_2 + 2\pi)] \tag{2}$$

$$= -\sqrt{(r_1 r_2)}\, [\cos \tfrac{1}{2}(\theta_1 + \theta_2) + i \sin \tfrac{1}{2}(\theta_1 + \theta_2)], \tag{3}$$

so the function enters upon a new branch. When P has made two circuits of the contour, θ_1 becomes $\theta_1 + 4\pi$, and if this value is inserted in (1) the function has its original value.

For case (b) each branch point is enclosed by a circle. The parallel lines are very close to each other, so that for one circuit at $z = 0$ and one at $z = a$, both θ_1 and θ_2 increase by 2π, so $\theta_1 + \theta_2$ increases by 4π. Hence after one circuit of the complete contour

$$w = \sqrt{(r_1 r_2)}\, [\cos \tfrac{1}{2}(\theta_1 + \theta_2 + 4\pi) + i \sin \tfrac{1}{2}(\theta_1 + \theta_2 + 4\pi)], \tag{4}$$

which is identical with (1), so the function regains its original value.

If we imagine a barrier or branch cut from $z = 0$ to a on the positive real axis in Fig. 8 (a), for any closed region not containing the barrier, the angle ranges $0 < \theta < 2\pi$ and $2\pi < \theta < 4\pi$

about O^* correspond to the first and second branches of $w = \sqrt{\{z(z-a)\}}$. In these ranges the function is analytic on both branches throughout the finite part of the z-plane. If both branch points are inside the region, during one turn round O, $(\theta_1 + \theta_2)$ increases by 4π, so by (1) the function resumes its original value. Accordingly, w is analytic and single-valued throughout the finite part of the z-plane, except on the barrier. $z = 0$, a are branch points, because there is no neighbourhood of either of them, throughout which w is analytic.

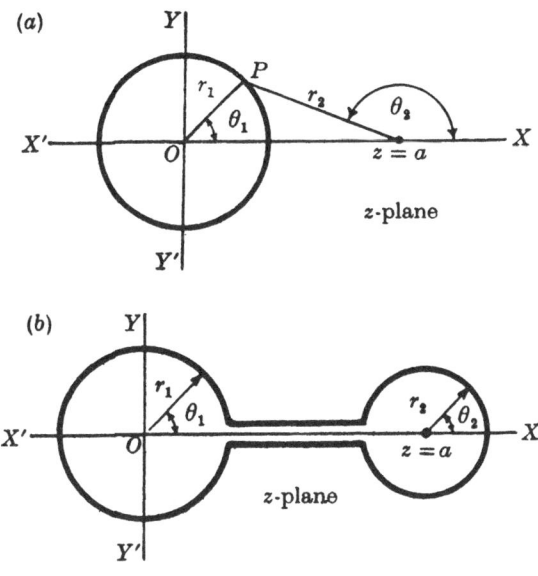

Fig. 8.

If $w = \sqrt{\{z(z-a)(z-b)(z-c)\}}$, there are four branch *points* at $z = 0$, a, b, c, respectively. By virtue of the square root, there are *two branches*, namely, $\pm\sqrt{\{z(z-a)(z-b)(z-c)\}}$. A single turn round a curve enclosing only one branch point brings w to the next branch, while after two turns w regains its initial value. If the index were $\frac{1}{3}$, there would still be four branch points, but three branches corresponding to each. On the second branch w starts with a phase $\frac{2}{3}\pi$, and on the third with phase $\frac{4}{3}\pi$ in relation to that on the first branch. For index $\frac{2}{3}$ the phase changes of w are $\frac{4}{3}\pi$, $\frac{8}{3}\pi$, and the function regains its initial value after three circuits of

---

* The radius vector r may be variable.

the branch point, the total phase change being $\frac{12}{3}\pi = 4\pi$. A new branch starts when phase z (initially zero, say) changes by 2π. The phase change in w depends upon the index.

1·64. The singularities of $1/[1+\sqrt{(az)}]$. With a real and >0, write $z = r\,e^{i\theta}$, and we get

$$w = 1/[1+\sqrt{(ar)}\,e^{\frac{1}{2}i\theta}], \tag{1}$$

where θ may have *any* value in the range $-\pi < \theta < \pi$, as illustrated in Fig. 9 (a). Suppose OA to rotate once, thereby increasing θ by 2π. To achieve this, we imagine OA to pass beneath the barrier. Then

$$w = 1/[1+\sqrt{(ar)}\,e^{i(\frac{1}{2}\theta+\pi)}] = 1/[1-\sqrt{(ar)}\,e^{\frac{1}{2}i\theta}] \tag{2}$$

$$= 1/[1-\sqrt{(az)}], \tag{3}$$

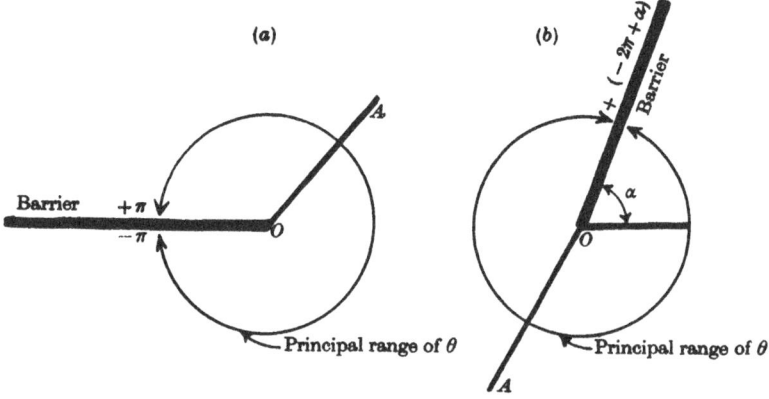

Fig. 9.

which differs from the original. Moreover, the function is now on another branch for which the angle range is $(-\pi+2\pi) < \theta < (\pi+2\pi)$, i.e. $\pi < \theta < 3\pi$. If OA rotates through 4π, we have $e^{\frac{1}{2}i(\theta+4\pi)} = e^{\frac{1}{2}i\theta}$, so after two circuits round the origin the function has returned to its original value, now being on the first branch. The above reasoning is valid wherever the barrier is situated, provided it radiates from the branch point O. In Fig. 9 (b) the angle ranges are

$$(-2\pi+\alpha) < \theta < \alpha \quad \text{for the first branch,}$$

and $\qquad\qquad \alpha < \theta < (\alpha+2\pi) \qquad \text{for the second branch.}$

In addition to a branch point at $z = 0$, there is also a singularity when the denominator of (1) vanishes. Thus we have

$$\sqrt{(ar)}\,e^{\frac{1}{2}i\theta} = -1, \tag{4}$$

which entails the reality of $e^{\frac{1}{2}i\theta}$. Then $\sqrt{(ar)} = 1$, and $e^{\frac{1}{2}i\theta} = -1$, so $\theta = 2\pi$. This lies in the angle range $\pi < \theta < 3\pi$, so the singularity occurs at $z = 1/a$ on the second branch. It appears to be a pole, and as shown in § 2·64, using Laurent's theorem this surmise is correct.

1·65. Finally, we consider the function $\log z$. Writing $z = r\, e^{i\theta}$, we get $\log z = \log_e r + i\theta$. When θ increases by 2π, $\log z$ increases by $2\pi i$; when θ increases by $2n\pi$, $\log z$ increases by $2n\pi i$, $n = 1, 2, 3, \ldots$, so in general $\log z = \log_e r + i(\theta + 2n\pi)$. Thus after each circuit of the origin, the function starts upon a new branch, of which there is an infinite number. To render the function analytic, as in § 1·543 we imagine a radial barrier or branch cut to start from O, so that each branch may be considered independently. The principal branch is usually taken to be $-\pi < \theta < \pi$ and the function is denoted by $\log z$ (*not* $\log_e z$). For z real > 0, i.e. on the positive real axis where $\theta = 0$, we get the Napierian or natural logarithm $\log_e x$. Although $\log z$ is not analytic on the negative real axis where $\theta = \pi$, by *convention* we regard its value there to be $\log_e |x| + i\pi$, i.e. the imaginary part of $\log_e (-x)$ is $i\pi$.

With the function $w = \log(z - a)$, if we move the origin to the point $z = a$ by writing $(\zeta + a)$ for z, we obtain $w = \log \zeta$. From what precedes this is seen to have a branch point at $\zeta = 0$, i.e. $z = a$. This is so, since $\log \zeta$ is not analytic throughout the neighbourhood of $\zeta = 0$.

1·7. Derivative of multivalued function. Consider a branch of the function where it is analytic throughout the finite part of the z-plane, except at singular points and on barriers. By § 1·33 it has a unique finite derivative at all 'ordinary' points of the region, and the derivative may be found by applying (4), (5), or (11) § 1·34. For example, take

$$w(z) = \log z = \log_e r + i(\theta + 2n\pi), \tag{1}$$

where $r = \sqrt{(x^2 + y^2)}$, $\theta = \tan^{-1}(y/x)$, with $0 < \theta < \pi$ for $y > 0$ and $-\pi < \theta < 0$ for $y < 0$. Then if $x \neq 0$, $y \neq 0$, w is analytic in $-\pi < \theta < \pi$, which is the angle range of the first branch. Also

$$w = u + iv, \tag{2}$$

where $u = \log_e \sqrt{(x^2 + y^2)}$, and $v = \tan^{-1}(y/x)$. Using (4) § 1·34 leads to

$$\frac{dw}{dz} = \frac{\partial}{\partial x} \log_e \sqrt{(x^2 + y^2)} + i \frac{\partial}{\partial x} \tan^{-1}(y/x) \tag{3}$$

$$= \frac{x - iy}{x^2 + y^2} = \frac{1}{x + iy} = \frac{1}{z}, \tag{4}$$

as in the case of z real. Usually when the Cauchy-Riemann conditions in § 1·34 are satisfied, the first derivative of a function may be found as in the case of the real variable.

II

INTEGRATION: CAUCHY'S THEOREM: TAYLOR'S AND LAURENT'S THEOREMS

2·11. Integration. We have seen already that the complex variable z can have any value in the z-plane which contains the real and imaginary axes $X'OX$, $Y'OY$. If a curve is drawn in the plane, it may be regarded as a contour. In Fig. 10 ABC is an open curve, while PQR is closed. Such curves may be used as the path of integration for functions of a complex variable. This will be understood more readily by reference to integration of the real variable, which is a particular case of a contour integral. For example, take $\int_0^a x^3\,dx$, then $y = x^3$ is the function,

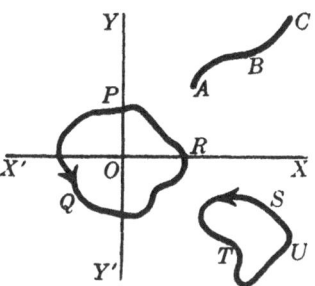

Fig. 10. Open and closed contours in the z-plane.

while the real axis from $x = 0$ to a is the contour. The function is given each value of x from 0 to a and the products $x^3\,dx$ are summed over this range. If the function $y = x^3$ is plotted above the x-axis in the usual way, the integration is represented geometrically by the area between the curve and the x-axis from 0 to a. In dealing with the complex variable it is well to be oblivious to any such interpretation, since there are two variables x and y, and the geometrical interpretation does not materialize. Integration of the complex variable is best regarded as a process of summation. It is known as Riemann integration after its leading exponent.

2·121. Suppose a function $f(z)$ is continuous for all values of z on the contour AB in Fig. 11.* Imagine the curve to be divided

* The contour may be a continuous open or closed curve, such that there are no singularities on it.

up into a large number of short arcs $z_0 z_1$, $z_1 z_2$, ... Take the value
of z at, say, the mid-point of each arc and find the corresponding
value of the function. Let the mid-point values be $z_{0m}, z_{1m}, ...$,
then the corresponding values of the function are $f(z_{0m}), f(z_{1m}), ...$
Multiplying $(z_1 - z_0)$, $(z_2 - z_1)$, ... by the appropriate values of
$f(z)$ at the mid-points of the respective arcs and adding, we obtain

$$S = (z_1 - z_0)f(z_{0m}) + (z_2 - z_1)f(z_{1m}) + ... + (z_n - z_{n-1})f(z_{n-1, m}). \quad (1)$$

Let $f(z) = u + iv$, where u, v are continuous functions of x, y for
all points on AB. If we write $z_0 = x_0 + iy_0$, $z_1 = x_1 + iy_1$, ... and
substitute into (1), we find ultimately, when the number of
divisions increases indefinitely, that (see [237])

$$\lim_{n \to \infty} S = \int_{AB} f(z)\, dz = \int_{AB} (u\, dx - v\, dy) + i \int_{AB} (u\, dy + v\, dx) \quad (2)$$

$$= \int_{AB} f(z)\,(dx + i\, dy). \quad (3)$$

This integral has been obtained by summation along the curve
AB, and its meaning is distinct from that of a real integral, whose
contour lies entirely on the real axis. In evaluating an integral
of this type, it may be expedient to transform to real integrals,
as shown in the next section.

2·122. Write $x = \xi(\theta)$, $y = \chi(\theta)$, where ξ, χ and their first
derivatives are continuous functions of the parameter θ, which
varies from θ_0 to θ_1 as z varies from A to B in Fig. 11. Then at any
point $z = x + iy$, we have

$$f(z) = u(x, y) + i\, v(x, y) = u[\xi(\theta), \chi(\theta)] + iv[\xi(\theta), \chi(\theta)] \quad (1)$$

$$= u(\theta) + iv(\theta), \quad (2)$$

u, v being continuous functions of θ. Substituting from (2) into
(3) § 2·121, with $dx = \xi'(\theta)\, d\theta$, $dy = \chi'(\theta)\, d\theta$, we obtain

$$\int_{AB} f(z)\, dz = \int_{\theta_0}^{\theta_1} [u(\theta) + iv(\theta)]\, [\xi'(\theta) + i\chi'(\theta)]\, d\theta, \quad (3)$$

the right-hand side depending on real integrals. This result may
be derived also from the left-hand side of (3) and (2) above by
writing $dz = [\xi'(\theta) + i\chi'(\theta)]\, d\theta$.

2·123. Example. Evaluate $\int_{C_1} \dfrac{dz}{z}$, where C_1 is a circle of radius r about the origin (Fig. 12).

Here we put $z = r\,e^{i\theta}$, r constant, so

$$\xi'(\theta) = -r\sin\theta, \quad \chi'(\theta) = r\cos\theta. \tag{1}$$

Also $\qquad f(z) = 1/z = (1/r)(\cos\theta - i\sin\theta),$ $\qquad\qquad$ (2)

giving $\quad u(\theta) = (1/r)\cos\theta,$ and $v(\theta) = -(1/r)\sin\theta.$ $\quad$ (3)

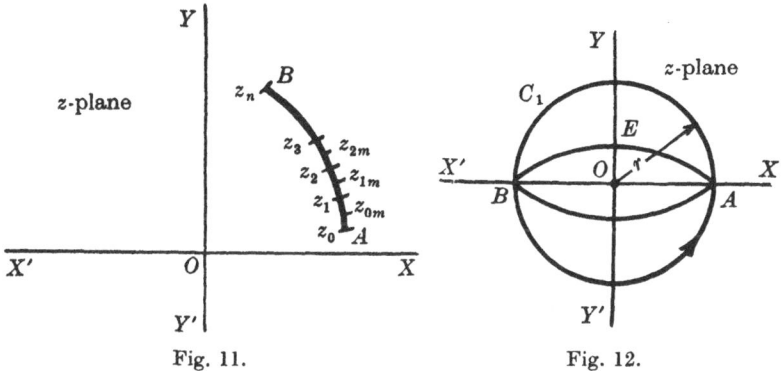

Fig. 11. $\qquad\qquad\qquad\qquad$ Fig. 12.

Substituting into (3) §2.122, we obtain

$$\int_{C_1} \frac{dz}{z} = ir \int_0^{2\pi} \frac{1}{r} (\cos\theta - i\sin\theta)(\cos\theta + i\sin\theta)\,d\theta \tag{4}$$

$$= i \int_0^{2\pi} d\theta = 2\pi i. \tag{5}$$

Hence $\qquad\qquad\qquad \dfrac{1}{2\pi i}\displaystyle\int_{C_1} \dfrac{dz}{z} = 1 \qquad\qquad\qquad$ (6)

for all *finite* $r > 0$, i.e. the result is independent of r.

If the curve C_1 were a circular arc subtending an angle $\alpha < 2\pi$ at its centre O, the value of the integral would be $\alpha/2\pi$. If it were a closed curve *not* containing the origin, the integral would be zero. As we shall see later, this latter result is due to the only singularity (pole at $z = 0$), being *outside* the curve C_1.

2·13. Consider the integral $\int_{C_1} \dfrac{dz}{z^2}$ taken round the same contour as in § 2·123. Writing $z = r\,e^{i\theta}$, we get

$$\int_{C_1} \frac{dz}{z^2} = \frac{i}{r}\int_0^{2\pi} e^{-i\theta}\,d\theta = \frac{i}{r}\int_0^{2\pi} (\cos\theta - i\sin\theta)\,d\theta = 0. \qquad (1)$$

The integral of the real part is positive from 0 to $\tfrac{1}{2}\pi$, and from $\tfrac{3}{2}\pi$ to 2π, but it is negative from $\tfrac{1}{2}\pi$ to $\tfrac{3}{2}\pi$. Since the two integrals are equal numerically, the sum is zero. A similar argument holds for the imaginary part. In general

$$\int_{C_1} \frac{dz}{z^n} = 0, \qquad (2)$$

provided n is a positive or a negative integer other than unity.

2·211. Cauchy's theorem. We now state, without proof (see [220, 237, 240, 243, 250]), a theorem of fundamental importance in the evaluation of contour integrals. If a function $f(z)$ is analytic and single-valued, within and upon a closed contour, the integral taken round the contour is zero. Alternatively, the value of the integral of a single-valued function taken once round any closed contour having no singularities of the function within or upon it is zero. For example, the contour STU in Fig. 10 does not contain the singularity of $w = 1/z$, so by Cauchy's theorem $\int_{STU} \dfrac{dz}{z} = 0$.

In Fig. 13 (a) there are two paths from P to Q, namely, PSQ and PRQ. If $f(z)$ is analytic on and within the contour, there is no singularity in the region, so by Cauchy's theorem

$$\int_{PSQ} f(z)\,dz + \int_{QRP} f(z)\,dz = 0. \qquad (1)$$

Now the first integral in (1) is finite, so the second must be equal but opposite. Since the positive direction is counter-clockwise, the integral along QRP is minus that along PRQ. Consequently

$$\int_{PSQ} f(z)\,dz = \int_{PRQ} f(z)\,dz, \qquad (2)$$

i.e. the value of the integral is independent of the path between the terminal points.

2·212. Morera's theorem. In a sense this is the converse of Cauchy's theorem. If $f(z)$ is continuous in a given region of the z-plane, and if $\int_C f(z)\,dz = 0$ for any *closed* contour C which lies within the region, then $f(z)$ is analytic within the region.

2·221. If we draw any curve between P and Q within or without the closed contour $PRQS$ in Fig. 13 (a), the integral along it from P to Q is independent of the path, provided no singularity of $f(z)$ is on or within two paths which form a closed region, i.e. the function must be analytic in the region. If $z = a$ in Fig. 13 (a) is a singularity within the closed contour $BDCE$, the integral of $f(z)$ from B to C is dependent upon whether the path lies above or below a. All paths above a give identical results, since no two of them enclose a. Similarly, all paths below a give identical results, but usually they differ from the foregoing.

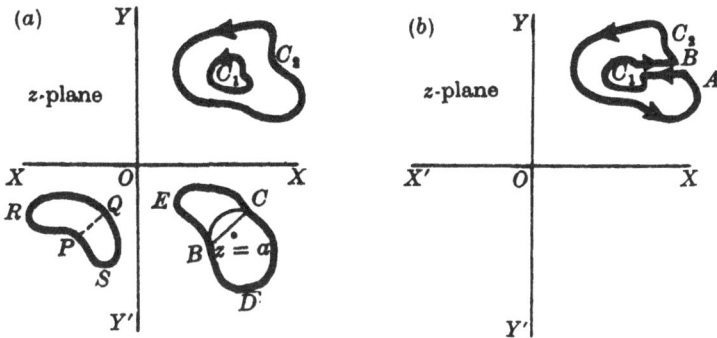

Fig. 13.

2·222. Example. Consider the integral $\int_{C_1} \dfrac{dz}{z}$ in § 2·123. In passing from $\theta = 0$ to π on a semicircle above the real axis (Fig. 12), the value of the integral is πi. For the semicircle below the axis, the limits are $\theta = 0$ to $-\pi$, so the value of the integral is $-\pi i$. The difference between these two values is due to the presence of a pole at the origin. By Cauchy's theorem *any* path from $+r$ to $-r$ above the real axis (AEB in Fig. 12) yields πi, and any similar path below (ADB) yields $-\pi i$. If both paths are described positively, the total result is $2\pi i$, and if described negatively, it is $-2\pi i$.

2·231. Theorem. If $f(z)$ and $f_1(z)$ are analytic in a region limited by a simple closed curve C, and $f(z) = df_1(z)/dz$, then

$$\int_{z_0}^{z_1} f(z)\, dz = f_1(z_1) - f_1(z_0),\tag{1}$$

where z_0 and z_1 are the terminal points of any path lying *within* C. The proof is given in [226]. Here integration may be effected as though the variable were real.

2·232. Examples. (a) $f(z) = z^n$ $(n \geqslant 1)$ is analytic everywhere in the finite part of the z-plane. Then $df_1(z)/dz = f(z)$, so $f_1(z) = z^{n+1}/(n+1)$, which is analytic also. Accordingly by (1) § 2·231, along any path in the finite part of the z-plane from z_0 to z_1,

$$\int_{z_0}^{z_1} z^n\, dz = \frac{z_1^{n+1} - z_0^{n+1}}{n+1}.\tag{1}$$

(b)
$$\int_1^e \frac{dz}{z} = [\log z]_1^e = 1.\tag{2}$$

The limits are both positive on the real axis, but the result is the same for any path between $z = 1$ and e, provided that it passes neither round nor through the origin, since $\log z$ has a branch point there. The value at (2) pertains to the principal branch of $\log z$, say $-\pi < \theta < \pi$ (see §§ 1·543, 1·65).

(c) We consider the contour of Fig. 14 where ϵ is the radius of a small semicircle at the origin. Moreover, the contour does *not* enclose O. Then

$$\int_{-a}^{a} \frac{dz}{z} = \int_{-a}^{-\epsilon} + \int_{\epsilon}^{a} + \text{integral round the semicircle}\tag{3}$$

$$= [\log z]_{-a}^{-\epsilon} + [\log z]_{\epsilon}^{a} - \pi i \quad \text{(from § 2·222)}\tag{4}$$

$$= \log(\epsilon/a) + \log(a/\epsilon) - \pi i\tag{5}$$

$$= -\pi i.\tag{6}$$

2·233. Example. Evaluate $\int_0^{z_0} z\, dz$, along the straight line $(0, z_0)$, where $z_0 = 3 + 4i$, by three different methods (see Fig. 15 (a)).

(1°) $\qquad \int_0^{z_0} z\,dz = \tfrac{1}{2}z_0^2 = \tfrac{1}{2}(3+4i)^2 = -3\cdot5+12i.$ $\qquad$ (1)

(2°) $\qquad \int_0^{z_0} z\,dz = \int_0^3 x\,dx + \int_0^4 (3+iy)\,d(3+iy)$

$$= \tfrac{9}{2} + i\int_0^4 (3+iy)\,dy = \tfrac{9}{2} - 8 + 12i$$

$$= -3\cdot5 + 12i. \qquad (2)$$

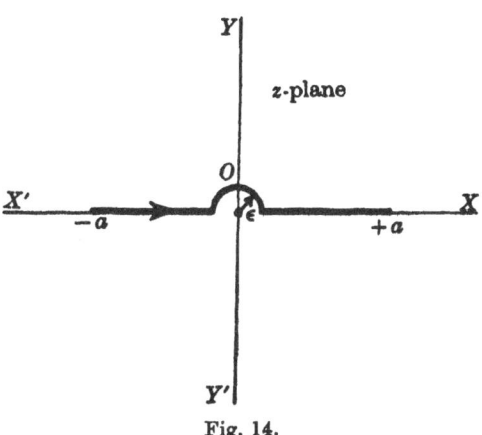

Fig. 14.

(3°) On Oz_0, Fig. 15 (a), put $z = re^{i\alpha}$, where r varies but α is constant.

Then $\qquad \int_0^{z_0} z\,dz = \int_0^5 re^{2i\alpha}\,dr = [\tfrac{1}{2}r^2 e^{2i\alpha}]_0^5,$

so $\qquad I = 12\cdot5[\cos 2\alpha + i\sin 2\alpha].$ $\qquad$ (3)

Now $\qquad \cos 2\alpha = 2\cos^2\alpha - 1 = \tfrac{18}{25} - 1 = -\tfrac{7}{25},$

and $\qquad \sin 2\alpha = 2\sin\alpha\cos\alpha = \tfrac{24}{25},$

so (3) becomes

$$I = 12\cdot5(-\tfrac{7}{25} + \tfrac{24}{25}i) = -3\cdot5 + 12i. \qquad (4)$$

2·234. Example. Show that in §2·233, if the path from O to z_0 is that illustrated in Fig. 15 (b), namely, Oz_1Bz_0, the value of the integral is unaltered.

On OY' we write $z = iy$, so

$$\int_0^{z_1} z\, dz = -\int_0^{-5} y\, dy = -12\cdot5. \tag{1}$$

On the circular arc $z_1 B z_0$, put $z = r\, e^{i\theta}$ and

$$\int_{z_1}^{z_0} z\, dz = i\int_{-\frac14\pi}^{\alpha} r^2\, e^{2i\theta}\, d\theta = \tfrac12 r^2 [e^{2i\theta}]_{-\frac14\pi}^{\alpha}$$

$$= \tfrac12 r^2[e^{2i\alpha}+1] = 12\cdot5(1+\cos 2\alpha + i\sin 2\alpha)$$

$$= 12\cdot5(\tfrac{18}{25}+\tfrac{24}{25}i) = 9+12i. \tag{2}$$

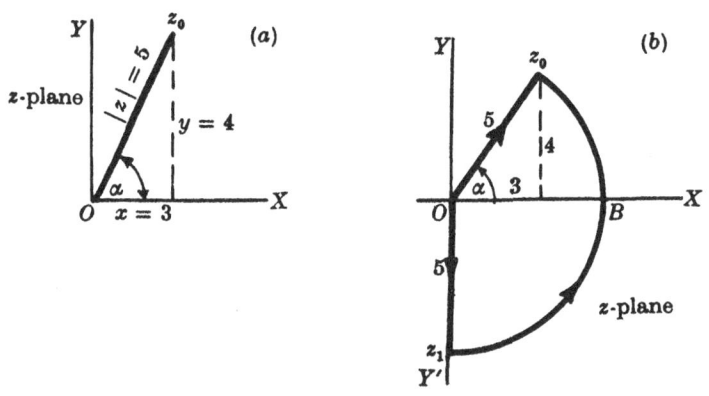

Fig. 15.

Hence from (1), (2)

$$\int_0^{z_0} z\, dz = -12\cdot5 + 9 + 12i = -3\cdot5 + 12i, \tag{3}$$

as before at (1) § 2·233.

2·24. Suppose that two simple closed curves are drawn in the z-plane, one enclosing the other completely, as shown by C_1, C_2 in the first quadrant of Fig. 13 (a). We shall now prove that if $f(z)$ is analytic in the ring-shaped space between the curves, and has no branch point within C_1,

$$\int_{C_2} f(z)\, dz - \int_{C_1} f(z)\, dz = 0, \tag{1}$$

both contours being described counter-clockwise, say. If C_1 contains no singularity, the function is analytic within C_1 and C_2, so each integral in (1) is zero. When C_1 contains one or more singularities which are not branch points, the artifice of Fig. 13 (b) is employed to prove the above theorem. The contours C_1, C_2 are connected by two parallel lines very close together, which necessitates gaps in C_1, C_2 at the junction points. Now C_2 is described positively and C_1 negatively, while if the parallel lines coalesce in the limit, the integrals along them are equal but opposite, since $f(z)$ is analytic and single-valued at all points thereon. Thus (1) follows.

2·25. Example. As a simple instance of this theorem, take the case of $\int \dfrac{dz}{z}$. It was shown in § 2·123 that for a circular contour with the origin as centre, the integral is $2\pi i$, whatever the radius. Hence if C_1, C_2 are concentric circles of radii r_1, r_2 about O, theorem (1) § 2·24 is satisfied.

2·31. Value of $f(z)$ at $z = a$ expressed as a contour integral. C is a closed contour within and upon which the function $f(z)$ is analytic, i.e. there are no singularities in the region. If a be any point within the contour, such that $f(a) \neq 0$, we can in effect create a singularity thereat by forming the new function

$$g(z) = f(z)/(z-a).$$

This function is analytic at all points in the region except at $z = a$, where it has a simple pole. With a as centre describe a small circle δ within C. Then $g(z)$ is analytic throughout the ring-shaped space between the two contours, so we can apply (1) § 2·24. Thus

$$\int_C \frac{f(z)\,dz}{(z-a)} = \int_\delta \frac{f(z)\,dz}{(z-a)}, \tag{1}$$

both contours being described positively. But by (2) § 1·31

$$f(z) = f(a) + (z-a)\,f'(a) + \lambda(z-a),$$

so

$$\int_C \frac{f(z)\,dz}{(z-a)} = f(a)\int_\delta \frac{dz}{(z-a)} + f'(a)\int_\delta dz + \int_\delta \lambda\,dz. \tag{2}$$

In the first integral on the right-hand side of (2), move the origin to $z = a$ by writing $(z-a) = \zeta$, $dz = d\zeta$. Then we get

$$\int \frac{d\zeta}{\zeta} = 2\pi i,$$

by (5) § 2·123. In the second integral write $z = r\,e^{i\theta}$, and

$$\int_\delta dz = i\int_0^{2\pi} r\,e^{i\theta}\,d\theta = 0. \tag{3}$$

In the third integral in (2) if η is the greatest value of $|\lambda|$ on, or inside the small circle, then by Appendix 1, since $|\lambda| < \eta$ when $|z-a| < r$,

$$\left|\int_\delta \lambda\,dz\right| \leqslant 2\pi r\eta \to 0 \tag{4}$$

in the limit when $r \to 0$. By hypothesis $f'(a)$ is finite, since a is not a singular point of $f(z)$, so that only the first term on the right-hand side of (2), namely, $2\pi i f(a)$, remains. Thus

$$f(a) = \frac{1}{2\pi i}\int_C \frac{f(z)\,dz}{(z-a)}. \tag{5}$$

(5) is known as Cauchy's integral formula. It gives the value of a function $f(z)$ at any point $z = a$ within a closed contour, where it is analytic, in terms of an integral taken round the contour.

2·32. Extension of (5) § 2·31 to a ring-shaped space.

This formula can be applied to a ring-shaped space between two closed contours, the function $f(z)$ being analytic in the region, but may have poles and/or essential singularities within the inner contour.

In Fig. 16 (a) let a be a point in the ring-shaped space between C_1 and C_2. We shall now show that when the two contours are described positively

$$f(a) = \frac{1}{2\pi i}\int_{C_2} \frac{f(z)\,dz}{(z-a)} - \frac{1}{2\pi i}\int_{C_1} \frac{f(z)\,dz}{(z-a)}. \tag{1}$$

Draw the contours as shown in Fig. 16 (b), where C_1 and C_2 are connected by parallel lines and two small semicircles δ of radius ϵ. Then the contour starting from A and described in the direction indicated by the arrows is a closed one having no singularities within or upon it. By Cauchy's theorem the integral round the contour is zero. In the limit when the connecting link between C_1 and C_2 consists of two straight lines and a circle at a, the integrals along the lines neutralize, because $f(z)/(z-a)$ is single valued and they are described in opposite directions. Thus we have

$$\frac{1}{2\pi i}\int_{C_1}\frac{f(z)\,dz}{(z-a)}-\frac{1}{2\pi i}\int_{C_1}\frac{f(z)\,dz}{(z-a)}-\frac{1}{2\pi i}\int_{s}\frac{f(z)\,dz}{(z-a)}=0. \qquad (2)$$

But from (5) § 2·31 the third integral is $f(a)$, so

$$f(a)=\frac{1}{2\pi i}\int_{C_2}\frac{f(z)\,dz}{(z-a)}-\frac{1}{2\pi i}\int_{C_1}\frac{f(z)\,dz}{(z-a)}. \qquad (3)$$

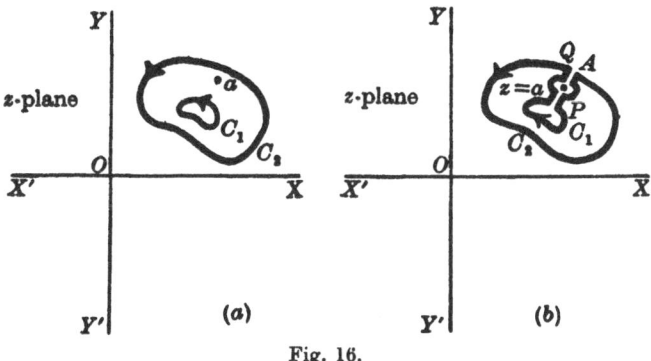

Fig. 16.

The integrals along the semicircles do not neutralize each other, since $f(z)/(z-a)$ differs in each. This will be realized if we move the origin to $z=a$ by writing $(z-a)=\zeta$ and put $\zeta=r\,e^{i\theta}$ on the semicircles. Then phase ζ varies continuously over the range 0 to 2π. Thus $z=a+r\,e^{i\theta}$, and since $r\,e^{i\theta}$ varies on the semicircles, so also does z. The radii of the semicircles are very small, but they are retained, since it is not permissible to integrate *through* a singularity, for the function is not analytic thereat, and Cauchy's theorem is inapplicable. It should be observed that any poles within C_1 are *not enclosed* by the contour shown in Fig. 16 (b).

2·41. Derivatives of $f(z)$ when $z=a$. We have already defined the first derivative of $f(z)$ to be

$$f'(z)=\lim_{|h|\to 0}\frac{f(z+h)-f(z)}{h}$$

(at (4) § 1·31).

Let C be a closed contour within and upon which $f(z)$ is analytic. Then from above, if a be any point inside the contour,

$$f'(a) = \lim_{|h| \to 0} \frac{f(a+h) - f(a)}{h}. \tag{1}$$

Now by (5) § 2·31

$$f(a) = \frac{1}{2\pi i} \int_C \frac{f(z)\,dz}{(z-a)} \quad \text{and} \quad f(a+h) = \frac{1}{2\pi i} \int_C \frac{f(z)\,dz}{(z-a-h)}.$$

Substituting these integrals in (1), we have

$$f'(a) = \lim_{|h| \to 0} \frac{1/h}{2\pi i} \int_C \left[\frac{1}{z-a-h} - \frac{1}{z-a} \right] f(z)\,dz$$

$$= \lim_{|h| \to 0} \frac{1}{2\pi i} \int_C \frac{f(z)\,dz}{(z-a-h)(z-a)} \tag{2}$$

$$= \frac{1}{2\pi i} \int_C \frac{f(z)\,dz}{(z-a)^2}. \tag{3}$$

In like manner, by taking

$$f''(a) = \lim_{|h| \to 0} \frac{f'(a+h) - f'(a)}{h},$$

we find that

$$f''(a) = \frac{2!}{2\pi i} \int_C \frac{f(z)\,dz}{(z-a)^3}, \tag{4}$$

and in general that

$$f^{(n)}(a) = \frac{n!}{2\pi i} \int_C \frac{f(z)\,dz}{(z-a)^{n+1}}, \tag{5}$$

n being a positive integer.

Since $f(z)$ is analytic at any point $z = a$ within C, by § 1·33 $f'(z)$ exists. Also we have shown that $f''(z)$ exists, so $f'(z)$ is analytic. Similarly $f''(z), \ldots, f^{(n)}(z)$ are analytic.

2·42. In deducing (3) § 2·41 from (2) § 2·41 we assumed tacitly that the limit of the latter integral when $|h| \to 0$ is identical with the integral of the limit of the integrand, this being (3). To corroborate this, write

$\dfrac{1}{(z-a-h)}$ in the form $\dfrac{1}{(z-a)} \left[1 + \dfrac{h}{(z-a-h)} \right]$ and (2) § 2·41 becomes

$$\frac{1}{2\pi i} \int_C \frac{f(z)\,dz}{(z-a)^2} + \lim_{|h| \to 0} \frac{h}{2\pi i} \int_C \frac{f(z)\,dz}{(z-a)^2(z-a-h)}.$$

Now the second integrand is finite at all points of the contour whose length is finite, so the integral exists. In the limit, therefore, when $|h| \to 0$, the second member vanishes and leaves the result at (3) § 2·41.

2·51. Taylor's theorem. In Fig. 17 (a), a is a point in the z-plane at which $f(z)$ is analytic, while b is the singularity of $f(z)$ nearest to a. With centre a and a radius just less than $|b-a|$, describe a circle C which excludes the singularity at b. Then $f(z)$ is analytic at all points upon and within this circle C. If $(a+h)$ is any point within C, then by (5) § 2·31 we have

$$f(a+h) = \frac{1}{2\pi i} \int_C \frac{f(z)\,dz}{(z-a-h)}. \tag{1}$$

Now
$$\frac{1}{(z-a-h)} = \frac{1}{(z-a)[1-h/(z-a)]}$$

$$= \frac{1}{z-a} + \frac{h}{(z-a)^2} + \frac{h^2}{(z-a)^3} + \cdots$$

$$+ \frac{h^n}{(z-a)^{n+1}} + \frac{h^{n+1}}{(z-a)^{n+1}(z-a-h)}, \tag{2}$$

provided $|h/(z-a)| < 1$; and this is so because z is on C while $(a+h)$ is within it.

Substituting from (2) into (1), we obtain

$$f(a+h) = \frac{1}{2\pi i} \int_C f(z)\,dz\left[\frac{1}{z-a} + \frac{h}{(z-a)^2} + \frac{h^2}{(z-a)^3} + \cdots\right.$$

$$\left. + \frac{h^n}{(z-a)^{n+1}} + \frac{h^{n+1}}{(z-a)^{n+1}(z-a-h)}\right]. \tag{3}$$

Using (5) § 2·31 and (5) § 2·41, (3) may be written

$$f(a+h) = f(a) + hf'(a) + \frac{h^2}{2!}f''(a) + \cdots$$

$$+ \frac{h^n}{n!}f^{(n)}(a) + \frac{h^{n+1}}{2\pi i}\int_C \frac{f(z)\,dz}{(z-a)^{n+1}(z-a-h)}. \tag{4}$$

We shall now show that the integral in (4) tends to zero as n tends to infinity. Since $f(z)/(z-a-h)$ is analytic on C, it follows that M, the maximum value of $|f(z)/(z-a-h)|$, is finite. Now

z is any point on C, so let $R = |z-a|$ be the distance of a from C. Then by Appendix 1

$$\left| \frac{h^{n+1}}{2\pi i} \int_C \frac{f(z)\,dz}{(z-a)^{n+1}(z-a-h)} \right|$$

$$\leqslant \frac{|h|^{n+1}}{2\pi} \frac{M \cdot 2\pi R}{R^{n+1}} = RM \left| \frac{h}{R} \right|^{n+1}. \quad (5)$$

Since $|h/R| < 1$, the right-hand side of (5) $\to 0$ as $n \to \infty$, so the integral in (4) does likewise.

Accordingly, (4) may be expressed in the form

$$f(a+h) = f(a) + hf'(a) + \frac{h^2}{2!} f''(a) + \ldots + \frac{h^n}{n!} f^{(n)}(a) + R_n, \quad (6)$$

where R_n, the remainder after n terms, tends to zero as $n \to \infty$. Series (6) is identical in form with Taylor's series for a real variable. In the theory of the complex variable it is preferable to use an alternative form, which is obtained by writing

$$z = a+h \quad \text{or} \quad h = z-a$$

in (6). Then

$$f(z) = f(a) + (z-a)f'(a) + \frac{(z-a)^2}{2!} f''(a) + \ldots. \quad (7)$$

The series (7) is the Taylor expansion of the function $f(z)$ in the region of the point a which must lie within the contour C, whose radius $R \simeq |b-a|$ is limited by the distance of the nearest singularity at $z = b$.

2·52. Liouville's theorem.

An integral function whose modulus is always less than some real positive number M is a constant.

An integral function is one which is analytic everywhere in the finite part of the z-plane, so Taylor's theorem may be applied. By (5) §2·41, $f^{(n)}(a)/n!$, the coefficient of $(z-a)^n$ in (7) § 2·51, is given by

$$\frac{1}{2\pi i} \int_C \frac{f(z)\,dz}{(z-a)^{n+1}} = a_n. \quad (1)$$

Let M_1 be the greatest value of $f(z)$ on a circle C centre a and radius $R = |z-a|$. Then by Appendix 1

$$|a_n| \leqslant \frac{1}{2\pi} \frac{M_1 2\pi R}{R^{n+1}} = \frac{M_1}{R^n}. \quad (2)$$

By hypothesis $M_1 < M$, and since R may be as large as we please, by letting $R \to \infty$, $a_n = 0$, $n \geqslant 1$. Thus (7) § 2·51 reduces to

$$f(z) = f(a), \quad \text{a constant.} \tag{3}$$

This theorem implies that every function, except a constant, has one or more singularities in the entire z-plane. For instance, the integral function $w = z^3$ has no singularity in the finite part of the plane, but by (2) §§ 1·4 and 1·521, it has a pole of order three at infinity.

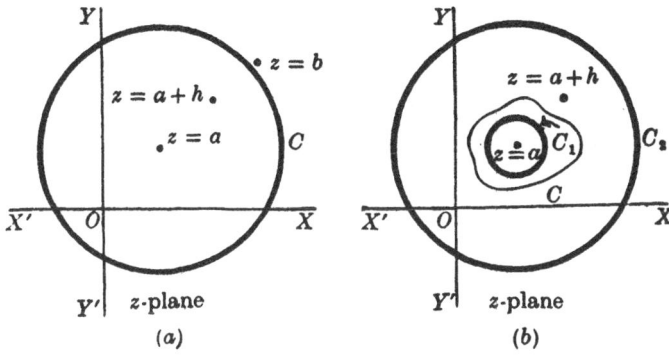

Fig. 17.

2·61. Laurent's theorem. Referring to Fig. 17 (b), C_1 and C_2 are two concentric circles described with a as centre. $(a+h)$ is any point in the ring-shaped space. $f(z)$ is analytic within the ring and on C_1, C_2. Then by (3) § 2·32

$$f(a+h) = \frac{1}{2\pi i} \int_{C_2} \frac{f(z)\, dz}{(z-a-h)} - \frac{1}{2\pi i} \int_{C_1} \frac{f(z)\, dz}{(z-a-h)}, \tag{1}$$

both contours being described positively. The first integral can be expressed in the form at (3) § 2·51, so we proceed to express the second in a suitable way. In this integral z is any point *on* C_1, and by reference to Fig. 17 (b) it is seen that since $(a+h)$ is outside C_1, $|z-a| < |h|$, or $\left|\dfrac{z-a}{h}\right| < 1$. Hence

$$-\frac{1}{(z-a-h)} = \frac{1}{h}\left(\frac{1}{1-(z-a)/h}\right)$$

$$= \frac{1}{h} + \frac{z-a}{h^2} + \frac{(z-a)^2}{h^3} + \ldots - \frac{(z-a)^{n+1}}{(z-a-h)\, h^{n+1}}. \tag{2}$$

Substituting from (3) § 2·51 and (2) into (1), the latter becomes

$$f(a+h) = \frac{1}{2\pi i}\int_{C_1} f(z)\,dz\left[\frac{1}{(z-a)} + \frac{h}{(z-a)^2} + \frac{h^2}{(z-a)^3} + \dots \right.$$
$$\left. + \frac{h^{n+1}}{(z-a)^{n+1}\,(z-a-h)}\right]$$
$$+ \frac{1}{2\pi i}\int_{C_1} f(z)\,dz\left[\frac{1}{h} + \frac{(z-a)}{h^2} + \frac{(z-a)^2}{h^3} + \dots\right]$$
$$- \frac{1}{2\pi i}\int_{C_1} \frac{f(z)\,(z-a)^{n+1}\,dz}{(z-a-h)\,h^{n+1}}. \tag{3}$$

We showed in § 2·51 that the last integral round C_2 tends to zero as $n \to \infty$. Taking the third integral in (3), the maximum value of

$$\left|\frac{f(z)}{(z-a-h)}\right|$$

is finite, since z is on C_1, and there are no singularities on the contour. The part $\left|\dfrac{z-a}{h}\right| < 1$, as shown previously, so $\left|\dfrac{z-a}{h}\right|^{n+1} \to 0$ as $n \to \infty$. Hence

$$\left|\int_{C_1} \frac{f(z)\,(z-a)^{n+1}\,dz}{(z-a-h)\,h^{n+1}}\right| \to 0, \quad n \to \infty,$$

and (3) may be written in the form

$$f(a+h) = c_0 + c_1 h + c_2 h^2 + \dots + \frac{d_1}{h} + \frac{d_2}{h^2} + \dots, \tag{4}$$

the coefficients $c_0, c_1, \dots d_1, d_2, \dots$ being obtained from the integrals

$$c_n = \frac{1}{2\pi i}\int_{C_1} \frac{f(z)\,dz}{(z-a)^{n+1}}, \tag{5}$$

$$d_n = \frac{1}{2\pi i}\int_{C_1} f(z)\,(z-a)^{n-1}\,dz. \tag{6}$$

The expansion (4) is known as Laurent's theorem. It can be put in a more convenient form by substituting $h = z - a$, which gives

$$f(z) = a_0 + a_1(z-a) + a_2(z-a)^2 + \dots + a_n(z-a)^n + \dots$$
$$+ \frac{b_1}{(z-a)} + \frac{b_2}{(z-a)^2} + \dots + \frac{b_n}{(z-a)^n} + \dots. \tag{7}$$

Since $h = (z=a)$, $z = (a+h)$, which means that z is *now* any point within the ring-shaped space between C_1 and C_2, where $f(z)$ is analytic. Formula (7) is the Laurent expansion of $f(z)$ at the point $z = (a+h)$ within the ring. The coefficients a_n, b_n are obtained from (5), (6), on replacing c_n, d_n, z by a_n, b_n, ζ, respectively. Here ζ is the variable *on* the contours, whereas z lies inside the ring. When $f(z)$ has a simple pole at $z = a$, there is only one term in inverse powers of $(z-a)$, namely, $b_1/(z-a)$. In general, if the pole is of the nth order, there are n such terms, the last being $b_n/(z-a)^n$; but any or all of $b_1, \ldots, b_{n-1}$ may be zero.

When $b_1 \ldots b_n$ are all zero, but $a_n \neq 0$, the function has a zero of order n at $z = a$, but there is no pole.

2·62. Application of Laurent's theorem.

Suppose a function $f(z)$ is analytic and single-valued in a certain region of the z-plane, excepting the point a, where it has a singularity. Since $f(z)$ is single-valued, the singularity will be either an essential one or merely a pole of finite order. Describe two concentric circles C_1, C_2 with a as centre (Fig. 17 (*b*)). Then $f(z)$ is analytic in the ring-shaped space between C_1, C_2, so it can be expanded at any point within this region in a Laurent series. Thus we obtain series (7) § 2·61. The sum of the terms in inverse powers of $(z-a)$ is called the principal part of the Laurent expansion near the singular point a. In their absence $f(z)$ would be analytic at a, since that part of (7) § 2·61 expressed in rising powers of $(z-a)$ is itself analytic.

Definition of pole. If m, the highest index of the inverse powers of $(z-a)$ in (7) § 2·61 is *finite*, $f(z)$ is said to have a pole of order m at $z = a$. Then

$$f(z) = \sum_{n=0}^{\infty} a_n(z-a)^n + \sum_{n=1}^{m} \frac{b_n}{(z-a)^n}. \tag{1}$$

The coefficient b_1 is the *residue* (see § 2·63) at the pole. The combined series in (1) is analytic, and converges for all values of z within C, except $z = a$.

Definition of essential singularity. If the series in *inverse* powers of $(z-a)$ in (7) § 2·61 does *not* terminate, $f(z)$ is said to have an essential singularity at $z = a$. Then

$$f(z) = \sum_{n=0}^{\infty} a_n(z-a)^n + \sum_{n=1}^{\infty} \frac{b_n}{(z-a)^n}. \tag{2}$$

The coefficient b_1 is the *residue* at the singularity (see § 2·63). The remarks with reference to the convergence of (1) apply here also.

2·63. Application to integration. If the Laurent expansion of a function $f(z)$ single-valued near a pole $z = a$ is known, the value of the integral taken round a contour enclosing this singularity alone can be found by aid of (7) § 2·61. For, moving the origin to the point a by writing $\zeta = z - a$ and $\phi(\zeta) = \sum_{n=0}^{\infty} a_n \zeta^n$, we have for a pole of order m,

$$f(\zeta + a) = \phi(\zeta) + b_1/\zeta + b_2/\zeta^2 + \ldots + b_m/\zeta^m, \tag{1}$$

where ζ is any point on a closed contour C which encircles C_1 and lies in the ring-shaped space, as shown in Fig. 17 (*b*). $\phi(\zeta)$ is analytic in C, so that $\int_C \phi(\zeta)\, d\zeta = 0$. Thus

$$\int_C f(\zeta + a)\, d\zeta = \int_C f(z)\, dz = \int_C \left[\frac{b_1}{\zeta} + \frac{b_2}{\zeta^2} + \ldots + \frac{b_m}{\zeta^m} \right] d\zeta \tag{2}$$

$$= 2\pi i b_1, \tag{3}$$

by § 2·122, since all integrals except the first vanish by § 2·13. Being the only contribution from (1), b_1 is called the *residue* at the pole $z = a$. From (3)

$$\int_C f(\zeta + a)\, d\zeta = \int_C f(z)\, dz$$

is $2\pi i$ times the residue at the pole, or $2\pi i$ times the coefficient of $1/\zeta = 1/(z-a)$. Thus if the Laurent expansion of a function $f(z)$ is known near a pole $z = a$, the residue is the coefficient of $1/(z-a)$, and the integral round any contour enclosing only this singularity is obtained on multiplying by $2\pi i$. If the singularity is an essential one, then by (7) § 2·61 the result (3) above is valid.

2·64. Continuation of § 1·64. We found that the function

$$w(z) = 1/[1 + \sqrt{(az)}] \quad (a \text{ real} > 0) \tag{1}$$

has a singularity at $z = 1/a$ on the second branch. We shall now show that it is a simple pole. At the singularity, $\sqrt{(ar)} = 1$, and by § 1·64 the phase angle for z real > 0 on the second branch is $\theta = 2\pi$, so the pole occurs at $z = e^{2\pi i}/a$. Thus

$$r = 1/a \quad \text{and} \quad z = r e^{i\theta} = e^{2\pi i}/a. \tag{2}$$

Writing $\zeta = z - e^{2\pi i}/a$ in (1), thereby moving the origin to the pole, then with $|\zeta a| < 1$ and

$$\sqrt{(e^{2\pi i} + a\zeta)} = e^{\pi i}\sqrt{(1+a\zeta)},$$

we get

$$w = 1/[1 - \sqrt{(1+a\zeta)}] \tag{3}$$

$$= -1/(\tfrac{1}{2}a\zeta - \tfrac{1}{8}a^2\zeta^2 + \tfrac{1}{16}a^3\zeta^3 - \ldots) \tag{4}$$

$$= -(2/a\zeta)/(1 - \tfrac{1}{4}a\zeta + \tfrac{1}{8}a^2\zeta^2 - \ldots). \tag{5}$$

Now $\zeta \to 0$ as $z \to 1/a$, so (5) may then be written

$$w \simeq -(2/a\zeta)(1 + \tfrac{1}{4}a\zeta - \tfrac{1}{16}a^2\zeta^2 + \ldots) \tag{6}$$

$$= -(2/a\zeta) - (\tfrac{1}{2} - \tfrac{1}{4}a\zeta + \ldots), \tag{7}$$

which is the Laurent expansion near the singularity. It is, therefore, a simple pole with residue $-2/a$. Another procedure is given at (5)–(7) § 3·231.

The function $w = 1/[1 + \sqrt{(az)}]$ occurs below (3) in § 6·21. On the second branch, the pole at $z = 1/a$ gives the factor $e^{t/a}$, so the p.d. would increase indefinitely with increase in t, which is impossible. Thus the first branch must be used in obtaining the solution of the cable problem.

If $w = 1/[\sqrt{(az)} - 1]$, the reader should confirm that there is a branch point at the origin, and a simple pole at $z = 1/a$ on the *first* branch, but not on the second.

2·65. Example. Find the Laurent expansion of

$$f(z) = 1/(z-a)(z-b)^n \quad (n \geqslant 1,\ a \neq b \neq 0)$$

near each pole.

Move the origin to $z = a$ by writing $\zeta = (z-a)$, then

$$f(z) = 1/\zeta(\zeta+c)^n = (1/c^n\zeta)/(1+\zeta/c)^n, \tag{1}$$

where $c = (a-b)$. Thus if $|\zeta/c| < 1$,

$$f(z) = \frac{1}{c^n\zeta}\left[1 - \frac{n\zeta}{c} + \frac{n(n+1)}{2!}\frac{\zeta^2}{c^2} - \ldots\right] \tag{2}$$

$$= \left[-\frac{n}{c^{n+1}} + \frac{n(n+1)\zeta}{2!\,c^{n+2}} - \ldots\right] + \frac{1}{c^n\zeta}, \tag{3}$$

which is the Laurent expansion near the simple pole at $z = a$. The residue is $1/c^n = 1/(a-b)^n$.

For the pole at $z = b$, write $\zeta = (z - b)$ and proceed as shown above. Then we get

$$f(z) = -\left(\frac{1}{c^{n+1}} + \frac{\zeta}{c^{n+2}} + \frac{\zeta^2}{c^{n+3}} + \ldots\right) - \left(\frac{1}{c^n \zeta} + \frac{1}{c^{n-1}\zeta^2} + \ldots + \frac{1}{c\zeta^n}\right). \quad (4)$$

The second series is the principal part of the expansion near the nth order pole at $z = b$, at which the residue is

$$-1/c^n = -1/(a-b)^n.$$

Hence the sum of the residues at the two poles is zero. More generally, by aid of §4·63 it may be shown that if $w(z) = f(z)/g(z)$ is a rational function, the order of whose denominator exceeds that of the numerator by at least two, the sum of the residues at all the poles* is zero.

2·66. Example. Show that $w = 1/\log z$ has a simple pole at $z = 1$. Write $z = (\zeta + 1)$, and near the pole where $|\zeta| \ll 1$ we get

$$w = 1/\log(\zeta + 1) = 1/(\zeta - \tfrac{1}{2}\zeta^2 + \tfrac{1}{3}\zeta^3 - \tfrac{1}{4}\zeta^4 + \ldots) \quad (1)$$

$$= 1/\zeta(1 - \tfrac{1}{2}\zeta + \tfrac{1}{3}\zeta^2 - \tfrac{1}{4}\zeta^3 + \ldots) = (1/\zeta)(1 + \tfrac{1}{2}\zeta - \tfrac{1}{12}\zeta^2 + \tfrac{1}{24}\zeta^3 - \ldots) \quad (2)$$

$$= (\tfrac{1}{2} - \tfrac{1}{12}\zeta + \tfrac{1}{24}\zeta^2 - \ldots) + \frac{1}{\zeta}, \quad (3)$$

which is the Laurent expansion near the pole $\zeta = 0$, i.e. $z = 1$. The residue there is the coefficient of $1/\zeta$, namely, unity. Hence if C is a contour which surrounds the point $z = 1$, but neither encloses nor passes through the branch point of w at the origin, then

$$\frac{1}{2\pi i} \int_C \frac{dz}{\log z} = 1. \quad (4)$$

2·67. Plurality of simple poles. In this case the Laurent expansion near each pole will have only one term of the type $b_1/(z-a)$. For instance, $w = 1/\sinh z$ has simple poles when $z = in\pi$, n integral. By (3) §1·521, $(1 - z/in\pi)$ is a non-repeated factor of $\sinh z$. To obtain the Laurent expansion near the nth pole, write $z = \zeta + in\pi$, and we get

$$\frac{1}{\sinh z} = \frac{1}{\sinh(\zeta + in\pi)} = \frac{(-1)^n}{\sinh \zeta} \quad (1)$$

$$= \frac{(-1)^n}{\zeta\left(1 + \dfrac{\zeta^2}{3!} + \dfrac{\zeta^4}{5!} + \ldots\right)}. \quad (2)$$

* They are finite in number and occur in the finite part of the z-plane.

Since $\zeta \to 0$ as $z \to in\pi$, (2) may be written

$$\frac{(-1)^n}{\sinh \zeta} = (-1)^n \left(\frac{1}{\zeta} - \frac{\zeta}{6} + \frac{7\zeta^3}{360} - \cdots \right), \qquad (3)$$

which is the required expansion. The residue at the pole is $(-1)^n$.

2·68. Discrimination of singularity. The function

$$w = (\log z)/(z-a)$$

has a branch point at the origin due to $\log z$, and a pole at $z = a$, $a \neq 0$. When $a = 0$, there is a branch point at the origin, this being the only singularity. For by Laurent's theorem a function is analytic in the neighbourhood of a pole, whereas by § 1·65, $\log z$ is not analytic in the neighbourhood of the origin, and the function has no Laurent expansion. In like manner it may be shown that $\sqrt{z}\, e^{1/z}$ has a branch point at the origin but no essential singularity there. The function $w = 1/\{(z-1)\log z\}$ has a branch point at $z = 0$, and a double pole at $z = 1$. The existence of the former is evident from § 1·65. To elucidate the double pole, write $(z-1) = \zeta$, and we get $w = 1/\{\zeta \log(1+\zeta)\}$. When $\zeta \to 0$, $1/\log(1+\zeta) \to 1/\zeta$, which is the principal part of the Laurent expansion corresponding to $1/\log(1+\zeta)$ near the pole $\zeta = 0$. Hence near $\zeta = 0$ we may write $w \simeq 1/\zeta^2$, so a double pole occurs there, i.e. at $z = 1$.

By aid of § 1·542 it may be shown that $e^{\sqrt{z}}/z$ has a branch point at the origin. Expanding, we obtain

$$\frac{e^{\sqrt{z}}}{z} = \frac{1}{z} + \frac{1}{\sqrt{z}} + \frac{1}{2!} + \frac{\sqrt{z}}{3!} + \cdots, \qquad (1)$$

which is *not* a Laurent expansion owing to presence of terms having fractional indices. The function $e^{\sqrt{z}}/(z-a)(z-b)$ has a branch point at the origin, and simple poles at $z = a, b$, if $a, b \neq 0$. Moreover, the type of singularity at a point is determined by the behaviour of the function in its neighbourhood. To summarize:

(i) If a function has a branch point at $z = a$, it has no other type of singularity there;

(ii) If a function has an essential singularity at $z = a$, it has no pole there.

2·711. Analytical continuation. Let $f_1(z)$ be a function of z which is analytic inside C_1 (Fig. 18), but neither upon it nor outside it, and let $f_2(z)$ be another function of z which is analytic within C_2.* Then if $f_1(z) = f_2(z)$ for all points within C_1, $f_2(z)$ is said to be the analytical continuation† of $f_1(z)$ outside C_1. As an example consider

$$f_1(z) = \frac{1}{a} \left[1 + \frac{z}{a} + \frac{z^2}{a^2} + \cdots + \frac{z^n}{a^n} + \cdots \right], \qquad (1)$$

* Since C_1 lies wholly within C_2, $f_2(z)$ is analytic within C_1.
† See [237, 250] for additional information.

the series being absolutely convergent if $|z/a| < 1$, i.e. $|z| < |a|$. Then $f_1(z)$ is analytic everywhere within a circle of radius $r < |a|$, whose centre is the origin. Within this circle (1) can also be represented by the function

$$f_2(z) = 1/(a-z), \qquad (2)$$

for (1) is the expansion of (2) when $|z/a| < 1$. Now both $f_1(z)$ and $f_2(z)$ are analytic and equal within the circle of radius $|a|$, while $f_2(z)$ is analytic at all points in the z-plane outside, and at all points on the circle except $z = a$. It follows, therefore, that $f_2(z) = 1/(a-z)$ is the analytical continuation of the series $f_1(z)$ outside the circle of radius $|a|$.

2·712. Circle of convergence. Let

$$f_1(z) = a_0 + a_1 z + a_2 z^2 + \ldots + a_n z^n + \ldots \qquad (1)$$

be a power series representing a function, the coefficients a_m being unrestricted numbers. Writing $A_m = |a_m|$, $R = |z|$, from (1) we obtain

$$S = A_0 + A_1 R + A_2 R^2 + \ldots + A_n R^n + \ldots. \qquad (2)$$

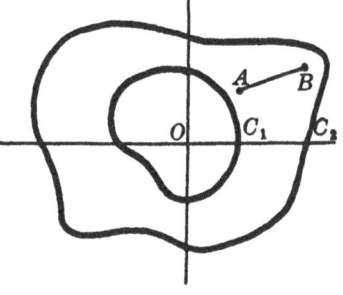

Then it may be shown that a positive number N exists such that (2) converges whenever $R < N$, but diverges whenever $R > N$. Since (1) converges absolutely when $|z| < R$, it does so for all z within a circle of radius R about the origin, and diverges for all z outside the circle. For this reason it is called the *circle of convergence* of the series (1). *On* the circle, (1) may converge or diverge, according to the series in question (proof in [226, 250]).

Fig. 18.

If (1) is differentiated term by term any number of times, each series so obtained is absolutely convergent, and has the same circle of convergence as the original. Hence every power series and its derivatives represent analytic functions at all points within the circle. When the latter is finite, there may be another function $f_2(z)$ which is the analytical continuation of $f_1(z)$ outside its circle of convergence.

If the circle includes the whole of the finite part of the z-plane, $f_1(z)$ is analytic everywhere except infinity. Then $f_1(z)$ is defined to be an *integral* function, e.g. (1) § 1·4, e^z, sinh z, cosh z, which have no singularities in the finite part of the z-plane.

2·72. Theorem. The following theorem in analytical continuation is given without proof. Referring to Fig. 18, if $f_1(z) = f_2(z)$ at all points of any line AB whose length > 0, and which lies within a contour C_2 throughout which *both* $f_1(z)$ and $f_2(z)$ are analytic, then the two functions are equal for all points within C_2. In practice it happens frequently that the line of equality is the x-axis or a portion of it (proof in [237, 250]).

2·73. Example. Show that $\sec^2 z = 1 + \tan^2 z$, for all finite values of z except singular points.

In this instance $f_1(z) = \sec^2 z$, while $f_2(z) = 1 + \tan^2 z$. Now $f_1(x) = f_2(x)$ at all points of the real axis, except the singular points $x = \frac{1}{2}(2n+1)\pi$, n being an integer, where they are both infinite and, therefore, undefined. Also $f_1(z)$ and $f_2(z)$ are analytic for all values of z, except the singular points aforesaid, so by the principle of analytical continuation

$$\sec^2 z = 1 + \tan^2 z$$

for all points at which both sides are analytic.

2·74. Example. Establish the validity of the binomial expansion for $(1+z)^\nu$ when $|z| < 1$, ν real and non-integral.

On the real axis when $x < 1$, the series

$$f_1(x) = 1 + \nu x + \frac{\nu(\nu-1)}{2!}x^2 + \frac{\nu(\nu-1)(\nu-2)}{3!}x^3 + \ldots \tag{1}$$

is convergent and equal to

$$(1+x)^\nu = f_2(x). \tag{2}$$

Now the function

$$f_1(z) = 1 + \nu z + \frac{\nu(\nu-1)}{2!}z^2 + \ldots \tag{3}$$

is analytic at all points within the circle of radius $|z| = 1$, as also is the function $f_2(z) = (1+z)^\nu$. Hence

$$(1+z)^\nu = 1 + \nu z + \frac{\nu(\nu-1)}{2!}z^2 + \ldots \tag{4}$$

for all values of z within the unit circle. But $f_2(z)$ is analytic outside the unit circle, so that $(1+z)^\nu$ is the analytical continuation of the infinite series $f_1(z)$ outside the circle of its convergence.

III

CALCULUS OF RESIDUES

3·11. In §2·63 we showed that the integral round any simple pole of a function $f(z)$ at $z = a$ is $2\pi i$ times the residue or coefficient of $1/(z-a)$ in the Laurent expansion near the pole. Since functions of the complex variable used in technical applications usually have a plurality of poles, we require a formula for obtaining $\int f(z)\,dz$ round a contour which includes any specified number of or all the poles of the function. Referring to Fig. 19 (a), let $z = a, b, c$ represent those poles of $f(z)$ which are enclosed by the contour C_2. There may, of course, be other singularities outside the contour, but they are excluded from the integration. The first step is to draw a suitable new contour enclosing the three poles, this contour being equivalent to C_2. The contour is illustrated in Fig. 19 (b) at C_1. It consists of a network of small circular arcs joined by parallel lines in close proximity. Then the space between C_1 and C_2 contains no singularity, so $f(z)$ is analytic within the region and (1) § 2·24 may be applied. Thus

$$\int_{C_2} f(z)\,dz - \int_{C_1} f(z)\,dz = 0, \tag{1}$$

both contours being described positively. In the limit when the respective pairs of parallel lines are coincident, the circles round the poles being small but finite, the integral along C_1 is equal to the sum of the integrals round the three circles, since the contributions along the lines cancel owing to their being traversed in opposite directions, the function being single-valued. Hence (1) may be written

$$\int_{C_2} f(z)\,dz = \int_{\delta_1} f(z)\,dz + \int_{\delta_2} f(z)\,dz + \int_{\delta_3} f(z)\,dz. \tag{2}$$

In general, when there are n poles, we have

$$\int_{C_2} f(z)\,dz = \sum_{m=1}^{n} \int_{\delta_m} f(z)\,dz = 2\pi i \Sigma \text{ Residues at poles}, \tag{3}$$

by § 2·63. This relationship is known as the theorem of residues.

3·12. Modification of the contour may be considered with reference to Fig. 19 (c), where C_1 and C_2 are drawn so that the combined contour contains none of the poles of $f(z)$. By Cauchy's theorem the integral round $C_1 + C_2$ is zero. In the limit when the parallel lines coalesce, the integral round the three circles plus that round C_2 is zero. But C_2 is described negatively, so that when *both* paths are described positively, the integral round C_2 is equal to the sum of the integrals round the poles, and (2) §3·11 is reproduced.

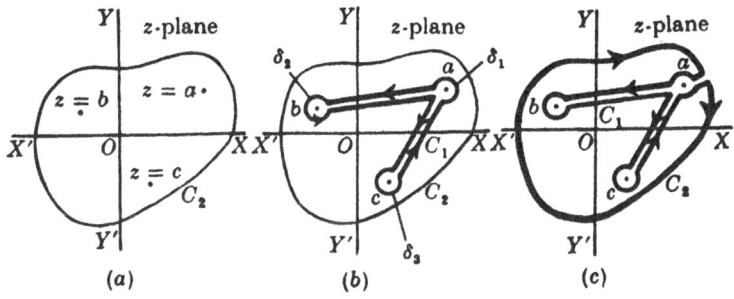

Fig. 19. (a) Contour enclosing poles at $z = a, b, c$. (b), (c) Contours C_1 round poles at $z = a, b, c$ are equivalent to the contour in (a). The separation between the parallel lines $\to 0$.

3·21. Evaluation of residues. The residues of a function of z at its poles can be evaluated in various ways, of which the simplest is treated first. If the Laurent expansion near each pole $z = a, b, \ldots$ is obtained easily, the residue is the coefficient of $1/(z-a)$, $1/(z-b)$, ..., as shown in § 2·63. Evaluation may be effected, without obtaining the expansion, by using the following procedure. Let the function be expressed in the form

$$w = f(z)/g(z), \tag{1}$$

$f(z)$ having no poles, though it may have other singularities, while $f(z)$ and $g(z)$ are prime to each other. Suppose that $g(z)$ can be factorized readily, then for a simple pole at $z = a$, (1) may be written

$$w = f(z)/(z-a)\,\xi(z), \tag{2}$$

where $g(z) = (z-a)\,\xi(z)$. Although we do not know the Laurent expansion of (1) near the pole, it exists nevertheless, provided w is analytic in a ring-shaped space enclosing a. Hence by (7) § 2·61

$$f(z)/g(z) = f(z)/(z-a)\,\xi(z)$$
$$= a_0 + a_1(z-a) + a_2(z-a)^2 + \ldots + b_1/(z-a), \qquad (3)$$

the expansion in inverse powers of $(z-a)$ terminating here, since the singularity is a simple pole. Multiplying both sides of (3) by $(z-a)$, and then substituting $z = a$, we get

$$b_1 = [(z-a)f(z)/g(z)]_{z=a} = f(a)/\xi(a), \qquad (4)$$

this being the residue of w at the pole $z = a$. Thus from (3) § 3·11 and (4) above

$$\int_C f(z)\frac{dz}{g(z)} = 2\pi i \frac{f(a)}{\xi(a)}, \qquad (5)$$

where C is a contour which encloses the singularity at a alone. If $g(z)$ has a number of factors of the form $(z-a)$, some of which may be repeated, thereby introducing multiple poles, the residue at each *simple* pole is found as shown above. Thus if

$$w = f(z)/(z-a)\,(z-b)\,(z-c)^2\,(z-d)^3, \qquad (6)$$

the residue at the simple pole $z = b$ is

$$\text{Res.} = f(b)/(b-a)\,(b-c)^2\,(b-d)^3. \qquad (7)$$

The procedure is to multiply by the factor in the denominator causing the singularity, namely $(z-b)$, thereby removing the pole—in effect—and put $z = b$ in the remainder. The residue at the double pole $z = c$ and at the triple pole $z = d$ cannot be evaluated by the foregoing method. The necessary technique is given in § 3·241.

3·22. Example. Find the residues of $f(z) = e^z/(z^2 + a^2)$ at its poles.

The function may be written

$$f(z) = e^z/(z-ia)\,(z+ia), \qquad (1)$$

so there are two simple poles on the imaginary axis at the points $z = +ia$ and $-ia$. The residue at ia is found on multiplying (1) by $(z-ia)$, and putting $z = ia$ into the remainder, namely,

$$e^z/(z+ia),$$

which gives $$\text{Res.}_1 = e^{ia}/2ia. \tag{2}$$

For the pole at $z = -ia$, change i to $-i$ in (2), and we get

$$\text{Res.}_2 = -e^{-ia}/2ia. \tag{3}$$

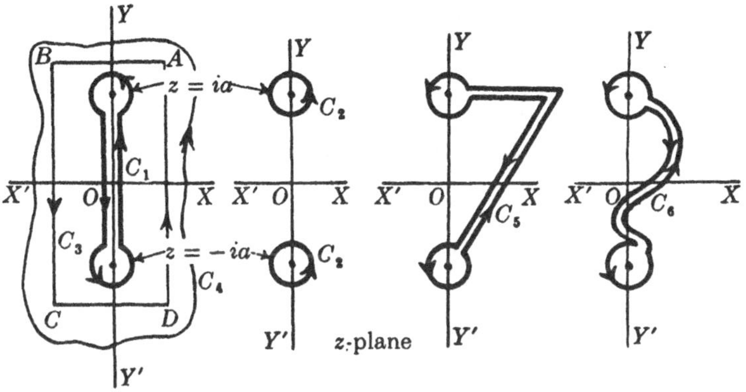

Fig. 20. Equivalent contours round two poles.

From (2) and (3) the sum of the residues

$$\Sigma \text{Res.} = \frac{1}{a}\left[\frac{e^{ia}-e^{-ia}}{2i}\right] = \frac{\sin a}{a}. \tag{4}$$

Hence by (3) § 3·11 $$\int_C \frac{e^z\,dz}{z^2+a^2} = 2\pi i\,\frac{\sin a}{a}, \tag{5}$$

or $$\frac{1}{2\pi i}\int_C \frac{e^z\,dz}{z^2+a^2} = \frac{\sin a}{a}, \tag{6}$$

where C is *any* closed contour which encircles both poles. This aspect is illustrated in Fig. 20, where the six contours $C_1, ..., C_6$ are all equivalent. In C_1 the integrals along the parallel lines (assumed coincident in the limit) neutralize each other, since they are equal but opposite, $f(z)$ being single-valued. Thus the effective part of the contour consists of two circles round the poles as shown by C_2, C_2. For C_3 the integrals along the parallel lines do not neutralize one another, since they are a finite distance apart, i.e. the values of z at A and B are different, as also are those at C and D.

3·231. Residues at simple poles of $f(z)/g(z)$, when $g(z)$ cannot be factorized readily. Let $w = f(z)/g(z)$ be a function having a limited number of simple poles* arising solely from the zeros of $g(z)$, $f(z)$ and $g(z)$ being prime to each other. Although $g(z)$ may not be factorizable in the usual way, w may be expressed near each pole in a Laurent series. By (4) § 3·21 the residue at the simple pole $z = a$, is

$$\text{Res.} = [(z-a)f(z)/g(z)]_{z=a}. \tag{1}$$

Since $z = a$ is a zero of $g(z)$, $g(a) = 0$, so (1) takes the indeterminate form $0/0$. Now by (3) § 1·31

$$g'(a) = \lim_{z \to a} \left[\frac{g(z) - g(a)}{z - a} \right] = \lim_{z \to a} \frac{g(z)}{(z-a)}, \tag{2}$$

since $g(a) = 0$. Inverting (2) and substituting into (1), yields

$$\text{Res.} = f(a)/g'(a). \tag{3}$$

Alternatively, (3) may be obtained by differentiating the numerator and denominator of (1) independently with respect to z, and putting $z = a$.

In general, if the *simple* poles of $f(z)/g(z)$ due to $g(z)$ are $a_1, a_2, ..., a_n$, the sum of the residues thereat is

$$\Sigma \text{Res.} = \sum_{m=1}^{n} \left[\frac{f(z)}{g'(z)} \right]_{z=a_m} = \sum_{m=1}^{n} \frac{f(a_m)}{g'(a_m)}. \tag{4}$$

Example. We apply (4) to (1) § 2·64 which has a branch point at the origin, as well as a simple pole. Then

$$\text{Res.} = 1 \bigg/ \frac{d}{dz} [\sqrt{(az)} + 1]_{z=e^{2\pi i/a}}. \tag{5}$$

$$= [2\sqrt{(z/a)}]_{z=e^{2\pi i/a}} \tag{6}$$

$$= 2e^{\pi i}/a = -2/a, \tag{7}$$

---

* There may be other singularities, but they do not affect the present discussion.

as below (7) §2·64. In (6) $e^{2\pi i}/a$ cannot be written $1/a$, since z must be treated as a complex number in virtue of the index $\frac{1}{2}$. Otherwise we should have $\sqrt{(z/a)} = \sqrt{(1/a^2)} = \pm 1/a$, which is ambiguous.

3·232. Sometimes the function takes the form

$$w = f(z)/zg(z),$$

where the numerator and denominator are prime to each other, while $g(z)$ has no zero at $z = 0$ and cannot be factorized readily. In addition to the poles arising from $1/g(z)$, there is a simple pole at the origin due to the factor $1/z$. By (4) §3·21 the residue thereat is

$$[f(z)/g(z)]_{z=0} = f(0)/g(0). \tag{1}$$

If $z = a_m$ be the general pole of $1/g(z)$, then provided $a_m \neq 0$, the residue is, by (4) §3·21,

$$\text{Res.} = [(z - a_m)f(z)/zg(z)]_{z=a_m}, \tag{2}$$

which assumes the indeterminate form $0/0$ when $z = a_m$. By §3·231

$$\text{Res.} = f(a_m)/a_m g'(a_m). \tag{3}$$

Hence from (1) and (3) the sum of the residues at all the simple poles is

$$\Sigma \text{Res.} = \left[\frac{f(z)}{g(z)}\right]_{z=0} + \sum_{m=1}^{n} \left[f(z) \Big/ z\frac{d}{dz}g(z)\right]_{z=a_m} \tag{4}$$

$$= \frac{f(0)}{g(0)} + \sum_{m=1}^{n} \frac{f(a_m)}{a_m g'(a_m)}. \tag{5}$$

If C is a finite contour enclosing all the poles of $f(z)/zg(z)$, then by (3) §3·11

$$\frac{1}{2\pi i} \int_C \frac{f(z)\,dz}{zg(z)} = \frac{f(0)}{g(0)} + \sum_{m=1}^{n} \frac{f(a_m)}{a_m g'(a_m)}, \tag{6}$$

it being understood that $1/g(z)$ has simple poles only. The result (6) is sometimes called the partial fraction rule, for it is equivalent to expanding in partial fractions and evaluating the residue at each pole by inspection, i.e. taking the coefficient of $1/(z - a_m)$.

If $z = 0$ is a zero of $g(z)$, w has a double pole at the origin, and the preceding formulae are invalid. When

$$w = f(z)/z^n g(z),$$

$n \geqslant 1$, and $g(z)$ has non-repeated zeros, the sum of *the residues arising from the poles of* $1/g(z)$ *alone*, $z = 0$ not being a pole thereof, is from (2) above

$$\Sigma \operatorname{Res.} = \sum_{m=1}^{s} \left[\frac{f(z)}{z^n g'(z)} \right]_{z=a_m} = \sum_{m=1}^{s} \frac{f(a_m)}{a_m^n g'(a_m)}. \tag{7}$$

The evaluation of the residue at a multiple pole is treated in § 3·241.

If w takes the form $w = f(z)/h(z) g(z)$, and the *simple* poles arising from $h(z)$ are not common to $g(z)$, the sum of the residues at all these poles can easily be shown to be

$$\Sigma \operatorname{Res.} = \sum_{m=1}^{n} \frac{f(a_m)}{h(a_m) g'(a_m)} + \sum_{r=1}^{s} \frac{f(b_r)}{h'(b_r) g(b_r)}, \tag{8}$$

where a_m and b_r are the simple poles due to $g(z)$ and $h(z)$ respectively.

3·233. Example. Find the sum of the residues of $e^z/\sin mz$ at the first $(s+1)$ poles on the negative real axis.

The poles, all of which are simple, occur when

$$\sin mz = 0, \quad \text{or} \quad z = -n\pi/m,$$

n being a positive integer. Thus from (4) § 3·231, we have

$$\Sigma \operatorname{Res.} = \sum_{n=0}^{s} \left[\frac{e^z}{m \cos mz} \right]_{z=-n\pi/m}$$

$$= \frac{1}{m} \sum_{n=0}^{s} (-1)^n e^{-n\pi/m}, \tag{1}$$

since $\cos mz = (-1)^n$ when $z = -n\pi/m$. Thus

$$\Sigma \operatorname{Res.} = \frac{[1 - e^{-\pi/m} + e^{-2\pi/m} - \dots + (-1)^s e^{-s\pi/m}]}{m}$$

$$= \frac{1 + (-1)^s e^{-(s+1)\pi/m}}{m(1 + e^{-\pi/m})}. \tag{2}$$

3·234. Example. Find the sum of the residues of

$$e^z/z \cosh mz$$

at the origin, and at the first s poles on each side of it.

The poles arising from the zeros of $\cosh mz$ occur at the points $z = -i(n+\frac{1}{2})\pi/m$, n integral, and are of unit order. Since $1/\cosh mz$ has no pole at the origin, by (5) § 3·232 we have

$$\Sigma \text{Res.} = 1 + \sum_{n=-s}^{s-1} \left[\frac{e^z}{mz \sinh mz} \right]_{z=-(n+\frac{1}{2})\,\pi i/m}$$

$$= 1 + \frac{2}{\pi} \sum_{n=-s}^{s-1} \left[\frac{e^{-(n+\frac{1}{2})\pi i/m}}{(2n+1)\sin(n+\frac{1}{2})\pi} \right]$$

$$= 1 + \frac{2}{\pi} \sum_{n=-s}^{s-1} (-1)^n \left[\frac{e^{-(n+\frac{1}{2})\pi i/m}}{(2n+1)} \right]$$

$$= 1 + \frac{4}{\pi} \left(\cos\alpha - \tfrac{1}{3}\cos 3\alpha + \dots + \frac{(-1)^{s-1}}{2s-1}\cos(2s-1)\alpha \right),$$

$$\tag{1}$$

where $\alpha = \pi/2m$.

3·235. Example. Find the residue of

$$(1°)\ w = e^z/z \sin mz; \quad (2°)\ w = z\,e^z/\sin mz$$

at the origin.

(1°) Both terms of the denominator of (1°) vanish when $z = 0$, and there is a double pole. Expanding both numerator and denominator, we get

$$\left(1 + z + \frac{z^2}{2!} + \dots \right) \Big/ mz^2 \left(1 - \frac{m^2 z^2}{3!} + \frac{m^4 z^4}{5!} - \dots \right). \tag{1}$$

Applying the binomial theorem to the denominator of (1) when z is very small, we have

$$\left(1 + z + \frac{z^2}{2!} + \dots \right) \left(1 + \frac{m^2 z^2}{3!} - \dots \right) \Big/ mz^2. \tag{2}$$

Then by § 2·63 the residue at the pole is the coefficient of $1/z$, namely, $1/m$. It follows from (3) § 3·11 that if C be a contour enclosing just the double pole at the origin, then

$$\frac{1}{2\pi i} \int_C \frac{e^z\,dz}{z \sin mz} = \frac{1}{m}. \tag{3}$$

The poles nearest to O occur at $z = \pm\pi/m$, so the radius of a circular contour C about the origin would have to be less than π/m. By drawing a closed contour C_1 which crosses the real axis between two poles, we have by (3) §3·11

$$\frac{1}{2\pi i}\int_{C_1}\frac{e^z\,dz}{z\sin mz} = \Sigma \text{ residues at poles within } C_1. \qquad (4)$$

(2°) The function $ze^z/\sin mz$ has no pole at the origin, since near $z = 0$, $\sin mz \simeq mz$, giving $ze^z/\sin mz \simeq 1/m$, which is analytic at $z = 0$. Thus if C be a circular contour about O as centre, whose radius $< \pi/m$ (the distance of the poles nearest to $z = 0$), then it follows that

$$\frac{1}{2\pi i}\int_C \frac{z\,e^z\,dz}{\sin mz} = 0. \qquad (5)$$

3·236. Example. Show that the sum of the residues at the poles of $w = 1/\log(z^2 - 1)$ is zero.

Log $(z^2 - 1)$ vanishes when $z^2 - 1 = 1$, so $z = \pm\sqrt{2}$. It may be shown that w has a Laurent expansion of type (7) § 2·61 with $b_2 = b_3 = \ldots b_n = 0$, near each of these values of z, so there are two simple poles. By § 3·231

$$\text{Res.} = 1 \bigg/ \frac{d}{dz}[\log(z^2 - 1)]_{z = \pm\sqrt{2}} \qquad (1)$$

$$= \left[\frac{z^2 - 1}{2z}\right]_{z = \pm\sqrt{2}} = \pm 1/2^{\frac{3}{2}}, \qquad (2)$$

so the sum of the residues is zero. We may write

$$w = 1/[\log(z - 1) + \log(z + 1)], \qquad (3)$$

which reveals the presence of branch points at $z = \pm 1$. This may be confirmed by moving the origin to each point in turn, and showing that the function is multi-valued. The reader may check that $w = e^{1/z}/\log z$ has a branch point at the origin, and a simple pole at $z = 1$, with residue e. There is no essential singularity at the origin (see § 2·68).

3·237. Example. Find the sum of the residues at the poles of $e^{zt}/z\sinh z$. The denominator has a double zero at $z = 0$, and simple zeros at $z = \pm in\pi$, n any positive integer $\neq 0$. By (3) § 1·521, $\sinh z$ has an infinite number of zeros. To effect the summation we obtain a general expression for the residue at a pole, and then let $n \to \infty$.*

When z is small enough, by aid of the binomial theorem, we obtain

$$\frac{e^{zt}}{z\sinh z} = \left(\frac{t}{z} + \frac{1}{z^2}\right) + \frac{1}{2!}(t^2 - \tfrac{1}{3}) + \frac{z}{3!}(t^3 - t) + \ldots, \qquad (1)$$

* By § 1·523 there is a limit point of poles at infinity.

which is the Laurent expansion near the *double* pole $z = 0.$* The residue, being the coefficient of $1/z$, is t. For the simple poles at $z = \pm in\pi$, by (4) § 3·231

$$\text{Res.} = \left[e^{zt} \middle/ z \frac{d}{dz} \sinh z \right]_{z=\pm in\pi} \tag{2}$$

$$= \left[\frac{e^{zt}}{z \cosh z} \right]_{z=\pm in\pi} \tag{3}$$

$$= (-1)^n \frac{(e^{in\pi t} - e^{-in\pi t})}{in\pi} = 2(-1)^n \sin n\pi t / n\pi. \tag{4}$$

Then $$\Sigma \text{Res.} = t + 2 \sum_{n=1}^{\infty} (-1)^n \frac{\sin n\pi t}{n\pi}. \tag{5}$$

3·241. Evaluation of residue at multiple pole by differentiation. Suppose the function $w = f(z)/g(z)$ has a pole of order n at $z = a$ so that $g(z) = (z-a)^n \xi(z)$. Then by Laurent's theorem, near the point a we have

$$w = a_0 + a_1(z-a) + a_2(z-a)^2 + \dots$$
$$+ \frac{b_1}{(z-a)} + \frac{b_2}{(z-a)^2} + \dots + \frac{b_n}{(z-a)^n}, \tag{1}$$

this expansion terminating with $b_n/(z-a)^n$, where n is finite.

The residue at $z = a$ is b_1, the coefficient of $1/(z-a)$, but as the coefficients in the expansion are unknown, we proceed as follows. Multiply both sides of (1) by $(z-a)^n$, thereby—in effect —removing the pole of order n. This gives

$$(z-a)^n \frac{f(z)}{g(z)} = \frac{f(z)}{\xi(z)} = [a_0 + a_1(z-a) + a_2(z-a)^2 + \dots](z-a)^n$$
$$+ b_1(z-a)^{n-1} + b_2(z-a)^{n-2} + \dots + b_n. \tag{2}$$

Differentiating the right-hand side $(n-1)$ times, and putting $z = a$ gives $b_1(n-1)!$. Hence the residue at the multiple pole is

$$b_1 = \frac{1}{(n-1)!} \frac{d^{n-1}}{dz^{n-1}} \left[(z-a)^n \frac{f(z)}{g(z)} \right]_{z=a}. \tag{3}$$

---

* Hence there are two terms in the principal part of the expansion, namely, $t/z + 1/z^2$, the second term indicating the double pole.

If $g(z)$ takes the form, say, $z(1-e^{-z})$, before applying (3) the exponential must be expanded, giving $z^2\left(1-\dfrac{z}{2!}+\dfrac{z^2}{3!}-\ldots\right)$, thereby obtaining the form $(z-a)^n\,\xi(z)=g(z)$. Actually this case would be evaluated more readily by using the binomial theorem for $|z|$ small as in § 3·235.

When $b_1=0$, there is no residue. For instance $1/(z-a)^3$ has a triple pole at $z=a$, but there is no residue. Hence $\displaystyle\int_{C}\dfrac{dz}{(z-a)^3}=0$, where C is a contour enclosing the point $z=a$.

3·242. Example. Find the residue of $ze^z/(z-a)^3$ at the triple pole $z=a$ by two different methods.

(1°) By (3) § 3·241 the residue is

$$\frac{1}{2!}\frac{d^2}{dz^2}(z\,e^z)_{z=a}. \tag{1}$$

Now $\qquad \dfrac{d}{dz}(z\,e^z)=e^z(1+z)$ and $\dfrac{d^2}{dz^2}(z\,e^z)=e^z(2+z)$.

Putting $z=a$ gives the residue

$$e^a(1+\tfrac{1}{2}a). \tag{2}$$

(2°) Moving the origin to $z=a$ by writing $z=\zeta+a$, we get

$$e^a.e^\zeta\left(\frac{1}{\zeta^2}+\frac{a}{\zeta^3}\right)=e^a\left(\frac{1}{\zeta^2}+\frac{a}{\zeta^3}\right)\left(1+\zeta+\frac{\zeta^2}{2!}+\ldots\right). \tag{3}$$

Taking the coefficient of $1/\zeta$ in (3), the residue is

$$e^a(1+\tfrac{1}{2}a), \tag{4}$$

as before. Accordingly

$$\frac{1}{2\pi i}\int_{C}\frac{z\,e^z\,dz}{(z-a)^3}=e^a(1+\tfrac{1}{2}a), \tag{5}$$

where C is any contour, in the finite part of the z-plane, which encloses the point $z=a$.

3·3.   Evaluation of infinite real integrals. The theorem of residues may be used for this purpose as exemplified below.

1°.   Evaluate
$$\int_{-\infty}^{\infty} \frac{dx}{1+x^2}. *$$

Consider $I = \int_C \frac{dz}{1+z^2}$, where C is the contour $OABDO$ in Fig. 21a. The integrand has poles at $z = \pm i$, of which that at $z = i$ is inside the contour. On the large semicircle $1/(1+z^2) = O(1/z^2)$,† and with $z = Re^{i\theta}$

$$\int O\left(\frac{1}{z^2}\right) dz = i \int_0^\pi O\left(\frac{1}{R}\right) e^{-i\theta} d\theta \to 0 \quad \text{as} \quad R \to \infty, \tag{1}$$

so there is no contribution to the integral from the semicircle. Thus along DOA

$$I = \int_{-\infty}^{\infty} \frac{dx}{1+x^2} = 2\pi i \times \text{residue at pole} \tag{2}$$

$$= \frac{2\pi i}{2i} = \pi. \tag{3}$$

This result may be obtained also by using a semicircle below DOA and evaluating at the pole $z = -i$. Since $1/(1+x^2)$ is an even function of x, it follows that

$$\int_0^\infty \frac{dx}{1+x^2} = \tfrac{1}{2}\pi. \tag{4}$$

2°.   Evaluate
$$\int_0^\infty \frac{dx}{x^4+1}.$$

Consider $I = \int \frac{dz}{z^4+1}$ taken round the contour $OABDO$ in Fig. 21b

Since $1/(z^4+1) = O(1/z^4)$ on the large semicircle, as in 1° there is no contribution to the integral therefrom when $R \to \infty$. Now

$$(z^4+1) = (z^2+i)(z^2-i) = (z+e^{-\frac{1}{4}i\pi})(z-e^{-\frac{1}{4}i\pi})(z+e^{\frac{1}{4}i\pi})(z-e^{\frac{1}{4}i\pi}),$$

which will be evident from Fig. 21 (c). The poles within the contour are at $z = e^{\frac{1}{4}i\pi}$ and $-e^{-\frac{1}{4}i\pi}$. Then as $R \to \infty$

$$\int_{DOA} = \int_{-\infty}^{\infty} \frac{dx}{x^4+1} = 2\pi i \times \text{sum of residues at poles} \tag{5}$$

$$= 2\pi i \frac{(e^{\frac{1}{4}i\pi} + e^{-\frac{1}{4}i\pi})}{4i} = \pi \cos \tfrac{1}{4}\pi \tag{6}$$

$$= \frac{\pi}{\sqrt{2}}. \tag{7}$$

---

* This integral is chosen for illustration. In terms of the real variable it is $[\tan^{-1} x]_{-\infty}^{\infty} = \tfrac{1}{2}\pi + \tfrac{1}{2}\pi$.

† O signifies 'of order'.

Hence
$$\int_0^\infty \frac{dx}{x^4+1} = \frac{\pi}{2^{\frac{3}{2}}}, \tag{8}$$

since the integrand in (5) is an even function of x.

The integral may be evaluated also using the contour $OABO$ in Fig. 21 (b) which encloses the pole at $z = e^{\frac{1}{4}\pi i}$, the residue there being $e^{-\frac{3}{4}i\pi}/4i$. Thus by aid of Fig. 21 (c)

$$\int_{OABO} \frac{dz}{z^4+1} = \frac{2\pi i \, e^{-\frac{3}{4}i\pi}}{4i} = \frac{\pi(1-i)}{2^{\frac{3}{2}}}. \tag{9}$$

(a)

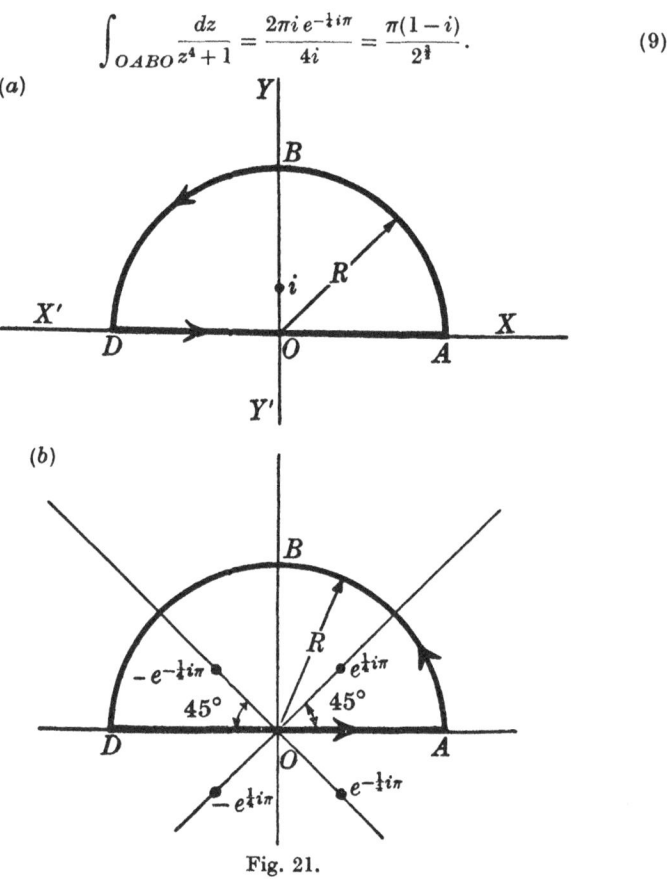

Fig. 21.

There is no contribution from AB as $R \to \infty$, so with $z = iy$ on BO and $z = x$ on OA, we have

$$i\int_\infty^0 \frac{dy}{y^4+1} + \int_0^\infty \frac{dx}{x^4+1} = \frac{\pi(1-i)}{2^{\frac{3}{2}}}, \tag{10}$$

and (8) follows immediately on equating real and imaginary parts.

3°. Evaluate $\displaystyle\int_0^\infty \frac{\sin ax\,dx}{x}$, a real > 0.

Consider $\displaystyle\int \frac{e^{iaz}\,dz}{z}$ taken round the contour $OABDO$ in Fig. 21 (d).

Writing $z = x + iy$, we have $iaz = a(ix - y)$, and since $y \geqslant 0$, $\lvert e^{iaz} \rvert \to 0$ on the large semicircle as $R \to \infty$, except at the points D, A where it is 1.

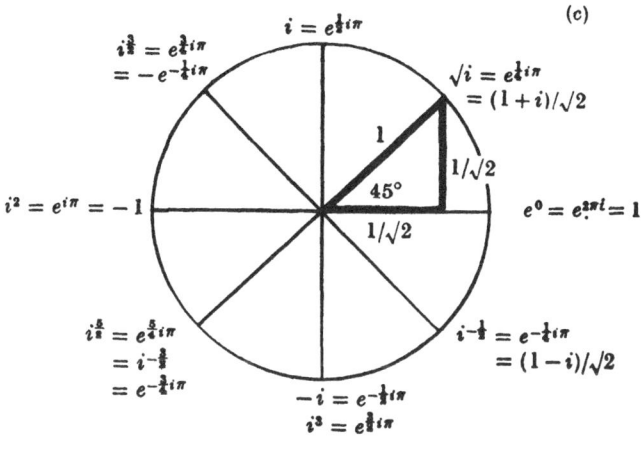

(c)

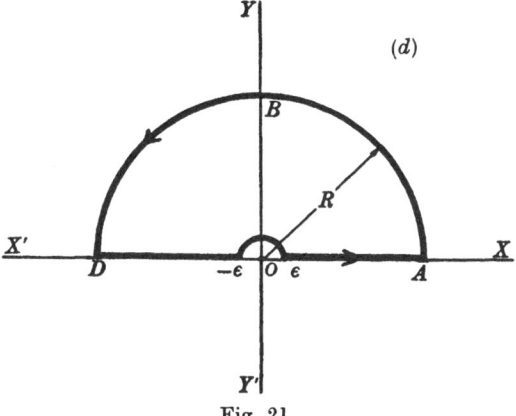

(d)

Fig. 21.

Thus the semicircle does not contribute to the value of the integral. The simple pole at the origin due to $1/z$ is outside the contour, so by Cauchy's theorem, the integral along DOA must be zero. Hence

$$\int_C \frac{e^{iaz}\,dz}{z} = \int_{-\infty}^{-\epsilon} \frac{e^{iax}\,dx}{x} + \int_{\epsilon}^{\infty} \frac{e^{iax}\,dx}{x} + i\int_{\pi}^{0} e^{ia\epsilon(\cos\theta + i\sin\theta)}\,d\theta = 0, \quad (11)$$

where the last integral is the contribution from the semicircle of radius ϵ. In the limit when $\epsilon \to 0$, (11) gives

$$\int_0^\infty \frac{e^{iax}\,dx}{x} - \int_0^\infty \frac{e^{-iax}\,dx}{x} = 2i\int_0^\infty \frac{\sin ax\,dx}{x} = -i\int_\pi^0 d\theta, \qquad (12)$$

so
$$\int_0^\infty \frac{\sin ax\,dx}{x} = \tfrac{1}{2}\pi. \qquad (13)$$

If the small semicircle were below the origin, the pole there would be within the contour. Then we should have

$$\int_{-\infty}^{-\epsilon} \frac{e^{iax}\,dx}{x} + \int_\epsilon^\infty \frac{e^{iax}\,dx}{x} + i\int_\pi^{2\pi} e^{ia\epsilon(\cos\theta + i\sin\theta)}\,d\theta = 2\pi i, \qquad (14)$$

since the residue at the pole $z = 0$ is unity. When $\epsilon \to 0$, the result at (13) is obtained, for the value of the third integral is πi. If $a < 0$, the value of integral (13) is $-\tfrac{1}{2}\pi$.

IV

THE BROMWICH CONTOUR: EQUIVALENT CONTOURS: EVALUATION OF INTEGRALS

4·1. The Bromwich contour. Hitherto we have used contours of ordinary type as illustrated in the various diagrams. In applying the complex variable to the solution of technical problems, the evaluation of the integrals obtained is effected on a contour specified by Bromwich. The contour in question, to

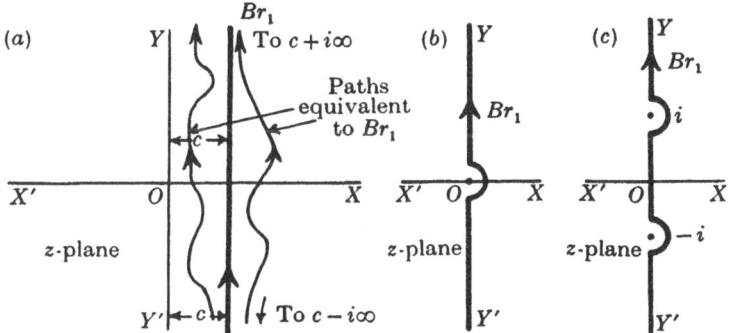

Fig. 22. Bromwich contour Br_1 and some equivalent paths.

which we shall append Bromwich's name, is shown in Fig. 22 (*a*), this being the basic form. As we shall see later, there are other contours equivalent to it, when certain conditions are fulfilled. The contour or path Br_1 indicated in Fig. 22 (*a*) is a straight line from $c - i\infty$ to $c + i\infty$, where $c \geqslant 0$ is a real number. c is usually constant, but it may vary throughout the length of the contour. The latter must be such that all the singularities of the function integrated are on its left, none being upon it. In the solution of problems pertaining to stable mechanisms, electrical circuits (with positive resistance*) and the like, the singularities of

---

* In an electrical system having an electronic device with a negative resistance, the singularities may be on the right of the imaginary axis.

the functions involved are either upon, or lie to the left of, the imaginary axis. In electrical circuit problems not involving wave motion, the singularities are usually poles. Where transmission lines, cables and acoustical problems are concerned, there is wave motion and branch points often occur. The type of integral with which we are concerned in the solution of technical problems is due to Bromwich and Wagner, who suggested it independently. It takes the form

$$I = \frac{1}{2\pi i} \int_{Br_1} \frac{e^{zt}\,\phi(z)\,dz}{z}, \tag{1}$$

where $\phi(z)$ is a function of z, all of whose singularities lie on the left of the path Br_1, and t is the time, which is always real and > 0.

4·21. Equivalent contours. We shall now indicate other paths of integration which are equivalent to Br_1 for integrals like (1) § 4·1. By applying Cauchy's theorem it may be shown that in the limit as $y \to \infty$, any path from $c - iy$ to $c + iy$ is equivalent to Br_1, provided that (a) all singularities of the integrand are on its left, and (b) $|e^{zt}\phi(z)/z| \to 0$ uniformly with respect to x^* as $y \to \infty$. The path might be a very large semicircle C of radius r on the right of Br_1. As $r \to \infty$ the integral round the semicircle counter-clockwise is equal to that along Br_1 from $c - iy$ to $c + iy$ as $y \to \infty$. This follows from the fact that the closed contour C and $c \pm iy$ contains no singularity, so the integral round it is zero. Hence the integral along $c - iy$ to $c + iy$, is equal but opposite to that taken counter-clockwise round C. This holds in the limit as r and $y \to \infty$.

The integral $\int \frac{e^{zt}\,dz}{z^3}$ has a triple pole at the origin. The contour may consist of the imaginary axis indented by a small semicircle on the right of the origin, as shown in Fig. 22 (b). The same path can be used for $\int \frac{e^{zt}\,dz}{\sqrt{z}}$, which has a branch point at O. For the integral

$$\int \frac{e^{zt}\,dz}{z^2 + 1} = \int \frac{e^{zt}\,dz}{(z + i)\,(z - i)}$$

* See first footnote to Appendix 4.

the path of Fig. 22 (c) would be equivalent to Br_1. It does not follow, of course, that because these particular paths are equivalent to Br_1, they are the best which can be chosen to evaluate the integral.

4·22. When all the singularities of the integrand are poles, and the order of the denominator exceeds that of the numerator by at least one*—apart from terms in the expansion of e^{zt}—Br_1 is equivalent to a circle or to any closed curve, or to a network of circles and pairs of parallel lines *enclosing* all the poles. This may be proved by taking a large circular arc of radius r on the left of $c \pm iy$ and considering the value of the integral round this arc as y and $r \to \infty$. Suppose the integral takes the form

$$\int \frac{e^{zt} z^n dz}{(z-a)(z-b)(z-c)^2 \dots},$$

where the highest power of z in the denominator exceeds n by at least unity $(t > 0)$. (For $t = 0$ the denominator should be of order z^{n+2}.) Then on the arc, $|a|$, $|b|$, $|c|$ may be neglected in comparison with $|z|$, which is very large, so the integral degenerates to $\int e^{zt} \frac{dz}{z^{m-n}}$, m and n being positive integers, $m > n$. It is proved in § 4·31 that as $r \to \infty$ the arc contributes nothing to the integral, so only the contribution along Br_1 remains. But the integral round the arc and Br_1 is equal to $2\pi i$ times the sum of the residues at the poles, and this is also equal to the integral round a circle or any closed curve containing all the poles. Consequently when $t > 0$, the path Br_1 is equivalent to the contours aforesaid.

4·23. Integration round a branch point. We have seen that the integral round any closed contour encircling only one pole of a function is independent of the size and shape of the contour. Also that after one circuit of the contour, the integrand or function returns to its initial value. When there is a branch point, after phase z has increased by 2π, the function does not

---

* One for $t > 0$ and two for $t = 0$.

resume its initial value. To avoid encroaching upon another branch, a barrier must be placed in a suitable position. The barrier is not to be crossed.

Consider the function $w = \sqrt{z}$, which has a branch point at the origin. We can integrate round O from $\theta = -\pi + \epsilon$ to $\pi - \epsilon$ as shown in Fig. 23, on a circle* of any radius r, but the result depends upon the magnitude of r. To demonstrate this, join the large and small solid circles by two lines parallel to and very close to the barrier. Then the closed contour so formed has no singularity within or upon it, and by Cauchy's theorem the integral round it is zero. Thus the integral along the parallel lines and round the small circle is equal to that round the large circle, both being taken positively. Consequently the latter integral exceeds that round the small circle by the contributions along the parallel lines. Since the contour encloses no singularity, the integral from A to B is independent of the path, so that the large solid circle may be replaced by the broken curve C.

The above result holds wherever the barrier is placed, provided it starts from the origin. For instance, it may make *any* angle with the real axis. After one circuit of the solid line contour of Fig. 23, the integrated function returns to its original value; for the origin is described once positively via the small circle and once negatively via the large one, phase z being unaltered.

For any closed contour associated with one or more branch points, the barrier must be drawn such that the function always remains on the same branch, and returns to its original value after z makes one complete circuit of the contour. In this way functions with fractional powers of z are made single-valued at all points of the z-plane, excluding the barrier. With $w = \sqrt{z}$, the positive value of the function is obtained if $-\pi < \theta < \pi$, since w remains on the first branch for all values of z.

4·24. Equivalent contour for function having a branch point.

If the integrand has a branch point, the integral round a semicircle in the second and third quadrants may $\rightarrow 0$ as its

---

* Strictly circular arc. The circle is approached as $\epsilon \rightarrow 0$.

radius $\to \infty$. When this is so, the equivalent contour may have a more convenient form than those illustrated in Fig. 22 (b), (c).

Take, for instance, the integral $\dfrac{1}{2\pi i}\displaystyle\int_{Br_1}\dfrac{e^{zt}\,dz}{z^{\nu+1}}$, when $R(\nu) > -1$, t being real and > 0. This has a branch point at the origin if ν is non-integral, so the equivalent contour should pass round O. As shown in § 4·31, the integrals along the arcs vanish as $r \to \infty$, so the equivalent contour can assume the form depicted in Fig. 24 (a) and marked Br_2. For the contour comprising Br_1, Br_2 and the arcs is closed and contains no singularity, so the integral round it is zero. Since the arcs contribute nothing,

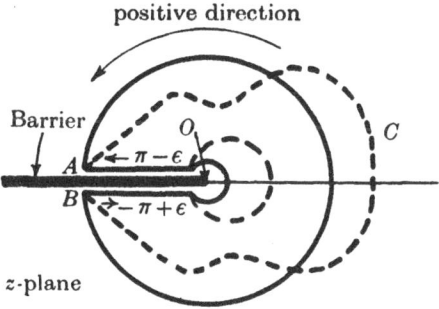

Fig. 23. Illustrating position of a 'barrier', when a branch point occurs.

provided $R(\nu) > -1$, the integral along Br_1 is equal to that along Br_2, both being described positively. The angle between the barrier and the positive real axis may have any value between $\tfrac{1}{2}\pi$ and $\tfrac{3}{2}\pi$. In general, if the arcs contribute nothing, and all the singularities of the integrand lie within Br_2, the two contours are equivalent $(t > 0)$. For ease of analytical manipulation, Br_2 would be drawn as a network of small circles round the poles and branch points, the circles being connected by pairs of parallel lines. When the only singularity is a branch point at the origin, the contour of Fig. 24 (b) is a suitable one, since integration along the lines parallel to the barrier can be effected in terms of the real variable x.

4·25. Example of integration round a branch point.

Evaluate $I = \dfrac{1}{2\pi i}\displaystyle\int_{Br_2} \dfrac{e^z\,dz}{\sqrt{z}}$, where Br_2 is the contour in Fig. 24 (b).

The integral may be taken in three parts, namely, (i) counter-clockwise round the circle of radius r from $-\pi$ to $+\pi$;* (ii) along the line below the barrier from $-\infty$ to $-r$; (iii) along the line above the barrier from $-r$ to $-\infty$.

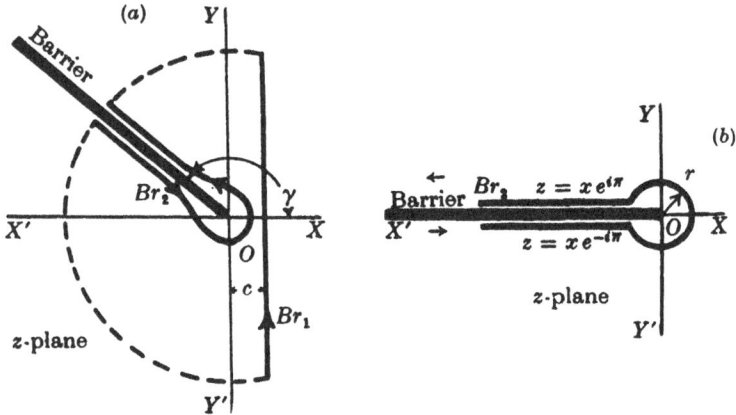

Fig. 24. The contour Br_2. In all diagrams of type (b), x signifies $|x|$.

(i) Write $z = r\,e^{i\theta}$ on the circle, and we get

$$I_1 = \frac{1}{2\pi i}\int_{-\pi}^{\pi} \frac{e^{r e^{i\theta}}\,d(r\,e^{i\theta})}{\sqrt{r}\,e^{\frac{1}{2}i\theta}} \tag{1}$$

$$= \frac{\sqrt{r}}{2\pi}\int_{-\pi}^{\pi} e^{r(\cos\theta + i\sin\theta) + \frac{1}{2}i\theta}\,d\theta, \tag{2}$$

which we shall leave in this form *pro tem.*

(ii) On the line below the barrier $z = x\,e^{-i\pi}$, where it is to be understood that x signifies $|x|$. Inserting this value of z into the integral gives

$$I_2 = \frac{1}{2\pi i}\int_{+\infty}^{+r} \frac{e^{x e^{-i\pi}}\,d(x\,e^{-i\pi})}{\sqrt{x\,e^{-\frac{1}{2}i\pi}}}, \tag{3}\dagger$$

$$= -\frac{1}{2\pi}\int_{\infty}^{r} e^{-x}x^{-\frac{1}{2}}\,dx = \frac{1}{2\pi}\int_{r}^{\infty} e^{-x}x^{-\frac{1}{2}}\,dx. \tag{4}$$

---

* In accordance with Fig. 23 the angle varies from $-\pi+\epsilon$ to $\pi-\epsilon$. In the limit $\epsilon\to 0$, and this is usually implied tacitly throughout the text.

† Both limits are positive since x is actually $|x|$.

(iii) On the line above the barrier $z = x e^{i\pi}$, so

$$I_3 = \frac{1}{2\pi i} \int_r^\infty \frac{e^{xe^{i\pi}} d(x e^{i\pi})}{\sqrt{x e^{\frac{1}{2}i\pi}}} \tag{5}$$

$$= \frac{1}{2\pi} \int_r^\infty e^{-x} x^{-\frac{1}{2}} dx. \tag{6}$$

Hence $\qquad I_2 + I_3 = \frac{1}{\pi} \int_r^\infty e^{-x} x^{-\frac{1}{2}} dx. \tag{7}$

When $r > 0$, the integral in (7) represents the incomplete gamma function with argument $\frac{1}{2}$. If we let $r \to 0$, (2) vanishes in the limit, and we get

$$I = I_1 + I_2 + I_3 = \frac{1}{\pi} \int_0^\infty e^{-x} x^{-\frac{1}{2}} dx = \frac{\sqrt{\pi}}{\pi} \tag{8}$$

$$= \frac{1}{\sqrt{\pi}}, \tag{9}$$

by (1) § 5·13. In evaluations of this type, it is usually expedient to let $r \to 0$. A case where $r = 1$ is given in § 7·45.

4·26. Integrand having two branch points. As an example take

$$\int_{Br_1} \frac{e^{zt} dz}{\sqrt{(z^2 + 1)}} = \int_{Br_1} \frac{e^{zt} dz}{\sqrt{\{(z+i)(z-i)\}}}. \tag{1}$$

The integrand has branch points at $z = \pm i$.

A short selection of contours equivalent to Br_1 is given in Fig. 25. Cases (a), (b), (d) are the most suitable for evaluation of integral (1). In (a) the barrier is on the imaginary axis, while in (b) there are two barriers each being parallel to the real axis. The equivalence of each path with Br_1 can be seen, for none of the contours completed via the broken lines encloses a singularity, while by § 4·31 the contributions from the arcs $\to 0$ as the radius $\to \infty$. Integration along (a) and (b) yield, respectively, the convergent and asymptotic series for $J_0(z)$, the Bessel function of the first kind of order zero. The former series is (2) § 5·32 with $\nu = 0$ and the latter is (8) § 5·351.

4·27. Integration round two branch points. 1°. 1. Consider the function $w = (z^2 - 1)^{\frac{1}{2}} = \{(z-1)(z+1)\}^{\frac{1}{2}}$ in relation to the contour in Fig. 26(b). There are two branch points at $z = \pm 1$, and the barrier is taken as the real axis from $z = -1$ to $+1$. The dumbbell contour comprises the lines $A_1 A_2$, $B_1 B_2$, parallel to and very near the real axis, together with two

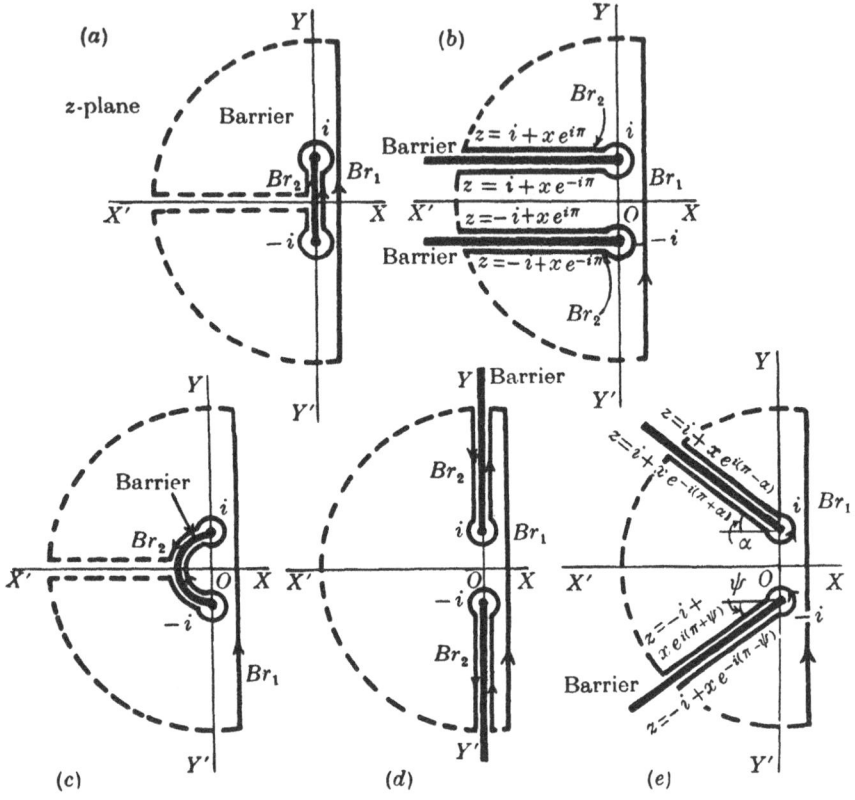

Fig. 25. Br_1 and equivalent contours when there are two branch points.

circular arcs (centres at the b.p.s.) whose radii we ultimately let $\rightarrow 0$. The function w is analytic on the contour and at all points in the finite part of the z-plane *except on the barrier*. Moreover w is single-valued at all points of the contour. So far as phase is concerned, we are at liberty to choose any convenient zero. In the present case it is expedient to regard the phase of z on $A_2 A_1$ to be zero.

We now consider what happens when the contour is traversed positively (counterclockwise). On $A_2 A_1$, if we let $r \rightarrow 0$, $z = x$ from $+1$ to -1.

To pass round the b.p. from A_1 to B_1, we move the origin to $z = -1$ by writing $(z+1) = \zeta$, $(z-1) = (\zeta-2)$, so that

$$w = \zeta^{\frac{1}{2}}(\zeta-2)^{\frac{1}{2}}. \tag{1}$$

With $\zeta = r\,e^{i\theta}$, (1) gives $\quad w = r^{\frac{1}{2}}e^{\frac{1}{2}i\theta}(r\,e^{i\theta}-2)^{\frac{1}{2}}.$ $\tag{2}$

At A_1, $\theta = 0$, so $\qquad\qquad w = r^{\frac{1}{2}}(r-2)^{\frac{1}{2}}.$ $\tag{3}$

At B_1, $\theta = 2\pi$, so by (2) $\quad w = r^{\frac{1}{2}}e^{i\pi}(r\,e^{2\pi i}-2)^{\frac{1}{2}}$

$$= -r^{\frac{1}{2}}(r-2)^{\frac{1}{2}}. \tag{4}$$

Thus *after* rounding the branch point, *w changes sign*; but since A_1, B_1 are on opposite sides of the barrier, no ambiguity occurs.

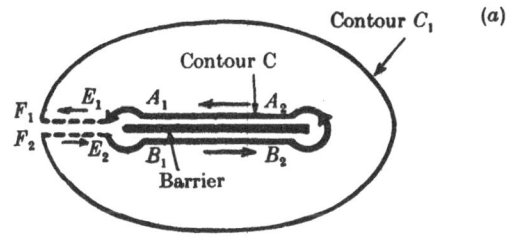

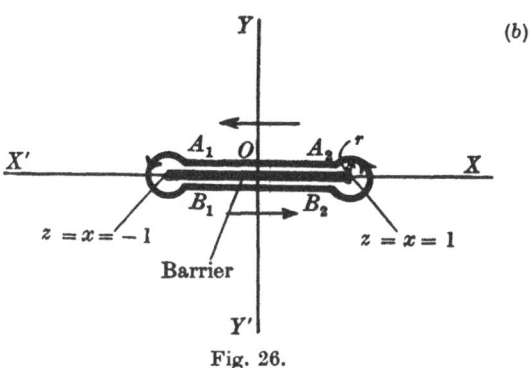

Fig. 26.

On B_1B_2 if $r \to 0$, $z = x\,e^{2\pi i} = x$ from -1 to $+1$. To pass round the b.p. from B_2 to A_2, we move the origin to $z = +1$ by writing $(z-1) = \zeta$, and $(z+1) = (\zeta+2)$, so that

$$w = \zeta^{\frac{1}{2}}(\zeta+2)^{\frac{1}{2}} = r^{\frac{1}{2}}e^{\frac{1}{2}i\theta}(r\,e^{i\theta}+2)^{\frac{1}{2}}. \tag{5}$$

At B_2, $\theta = 2\pi$, so $\qquad w = r^{\frac{1}{2}}e^{i\pi}(r\,e^{2\pi i}+2)^{\frac{1}{2}}$

$$= -r^{\frac{1}{2}}(r+2)^{\frac{1}{2}}, \tag{6}$$

which has the same sign as at B_1.

At A_2, $\theta = 0$ (by hypothesis), so by (5)

$$w = r^{\frac{1}{2}}(r+2)^{\frac{1}{2}}, \tag{7}$$

and, therefore, w *changes sign after* rounding the branch point. There is no ambiguity, since B_2, A_2 are on opposite sides of the barrier.

2. *Values of w on A_2A_1 and B_1B_2.*

On A_2A_1, if $r \to 0$, $z = x$ from $+1$ to -1, and

$$w = (x^2 - 1)^{\frac{1}{2}}, \tag{8}$$

since by (7) the sign of w is positive on A_2A_1. At $x = \pm 1$, $w = 0$. On B_2B_1, if $r \to 0$, $z = x\,e^{2\pi i} = x$, from -1 to $+1$, and

$$w = -(x^2 - 1)^{\frac{1}{2}}, \tag{9}$$

since by (4) the sign of w is minus on B_1B_2. At $x = \pm 1$, $w = 0$.

3. *Integration round contour.*

Suppose we evaluate

$$I = \int_C (z^2 - 1)^{\frac{1}{2}}\,dz, \tag{10}$$

the integral being taken round the contour of Fig. 26(b). We commence at A_2 and move counterclockwise. Then from (8) on A_2A_1, $w = (x^2 - 1)^{\frac{1}{2}}$, so

$$I_1 = \int_{+1}^{-1} (x^2 - 1)^{\frac{1}{2}}\,dx, \tag{11}$$

it being implied that $r \to 0$ to obtain the limits ± 1.

Similarly from B_1 to B_2 we get

$$I_2 = -\int_{-1}^{+1} (x^2 - 1)^{\frac{1}{2}}\,dx = \int_{+1}^{-1} (x^2 - 1)^{\frac{1}{2}}\,dx, \tag{12}$$

the minus being due to w changing sign after z has passed round the branch point.

Going round the small circle having centre $z = -1$, from (2) we have

$$I_3 = r^{\frac{1}{2}} \int_0^{2\pi} e^{\frac{1}{2}i\theta}\,(r\,e^{i\theta} - 2)^{\frac{1}{2}}\,d(r\,e^{i\theta}) \to 0 \tag{13}$$

as $r \to 0$, and similarly for the corresponding integral at $z = +1$. Hence the integral of w taken round the contour is

$$I = I_1 + I_2 = 2\int_{+1}^{-1} (x^2 - 1)^{\frac{1}{2}}\,dx$$

$$= 2i \int_{+1}^{-1} (1 - x^2)^{\frac{1}{2}}\,dx. \tag{14}$$

Write $x = \sin\phi$, $dx = \cos\phi$, $(1 - x^2)^{\frac{1}{2}} = \cos\phi$, and (14) becomes

$$I = 2i \int_{\frac{1}{2}\pi}^{-\frac{1}{2}\pi} \cos^2\phi\,d\phi = -\pi i. \tag{15}$$

4. Equivalent contour.

We shall now show that the value of the integral in §3 taken round any contour C_1 which encloses C, is equal to the value of the integral round C (Fig. 26 (a)).

Draw the parallel broken lines $F_1 E_1$, $F_2 E_2$, which ultimately coalesce. Then starting at A_2 and traversing the contour C and C_1 via the lines FE positively, the complete contour contains no singularity of w, so by Cauchy's theorem the integral round it is zero. Now when the lines at EF coalesce, the integral along $F_1 E_1$ is equal but opposite to that along $E_2 F_2$, so they cancel. Hence the integral counterclockwise round C is equal to that counterclockwise round C_1. The two contours are then said to be equivalent. If C_1 is a circle about O whose radius $R > 1$, by expanding binomially and integrating term by term, (15) is obtained.

$2°$. *Evaluation of* $I = \dfrac{1}{2\pi i} \displaystyle\int_C \dfrac{e^{zt}dz}{\sqrt{(z^2-1)}}$, *where* C *is the dumbbell-type contour of Fig.* 26 (b).*

By §4·26 this contour is equivalent to Br_1. Suppose that the phase of z along the line $A_2 A_1$ is zero,† then on $A_2 A_1$, $z = x$ from $+1$ to -1. Thus

$$I_1 = \frac{1}{2\pi i}\int_1^{-1} \frac{e^{zt}dx}{\sqrt{(x^2-1)}} = \frac{1}{2\pi}\int_{-1}^{1} \frac{e^{xt}dx}{\sqrt{(1-x^2)}}, \tag{1}$$

since on the contour $|x| \leqslant 1$. Passing round the branch point at $z = -1$, the phase of the integrand changes by π,‡ and on $B_1 B_2$, $z = x\,e^{2\pi i}$. Thus

$$I_2 = -\frac{1}{2\pi i}\int_{-1}^{1} \frac{e^{zt}dx}{\sqrt{(x^2-1)}} = \frac{1}{2\pi}\int_{-1}^{1} \frac{e^{xt}dx}{\sqrt{(1-x^2)}}. \tag{2}$$

Moving the origin to the point $z = -1$, we write $\zeta = z + 1$ or $z = \zeta - 1$, which gives

$$I_3 = \frac{e^{-t}}{2\pi i}\int \frac{e^{\zeta t}d\zeta}{\sqrt{[(\zeta-2)\,\zeta]}}. \tag{3}$$

* C is equivalent to Br_1. If the denominator is expanded binomially, the integral may be evaluated term by term on Br_1 using (3) §4·51, as at (2)–(5) §6·14.

† The phase of z may have any suitable value on $A_2 A_1$. In §4·25 it was taken as π, but it is convenient to regard the phase to be zero here. The reader may test that the same result is obtained by using π.

‡ Since there are two branch points, by §1·63 the phase change of the *integrand* is 2π, when the complete contour is traversed, but π when only one branch point has been rounded. The integrand in (4) may be written

$$\sqrt{r}\,e^{rt(\cos\theta + i\sin\theta)}\,(\cos\tfrac{1}{2}\theta + i\sin\tfrac{1}{2}\theta)/\sqrt{[r(\cos\theta + i\sin\theta) - 2]}.$$

When $\theta = 0$ this has the value $+\sqrt{r}\,e^{rt}/\sqrt{(r-2)}$, and for $\theta = 2\pi$ the value is $-\sqrt{r}\,e^{rt}/\sqrt{(r-2)}$; so the integrand changes sign in rounding the branch point at $z = -1$. Similarly for the branch point at $z = 1$, where the change is from $-$ to $+$.

On the small circle with $z = -1$ as centre, write $\zeta = r e^{i\theta}$, and we get

$$I_3 = \frac{e^{-t}}{2\pi} \int_{+\pi}^{-\pi} \frac{e^{rt(\cos\theta + i\sin\theta) + \frac{1}{2}i\theta} \sqrt{r}\, d\theta}{\sqrt{(r e^{i\theta} - 2)}}. \tag{4}$$

As $r \to 0$, I_3 tends to zero, so in the limit I_3 vanishes, as also does the integral corresponding to $z = 1$. Hence

$$I = I_1 + I_2 = \frac{1}{\pi} \int_{-1}^{1} \frac{e^{xt}\, dx}{\sqrt{(1 - x^2)}}. \tag{5}$$

To evaluate (5) write $x = \cos\theta$, $dx = -\sin\theta\, d\theta$, and $\sqrt{(1 - x^2)} = \sin\theta$, then

$$I = \frac{1}{\pi} \int_0^{\pi} e^{t\cos\theta}\, d\theta = \frac{1}{\pi} \int_0^{\pi} \sum_{r=0}^{\infty} \frac{(t\cos\theta)^r}{r!}\, d\theta. \tag{6}$$

Term-by-term integration is permissible in virtue of uniform convergence (see [215]), so

$$I = \frac{1}{\pi} \left[\pi + \pi \cdot \frac{1}{2} \cdot \frac{t^2}{2!} + \pi \cdot \frac{3}{4} \cdot \frac{1}{2} \cdot \frac{t^4}{4!} + \pi \cdot \frac{5}{6} \cdot \frac{3}{4} \cdot \frac{1}{2} \cdot \frac{t^6}{6!} + \cdots \right] \tag{7}$$

$$= 1 + \frac{t^2}{2^2} + \frac{t^4}{2^2 \cdot 4^2} + \frac{t^6}{2^2 \cdot 4^2 \cdot 6^2} + \cdots = \sum_{r=0}^{\infty} \frac{(\frac{1}{2}t)^{2r}}{(r!)^2}. \tag{8}$$

By (4) §5·33 with $\nu = 0$, (8) represents $I_0(t)$, the modified Bessel function of the first kind of order zero. Hence

$$\frac{1}{2\pi i} \int_C \frac{e^{zt}\, dz}{\sqrt{(z^2 - 1)}} = I_0(t), \tag{9}$$

as at (5) §5·34, with $\nu = 0$, $a = 1$, $z = t$, $\zeta = z$. (See note on p. 90.)

4·28. If $R(\nu) \leqslant -1$, $t > 0$, the integral $\int \frac{e^{zt}\, dz}{z^{\nu+1}}$ does not vanish round the arcs AC, DF in Fig. 27 (a) when $r \to \infty$, owing to contributions which occur in the vicinity of the imaginary axis. It does vanish, however, on the arcs AC, DE in Fig. 27 (b), provided the angles α and β are > 0, so the paths Br_2, Br_2' are then equivalent, since the closed contour $ACDEA$ contains no singularity. The angle of the barrier with the positive real axis lies between $\frac{1}{2}\pi + \alpha$ and $\frac{3}{2}\pi - \beta$. When the integrand differs from that given above, Br_2 and Br_2' are equivalent, provided the contribution from the arcs $\to 0$ as the radius $\to \infty$ and all the singularities lie within Br_2.

4·31. Theorem. If $t > 0$, and $|\phi(z)/z| \to 0$ uniformly on the arc ACF in Fig. 27 (a) as $|z| = r \to \infty$, then

$$I = \int_{ACF} \frac{e^{zt}\,\phi(z)\,dz}{z} = 0.$$

Proof. With $z = r\,e^{i\theta}$ on the arc BC, we have

$$I_1 = r \int_{\frac{1}{2}\pi}^{\pi} e^{rt\cos\theta + i(rt\sin\theta + \theta + \frac{1}{2}\pi)} \frac{\phi(r\,e^{i\theta})\,d\theta}{r\,e^{i\theta}}. \tag{1}$$

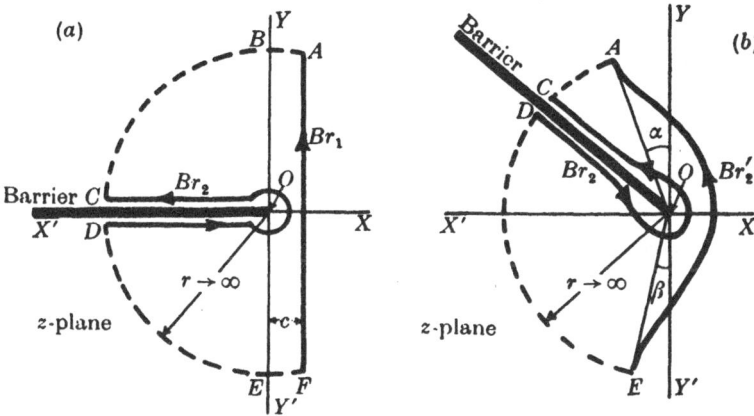

Fig. 27.

If $|\phi(r\,e^{i\theta})/r\,e^{i\theta}| < \epsilon$ when r is large,

$$|I_1| < \epsilon r \int_{\frac{1}{2}\pi}^{\pi} e^{rt\cos\theta}\,d\theta = \epsilon r \int_{0}^{\frac{1}{2}\pi} e^{-rt\sin\alpha}\,d\alpha, \tag{2}$$

where $\theta = \alpha + \frac{1}{2}\pi$. Now $\sin\alpha \geqslant 2\alpha/\pi$ in $0 \leqslant \alpha \leqslant \frac{1}{2}\pi$, so

$$|I_1| < \epsilon r \int_{0}^{\frac{1}{2}\pi} e^{-2rt\alpha/\pi}\,d\alpha = \frac{\pi\epsilon}{2t}(1 - e^{-rt}). \tag{3}$$

When $r \to \infty$, $\epsilon \to 0$, so $I_1 \to 0$, as also does the integral along DE,

The arc AB may be replaced by a line from B perpendicular to Br_1. Thus

$$I_2 = \int_{AB} \frac{e^{zt}\,\phi(z)\,dz}{z} = -\int_{0}^{c} \frac{e^{t(x+iy)}\,\phi(z)\,dx}{z}, \tag{4}$$

so

$$|I_2| < \int_{0}^{c} e^{tc}\left|\frac{\phi(z)}{z}\right|\,dx. \tag{5}$$

Since $|\phi(z)/z| < \epsilon$ on AB, by Appendix 1, $|I_2| < c\,e^{lc}\epsilon$, which $\to 0$ as $r \to \infty$ and $\epsilon \to 0$. A similar result is obtained for EF, and this completes the proof.

Equivalence of Br_1 and Br_2. When the above theorem holds, and all the singularities of the integrand are on the left of both contours, as stated in § 4·24, they are equivalent. For instance, if $\phi(z)/z = 1/z^{\nu+1}$, $R(\nu) > -1$, $|\phi(z)/z| = 1/r^{\nu+1} \to 0$ as $r \to \infty$, so Br_1 and Br_2 are equivalent.

When $R(\nu) \leqslant -1$, the contributions from the arcs AC, DF in Fig. 27 (a) do not vanish when $r \to \infty$, even if Br_1 begins and ends on the imaginary axis, the real axis being crossed at a distance c to the right of O. As shown in Appendix 7, the integral is convergent on Br_2 but divergent on Br_1.

4·32. The results in the last two paragraphs of § 4·31 may be summarized as follows:

$$\int_{Br_1} \frac{e^{zt}\,dz}{z^{\nu+1}} = \int_{Br_2} \frac{e^{zt}\,dz}{z^{\nu+1}}, \tag{1}$$

provided that $R(\nu) > -1$;

$$\int_{Br_1} \frac{e^{zt}\,dz}{z^{\nu+1}} \tag{2}$$

does not converge if $R(\nu) \leqslant -1$, while

$$\int_{Br_2} \frac{e^{zt}\,dz}{z^{\nu+1}} \tag{3}$$

is convergent for all values of ν.

4·41. Theorem:

$$\int_{Br_1} \frac{e^{zt}\phi(z)\,dz}{z} = 0, \quad \text{when } t < 0 \text{ and } \left|\frac{\phi(z)}{z}\right| \to 0$$

uniformly with respect to phase z as $|z| \to \infty$, $-\tfrac{1}{2}\pi \leqslant \text{phase } z \leqslant \tfrac{1}{2}\pi$. As in § 4·1 we take all the singularities of the integrand to lie on the left of Br_1. Then the latter can be replaced by a semicircle on its right whose radius $r \to \infty$, since the closed contour contains no singularity. Thus writing $z = r\,e^{i\theta}$, we have

$$I = r \int_{-\frac{1}{2}\pi}^{\frac{1}{2}\pi} e^{rt\cos\theta + i(\sin\theta + \theta + \frac{1}{2}\pi)} \frac{\phi(r\,e^{i\theta})}{r\,e^{i\theta}}\,d\theta. \tag{1}$$

In virtue of the above proviso, $\left|\dfrac{\phi(r\,e^{i\theta})}{r\,e^{i\theta}}\right| < \epsilon$, when r is large and $-\frac{1}{2}\pi \leqslant \theta \leqslant \frac{1}{2}\pi$. Then

$$|I| < \epsilon r \int_{-\frac{1}{2}\pi}^{\frac{1}{2}\pi} e^{rt\cos\theta}\,d\theta = 2\epsilon r \int_{0}^{\frac{1}{2}\pi} e^{rt\sin\theta}\,d\theta. \tag{2}$$

By aid of Appendix 2, since $t < 0$, (2) may be written

$$|I| < 2\epsilon r \int_{0}^{\frac{1}{2}\pi} e^{2rt\theta/\pi}\,d\theta = \frac{\pi\epsilon}{t}\,[e^{2rt\theta/\pi}]_{0}^{\frac{1}{2}\pi} \tag{3}$$

$$= -\frac{\pi\epsilon}{t}\,(1 - e^{rt}). \tag{4}$$

As $r \to \infty$, $\epsilon \to 0$ by hypothesis, and with $t < 0$,

$$|I| = 0, \tag{5}$$

so the required result is established. If

$$\phi(z)/z = 1/z^{\nu+1},$$

the above holds provided $R(\nu) > -1$.

4·42. Note on contours. Those used herein usually satisfy one of the following conditions:

(i) All singularities of $\phi(z)/z$ in the z-plane lie to the left of the contour if it is open like Br_1, Br_2;

(ii) $|e^{zt}\phi(z)/z|$ vanishes at both extremities of the contour, if open;

(iii) $e^{zt}\phi(z)/z$ regains its initial value after one circuit of the contour, if closed.

Strictly the complex integral of the form (1) §4·1 should be written

$$\lim_{y \to +\infty} \frac{1}{2\pi i} \int_{c-iy}^{c+iy} \frac{e^{zt}\,\phi(z)\,dz}{z}. \tag{1}$$

For convenience we shall not use this form, but it will be implied tacitly.

4·51. Example. Evaluate $I = \dfrac{1}{2\pi i} \displaystyle\int_{Br_1} \dfrac{e^{zt}\,dz}{z^{n+1}}$, n being a positive integer.

From § 4·22 the contour Br_1 is equivalent to a circle enclosing the origin, so the value of I is the residue at the pole, which is of order $(n+1)$. Expanding the exponential, we get

$$I = \frac{1}{2\pi i} \int_C \left[1 + zt + \frac{z^2 t^2}{2!} + \dots + \frac{z^n t^n}{n!} + \dots \right] \frac{dz}{z^{n+1}}. \tag{1}$$

Now, the integrand may be written

$$\frac{1}{z^{n+1}} + \frac{t}{z^n} + \dots + \frac{t^n}{n!\,z} + \frac{t^{n+1}}{(n+1)!} + \frac{z t^{n+2}}{(n+2)!} + \dots, \tag{2}$$

which is the Laurent expansion near the pole $z = 0$ (§ 2·61 et seq.). The residue is the coefficient of $1/z$ in the expansion, so

$$I = t^n/n!. \tag{3}$$

4·52. Example. Evaluate $I = \dfrac{1}{2\pi i} \displaystyle\int_{Br_1} \dfrac{e^{zt}\,dz}{(z-a)^{n+1}}$, n a positive integer.

Moving the origin to the pole at $z = a$ on the left of Br_1, we write $(z - a) = \zeta$, a transformation in accordance with Appendix 5. Thus

$$I = \frac{1}{2\pi i} \int_{Br_1} \frac{e^{(\zeta + a)t}\,d\zeta}{\zeta^{n+1}} = \frac{e^{at}}{2\pi i} \int_{Br_1} \frac{e^{\zeta t}\,d\zeta}{\zeta^{n+1}} \tag{1}$$

$$= e^{at} t^n/n!, \tag{2}$$

by § 4·51.

4·53. Example. Evaluate

$$I = \frac{1}{2\pi i} \int_{Br_1} \frac{e^{zt}\,dz}{(z+a)(z+b)(z+c)}. \tag{1}$$

Since all the singularities of the integrand are simple poles at $z = -a,\ -b,\ -c$, Br_1 is equivalent to a circle enclosing them.

Evaluating the residue at each pole by the method shown in § 3·21, we have

$$I = \Sigma\,\text{Residues} = \frac{e^{-at}}{(b-a)(c-a)} + \frac{e^{-bt}}{(a-b)(c-b)} + \frac{e^{-ct}}{(a-c)(b-c)}. \tag{2}$$

4·54. Example. Evaluate

$$I = \frac{1}{2\pi i} \int_{Br_1} \frac{e^{zt}\, dz}{z(1 - e^{-hz})}, \tag{1}$$

h real > 0.

The integrand has a double pole at the origin, simple poles on the imaginary axis at $z = 2n\pi i/h$, $n \neq 0$, and a limit point of poles at infinity. To evaluate the residue at the double pole, we expand the numerator and denominator, thereby obtaining

$$I_1 = \frac{1}{2\pi i} \int_{Br_1} \frac{\left(1 + zt + \dfrac{z^2 t^2}{2!} + \dots\right) dz}{hz^2\left(1 - \dfrac{hz}{2!} + \dfrac{h^2 z^2}{3!} - \dots\right)}. \tag{2}$$

When z is very small the denominator can be expanded by the binomial theorem, so (2) becomes

$$I_1 = \frac{1}{2\pi i} \int_{Br_1} \frac{\left(1 + zt + \dfrac{z^2 t^2}{2!} + \dots\right)\left(1 + \dfrac{hz}{2!} - \dots\right) dz}{hz^2}. \tag{3}$$

Selecting the coefficient of $1/z$, we obtain

$$I_1 = \left(\frac{t}{h} + \frac{1}{2}\right). \tag{4}$$

In previous examples the poles have been in the finite part of the z-plane, so by § 4·22, Br_1 could be replaced by any closed contour containing them all. In view of the limit point of simple poles, we take a large circular arc radius R on the left of $c \pm iy$, which intersects the imaginary axis in two points *between* the mth and $(m + 1)$th poles on either side of $X'OX$. Maintaining this condition, if the integral along the arc $\to 0$ as m and, therefore, $R \to \infty$, the value of (1) along Br_1—apart from the contribution due to the double pole at $z = 0$—is found by summing the residues as in § 3·237.

Br_1 may be taken as the imaginary axis indented at each pole by a small semicircle on its right, as illustrated in Fig. 28. Then we have to demonstrate that the value of (1) taken round the semicircle C_1 tends to zero as $R \to \infty$. For this purpose we use § 4·31, and therefore have to show that $1/|z(1-e^{-hz})| \to 0$ as $R \to \infty$. Now $z \neq 2n\pi i/h$ on C_1, since the contour passes *between* the poles, so $|1-e^{-hz}| > 0$; also $1/|z| = 1/R$. Hence the above condition is satisfied. Accordingly by § 3·237, if $t > 0$, the contribution to (1) from the simple poles,* is

$$I_2 = \sum_{n=-\infty}^{\infty}{}' \left[\frac{e^{zt}}{z\dfrac{d}{dz}(1-e^{-hz})} \right]_{z=2\pi ni/h} = \frac{1}{h} \sum_{n=-\infty}^{\infty}{}' \left(\frac{e^{zt}}{z\,e^{-hz}} \right)_{z=2\pi ni/h} \quad (5)$$

$$= \sum_{n=-\infty}^{\infty}{}' \frac{e^{2\pi nti/h}}{2\pi ni} = \frac{1}{\pi} \sum_{n=1}^{\infty} \frac{\sin(2\pi nt/h)}{n}. \quad (6†)$$

Hence from (4) and (6)

$$I = \frac{1}{2} + \frac{t}{h} + \frac{1}{\pi} \sum_{n=1}^{\infty} \frac{\sin(2\pi nt/h)}{n}. \quad (7)$$

4·61. Approximations. In practical applications of integrals like (1) § 4·53, it may be desirable to know the form of the function I when t is small, so we shall indicate the method of procedure in certain cases.

Consider the function $I = \dfrac{1}{2\pi i} \displaystyle\int_{Br_1} \frac{e^{zt}dz}{f(z)}$, where $f(z)$ is a polynomial of the nth degree, namely,

$$f(z) = a_n z^n + a_{n-1} z^{n-1} + \ldots + a_0, \quad (1)$$

n being a positive integer ≥ 1. Then $f(z)$ can be factorized and all the singularities are poles, so Br_1 is equivalent to a circle enclosing these poles. Write $\zeta = zt$, t being real and > 0. By Appendix 5 the contour in the ζ-plane is a circle of radius t times that in the z-plane. Consequently the integral becomes

$$I = \frac{(1/t)}{2\pi i} \int_C \frac{e^\zeta\, d\zeta}{a_n(\zeta/t)^n + a_{n-1}(\zeta/t)^{n-1} + \ldots + a_0} \quad (2)$$

$$= \frac{t^{n-1}}{2\pi i} \int_C \frac{e^\zeta d\zeta}{a_n \zeta^n + a_{n-1}\zeta^{n-1}t + \ldots + a_0 t^n}. \quad (3)$$

---

* Σ' signifies the omission of $n = 0$.

† (6) in bold type means that convergence of the series is dealt with in Appendix 9.

When t is small enough all the terms in the denominator of (3) can be neglected in comparison with $|a_n\zeta^n|$.* Hence (3) may be written

$$I \simeq \frac{(t^{n-1}/a_n)}{2\pi i} \int_C \frac{e^\zeta\, d\zeta}{\zeta^n} = \frac{t^{n-1}}{a_n(n-1)!},\qquad(4)$$

by (3) §4·51. This formula gives a first approximation to the integral when t is small.

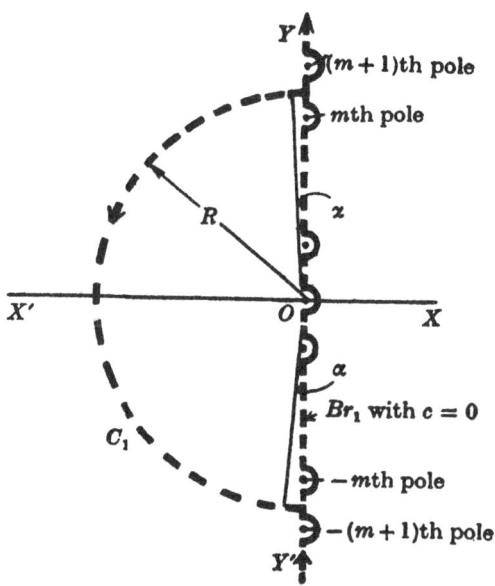

Fig. 28.

4·62. The technique adopted in ascertaining the form of say (1) §4·53, when t is small, is to take $|z|$ very large and neglect $|a|$, $|b|$, $|c|$, in comparison, thereby obtaining

$$I \simeq \frac{1}{2\pi i} \int_C \frac{e^{zt}\, dz}{z^3} = \frac{t^2}{2!},\qquad(1)$$

this being the main contribution to the integral when t is small. In (1) §4·54, it is necessary to evaluate the integral as shown there. The above procedure would yield unity.

---

* The radius of the circle in the z-plane and, therefore, in the ζ-plane can be as large as we please so that $|a_n\zeta^n|$ is the largest term.

4·63. Example. Prove that

$$\frac{1}{2\pi i}\int_{Br_1}\frac{e^{zt}dz}{(z-a)(z-b)(z-c)^m(z-d)^n\ldots}\to 0$$

as $t\to 0$, if all the singularities of the integrand are poles, finite in number.

From the analysis given in §§ 4·61, 4·62, if t is small the integral may be written

$$I\simeq\frac{1}{2\pi i}\int_{Br_1}\frac{e^{zt}dz}{z^r}=\frac{t^{r-1}}{(r-1)!},\tag{1}$$

where z^r is the product of all the factors in the denominator when $a,\ b,\ \ldots$ are neglected. Hence I vanishes with t.

4·64. When the integral takes the form $\dfrac{1}{2\pi i}\displaystyle\int_C\dfrac{e^{zt}f(z)}{g(z)}dz$, where $f(z)$ and $g(z)$ are polynomials, the former being of equal or higher degree than the latter, it is necessary to divide out or resolve into partial fractions before proceeding to obtain the form of the function when t is small. Thus in the case

$$I=\frac{1}{2\pi i}\int_C\frac{e^{zt}z^3\,dz}{z^2+1},$$

C being a contour enclosing the two poles at $z=\pm i$, we have

$$I=\frac{1}{2\pi i}\int_C e^{zt}\left[z-\frac{z}{z^2+1}\right]dz.\tag{1}$$

The first integral is zero, and the functional form of the second when t is small is given by taking z very large, so

$$-I\simeq\frac{1}{2\pi i}\int_C\frac{e^{zt}dz}{z}=1.\tag{2}$$

With Br_1 as the contour, the integral diverges unless $f(z)$ is of lower order than $g(z)$. This is considered in Appendix 7.

4·65. Value of $f(t)=\dfrac{1}{2\pi i}\displaystyle\int_{Br_1}\dfrac{e^{zt}\phi(z)\,dz}{z}$, when $\dfrac{\phi(z)}{z}$ has poles only and $t\to\infty$. When the singularities of $\phi(z)/z$ are poles to the left of the imaginary axis, their real parts are negative, and it may be shown that the contribution from a pole at $z=-a$ takes the form

$$(a_1+a_2t+\ldots+a_nt^{n-1})e^{-at},\tag{1}$$

n being the order of the pole. Consequently, as $t\to\infty$, $f(t)\to 0$. If the nth order pole is at the origin, $a=0$, and (1) becomes

$$(a_1+a_2t+\ldots+a_nt^{n-1}).\tag{2}$$

When $t \to \infty$, $f(t) \to \infty$, except when a simple pole occurs, and then $f(t) \to a_1$ as $t \to \infty$. If the pole is on the right of the imaginary axis the real part of z is positive, and the exponential in (1) becomes e^{at}, so $f(t) \to \infty$ as $t \to \infty$. When the pole is on the imaginary axis $f(t)$ oscillates with increasing amplitude as t increases, except for a simple pole, when the amplitude is a_1, being independent of t.

4·66. Example. What is the form of $\dfrac{1}{2\pi i}\displaystyle\int_{Br_1} \dfrac{e^{zt}\,dz}{z^2(z+a)^3}$ when t is large? By (1) § 4·65 we see that the contribution from the triple pole at $z = -a$ tends to zero as $t \to \infty$. That from the double pole at the origin, by (2) § 4·65, takes the form $a_2 t$. Since the pole at $z = -a$ can be disregarded, the main contribution to the integral occurs when z is small, and Br_1 is equivalent to a circle round O. Hence, with z small, we get

$$I \simeq \frac{1}{2\pi i}\int \frac{e^{zt}\,dz}{a^3 z^2} = \frac{t}{a^3}, \tag{1}$$

when t is large, so the function increases in proportion to t. In cases of this type having a pole at $z = 0$, the technique is to make z small and evaluate round a circle about O.

4·71. Theorem. If in Fig. 29 all the singularities of $\phi(z)/z$ are to the left of Br_1, and if $\dfrac{1}{2\pi i}\displaystyle\int_{BC_1A} \dfrac{\phi(z)\,dz}{z} \to 0$ as the radius of the circle $\to \infty$, the integral

$$\frac{1}{2\pi i}\int_{Br_1} \frac{e^{zt}\,\phi(z)\,dz}{z} = 0 \quad \text{when } t = 0.$$

Proof. Since the contour $Br_1 C_2$ encloses no singularity of the integrand, by Cauchy's theorem the integral along Br_1 is equal to that round C_2. By hypothesis the latter is zero, so the theorem is proved.

4·72. Example. Evaluate

$$\frac{1}{2\pi i}\int_{Br_1} \frac{e^{zt}\sinh r\sqrt{z}\,dz}{rz\sinh a\sqrt{z}} \quad (a > r > 0),$$

and show that
$$1 + 2\sum_{n=1}^{\infty}(-1)^n \frac{\sin n\pi r/a}{n\pi r/a} = 0.$$

Expanding the integrand without e^{zt} and letting $z \to 0$, yields $1/az$, so there is a pole at the origin. Thus when $z \to 0$, the integral becomes

$$I_1 = \frac{1}{2\pi i} \int_{Br_1} \frac{e^{zt}\,dz}{az} = \frac{1}{a}. \qquad (1)$$

There are also simple poles when $\sinh a \sqrt{z} = 0$, i.e. $ia \sqrt{z} = n\pi$ or $z = -n^2\pi^2/a^2$ ($n = 1, 2, 3, \ldots$ with a limit point at infinity). Only positive values of n are used, since negative values merely

z-plane

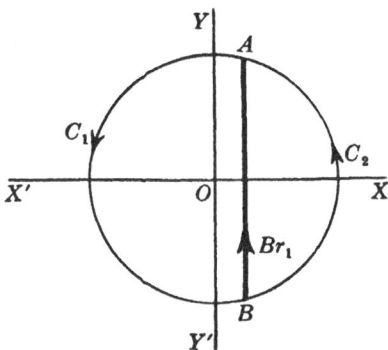

Fig. 29.

give repetitions due to the presence of n^2. Proceeding as in § 4·54, we have to show that the integral round a large semicircle radius R, on the left of Br_1 (taken as the imaginary axis indented on its right at the origin) which intersects the negative real axis *between* the mth and $(m+1)$th poles, vanishes as $R \to \infty$. By § 4·31 this is true if $|\sinh r \sqrt{z}/rz \sinh a \sqrt{z}| \to 0$ as $R \to \infty$ ($\frac{1}{2}\pi \leqslant \theta \leqslant \frac{3}{2}\pi$, $0 \leqslant r \leqslant a$).

Demonstration. When $r = a$,

$$|\phi(z)/z| = 1/a\,|z| \to 0 \quad \text{as} \quad |z| = R \to \infty. \qquad (2)$$

For the range $0 \leqslant r < a$, take $z = R e^{i\theta}$ on C_1 (see Fig. 28), then

$$\sqrt{z} = \sqrt{R}\,(\cos \tfrac{1}{2}\theta + i \sin \tfrac{1}{2}\theta),$$

with

$$r\sqrt{z} = p+iq, \quad a\sqrt{z} = p_1+iq_1, \quad p = r\sqrt{R}\cos\tfrac{1}{2}\theta,$$
$$q = r\sqrt{R}\sin\tfrac{1}{2}\theta, \quad p_1 = a\sqrt{R}\cos\tfrac{1}{2}\theta, \quad q_1 = a\sqrt{R}\sin\tfrac{1}{2}\theta.$$

Thus we obtain

$$\frac{\phi(z)}{z} = \frac{1}{rRe^{i\theta}}\left[\frac{\sinh p \cos q + i\cosh p \sin q}{\sinh p_1 \cos q_1 + i\cosh p_1 \sin q_1}\right], \tag{3}$$

so

$$\left|\frac{\phi(z)}{z}\right| = \frac{1}{rR}\sqrt{\left[\frac{\cosh 2p - \cos 2q}{\cosh 2p_1 - \cos 2q_1}\right]}. \tag{4}$$

On C_1, excluding the neighbourhood of $\theta = \pi$, $\cos\tfrac{1}{2}\theta \neq 0$, and (4) may be written asymptotically as $e^{-(a-r)\sqrt{R}\cos\frac{1}{2}\theta}/rR$, which vanishes when $R\to\infty$, $0 < r < a$.

In the neighbourhood of $\theta = \pi$, with R finite, $0 < r < a$, (4) is approximately

$$\frac{1}{rR}\sqrt{\left[\frac{1-\cos 2r\sqrt{R}}{1-\cos 2a\sqrt{R}}\right]}. \tag{5}$$

Since C_1 passes *between* consecutive poles on the negative real axis, $|\sinh a\sqrt{z}|$ and, therefore, $\sqrt{(1-\cos 2a\sqrt{R})} > 0$. Hence (5) vanishes as $R\to\infty$. With R finite, as $r\to 0$, $\tfrac{1}{2}\pi \leqslant \theta \leqslant \tfrac{3}{2}\pi$, (4) takes the indeterminate form $0/0$. Expanding the numerator, we get

$$\sqrt{2}\sqrt{[r^2R\cos^2\tfrac{1}{2}\theta + \tfrac{1}{3}r^4R^2\cos^4\tfrac{1}{2}\theta + \dots}$$
$$+ (r^2R\sin^2\tfrac{1}{2}\theta - \tfrac{1}{3}r^4R^2\sin^4\tfrac{1}{2}\theta + \dots)]. \tag{6}$$

When r is small enough, (6) is approximately $\sqrt{(2R)}r$, so by (4)

$$|\phi(z)/z| \simeq \sqrt{[2/(\cosh 2p_1 - \cos 2q_1)R]} \to 0 \tag{7}$$

as $R\to\infty$, since from above $\sqrt{(\cosh 2p_1 - \cos 2q_1)} > 0$. Hence we have proved that $|\phi(z)/z| \to 0$ on C_1 as $R\to\infty$, $0 \leqslant r \leqslant a$.

Evaluation at poles other than at $z = 0$. In virtue of the above demonstration, if $t > 0$, by §§ 3·237, 4·54

$$I_2 = \sum_{n=1}^{\infty}\left[\frac{e^{zt}\sin(ir\sqrt{z})}{rz\,d(\sin ia\sqrt{z})/dz}\right]_{z=-n^2\pi^2/a^2} \tag{8}$$

$$= 2\sum_{n=1}^{\infty}\left[\frac{e^{zt}\sin(ir\sqrt{z})}{r\{ia\sqrt{z}\cos(ia\sqrt{z})\}}\right]_{z=-n^2\pi^2/a^2,\ ia\sqrt{z}=n\pi} \tag{9}$$

$$= \frac{2}{a}\sum_{n=1}^{\infty}(-1)^n e^{-n^2\pi^2t/a^2}\frac{\sin(n\pi r/a)}{(n\pi r/a)}. \tag{10}$$

Hence by (1), (10)

$$\frac{1}{2\pi i}\int_{Br_1}\frac{e^{zt}\sinh r\sqrt{z}\,dz}{rz\sinh a\sqrt{z}} = \frac{1}{a}\left[1 + 2\sum_{n=1}^{\infty}(-1)^n e^{-n^2\pi^2 t/a^2}\frac{\sin(n\pi r/a)}{(n\pi r/a)}\right]. \tag{11}$$

Consideration of (11) *with* $t = 0$. From §4 Appendix 9, the 'Σ' part of (11) converges uniformly in $0 < r \leqslant a$, $0 \leqslant t \leqslant t_1$, while the integral does likewise. Thus (11) is valid for $t = 0$, in this r-range, and we get

$$\frac{1}{2\pi i}\int_{Br_1}\frac{\sinh r\sqrt{z}\,dz}{rz\sinh a\sqrt{z}} = \frac{1}{a}\left[1 + 2\sum_{n=1}^{\infty}(-1)^n\frac{\sin(n\pi r/a)}{(n\pi r/a)}\right]. \tag{12}$$

By §4·71, Br_1 is equivalent to $BC_2 A$ in Fig. 29 as its radius $R \to \infty$, and we shall study

$$I_3 = \int_{BC_2 A}\frac{\sinh r\sqrt{z}\,dz}{rz\sinh a\sqrt{z}}. \tag{13}$$

For large $|u| = |\sqrt{z}|$, if $R(u) > 0$, $\sinh u \sim \frac{1}{2}e^u$, so (13) may be written

$$I_3 \sim \int_{BC_2 A}\frac{e^{-(a-r)\sqrt{z}}\,dz}{rz} \to 0 \tag{14}$$

as $R \to \infty$, if $0 < r < a$. When $r \to 0$, (13) yields

$$\int_{BC_2 A}\frac{e^{-a\sqrt{z}}\,dz}{\sqrt{z}} \to 0 \tag{15}$$

as $R \to \infty$. Hence by §4·71, $I_3 = 0$, $0 \leqslant r < a$. It follows that the left-hand side of (12) vanishes in $0 \leqslant r < a$, and when $r = a$, both sides have the value $1/a$. The right-hand side does not hold for $r = 0$, so in the range $0 < r < a$,

$$1 + 2\left|\sum_{n=1}^{\infty}(-1)^n\frac{\sin(n\pi r/a)}{(n\pi r/a)}\right| = 0. \tag{16}$$

4·73. Example.* Evaluate

$$\frac{1}{2\pi i}\int_{Br_1}\frac{e^{zt}J_0(ir\sqrt{z})\,dz}{zJ_0(ia\sqrt{z})} \quad (0 < r < a). \tag{1}$$

* It is assumed that the reader is acquainted with, say, chapters I, II, reference [234]. See p. 216 for first ten zeros of $J_0(u)$.

When $z \to 0$ both of the Bessel functions $\to 1$, so the origin is a simple pole the contribution from which is unity. The remaining poles occur when $J_0(ia \sqrt{z}) = 0$, so $ia \sqrt{z} = \alpha_n$ or $z = -\alpha_n^2/a^2$, where $\alpha_1, \alpha_2, \ldots, \alpha_n, \ldots$ are the positive zeros of $J_0(u)$, and there is a limit point at infinity. As in §§ 4·54, 4·72, it may be demonstrated that the integral round a large semicircle on the left of Br_1 tends to zero as the radius $\to \infty$. Then

$$I_2 = \sum_{n=1}^{\infty} \left[\frac{e^{zt} J_0(ir \sqrt{z})}{z \dfrac{d}{dz} J_0(ia \sqrt{z})} \right]_{z=-\alpha_n^2/a^2}, \tag{2}$$

or

$$I_2 = -2 \sum \left[\frac{e^{zt} J_0(ir \sqrt{z})}{ia \sqrt{z} J_1(ia \sqrt{z})} \right]_{\substack{z=-\alpha_n^2/a^2 \\ ia\sqrt{z}=\alpha_n}}, \tag{3}$$

since

$$\frac{d}{dz} J_0(ia \sqrt{z}) = \frac{dJ_0(ia \sqrt{z})}{d(ia \sqrt{z})} \frac{d(ia \sqrt{z})}{dz} = -J_1(ia \sqrt{z}) \frac{d(ia \sqrt{z})}{dz};$$

so

$$I_2 = -2 \sum_{n=1}^{\infty} \frac{e^{-\alpha_n^2 t/a^2} J_0(\alpha_n r/a)}{\alpha_n J_1(\alpha_n)}. \tag{4}$$

Hence adding the contribution from the origin to (4), we find that

$$\frac{1}{2\pi i} \int_{Br_1} \frac{e^{zt} J_0(ir \sqrt{z}) \, dz}{z J_0(ia\sqrt{z})} = 1 - 2 \sum_{n=1}^{\infty} \frac{e^{-\alpha_n^2 t/a^2} J_0(\alpha_n r/a)}{\alpha_n J_1(\alpha_n)}. \tag{5}$$

Proceeding as in § 4·72, by aid of (5) Appendix 9, (5) above holds for $t = 0$, $0 < r < a$, so

$$\frac{1}{2\pi i} \int_{Br_1} \frac{J_0(ir \sqrt{z}) \, dz}{z J_0(ia \sqrt{z})} = 1 - 2 \sum_{n=1}^{\infty} \frac{J_0(\alpha_n r/a)}{\alpha_n J_1(\alpha_n)}. \tag{6}$$

Also the integral is zero in $0 < r < a$, but has the value unity for $r = a$. Hence

$$\sum_{n=1}^{\infty} \frac{J_0(\alpha_n r/a)}{\alpha_n J_1(\alpha_n)} = \frac{1}{2} \quad (0 < r < a). \tag{7}$$

4·74. In the case of

$$I = \frac{1}{2\pi i} \int_{Br_1} \frac{e^{zt} J_0(ir \sqrt{z}) \, dz}{z^2 J_0(ia \sqrt{z})} \quad (0 \leqslant r \leqslant a), \tag{1}$$

the residue at $z = 0$ may be found by expanding the integrand. We
obtain

$$(1 + zt + \ldots) \left(1 + \frac{r^2 z}{4} + \ldots\right) \Big/ z^2 \left(1 + \frac{a^2 z}{4} + \ldots\right). \tag{2}$$

When $z \to 0$, (2) can be written

$$(1 + zt + \ldots) \left(1 + \frac{z}{4}(r^2 - a^2) + \ldots\right) \Big/ z^2 \tag{3}$$

$$= \frac{1}{z^2}\{1 + z[t + (r^2 - a^2)/4] + \text{terms containing higher powers of } z\}. \tag{4}$$

The residue at the origin is the coefficient of $1/z$ in (4), namely, $t + (r^2 - a^2)/4$.
Proceeding as in § 4·73 we find that the contribution from the singularities
arising from the zeros of $J_0(ia\sqrt{z})$ is

$$+ 2a^2 \sum_{n=1}^{\infty} \frac{e^{-a_n^2 t/a^2} J_0(\alpha_n r/a)}{\alpha_n^3 J_1(\alpha_n)}. \tag{5}$$

Adding the residue at O to (5), we obtain

$$\frac{1}{2\pi i} \int_{Br_1} \frac{e^{zt} J_0(ir\sqrt{z})\, dz}{z^2 J_0(ia\sqrt{z})} = t + \tfrac{1}{4}(r^2 - a^2) + 2a^2 \sum_{n=1}^{\infty} \frac{e^{-a_n^2 t/a^2} J_0(\alpha_n r/a)}{\alpha_n^3 J_1(\alpha_n)}. \tag{6}$$

Proceeding as in §§ 4·72, 4·73, by aid of §§ 5, 6, Appendix 9, we find that
(6) is valid in $0 \leqslant r \leqslant a$, $0 \leqslant t \leqslant t_1$.

When $t = 0$ in (1), using the contour of Fig. 29, we have

$$\frac{1}{2\pi i} \int_{BC_2A} \frac{J_0(ir\sqrt{z})\, dz}{z^2 J_0(ia\sqrt{z})} = 0, \tag{7}$$

so it follows that

$$\sum_{n=1}^{\infty} \frac{J_0(\alpha_n r/a)}{\alpha_n^3 J_1(\alpha_n)} = \tfrac{1}{8}(1 - r^2/a^2) \quad (0 \leqslant r \leqslant a). \tag{8}$$

This result will be understood more readily when it is remembered that
$J_1(\alpha_1)$ is positive, $J_1(\alpha_2)$ is negative, and so on alternately.

Note on 2°, § 4·27. The integral may be evaluated also on a circular
contour C_1, centre O in Fig. 26 (b) and radius $R > 1$. Since $|z| > 1$ on C_1,
the integrand may be expanded binomially, and integrated term by
term using (3) § 4·51. Thus:

$$I = \frac{1}{2\pi i} \int_{C_1} \frac{e^{zt}\, dz}{z\sqrt{(1 - 1/z^2)}} = \frac{1}{2\pi i} \int_{C_1} \frac{e^{zt}\, dz}{z} \left(1 + \frac{1}{2z^2} + \frac{1.3}{2^2.2!} \cdot \frac{1}{z^4} + \frac{1.3.5}{2^3.3!} \cdot \frac{1}{z^6} + \ldots\right)$$

$$= \text{the result at (8), } 2°, \S 4·27.$$

V

GAMMA, ERROR AND BESSEL FUNCTIONS

5·111. The gamma function $\Gamma(z)$. This function was introduced by Euler, and his definition is given in § 5·13. We introduce it now by the following definition due to Weierstrass. For all real or complex values of z

$$\frac{1}{\Gamma(z)} = z\,e^{\gamma z} \prod_{m=1}^{\infty} \left\{ \left(1 + \frac{z}{m}\right) e^{-z/m} \right\}, \tag{1}$$

where Π signifies that the product of the terms is to be taken. γ is Euler's well-known constant defined by

$$\gamma = \lim_{m \to \infty} \left[1 + \frac{1}{2} + \frac{1}{3} + \dots + \frac{1}{m} - \log m \right] \simeq 0.5772. \tag{2}$$

Using (1) it can be shown that

$$\Gamma(z) = \lim_{m \to \infty} \frac{m!\, m^z}{z(z+1)(z+2)\dots(z+m)}. \tag{3}$$

Writing $(z+1)$ for z in (3), we get

$$\Gamma(z+1) = z \lim_{m \to \infty} \frac{m!\, m^z m}{z(z+1)(z+2)\dots(z+m)(z+m+1)} \tag{4}$$

$$= z\Gamma(z). \tag{5}$$

In like manner we have

$$\Gamma(z) = (z-1)\,\Gamma(z-1) = (z-1)(z-2)\,\Gamma(z-2), \text{ and so on;} \tag{6}$$

$$\Gamma(1-z) = -z\Gamma(-z) = (-z)(-z-1)\,\Gamma(-z-1), \text{ and so on.} \tag{7}$$

When $z = n$, a positive integer,

$$\Gamma(n+1) = n(n-1)(n-2)\dots 1 = n!. \tag{8}$$

5·112. We shall now prove that

$$\Gamma(z)\,\Gamma(1-z) = \pi/\sin \pi z. \tag{1}$$

To do so we use formula (1) in § 1·521, namely,

$$\sin z = z \prod_{m=1}^{\infty} \left(1 - \frac{z^2}{m^2 \pi^2}\right). \tag{2}$$

From (1) § 5·111

$$\frac{1}{\Gamma(z)}\frac{1}{(-z)\,\Gamma(-z)} = z\,e^{\gamma z}\prod_{m=1}^{\infty}\left\{\left(1+\frac{z}{m}\right)e^{-z/m}\right\}e^{-\gamma z}\prod_{m=1}^{\infty}\left\{\left(1-\frac{z}{m}\right)e^{z/m}\right\} \tag{3}$$

so $$\frac{1}{\Gamma(z)\,\Gamma(1-z)} = z\prod_{m=1}^{\infty}\left(1-\frac{z^2}{m^2}\right) = \frac{\sin \pi z}{\pi}, \tag{4}$$

from (2), whence $\Gamma(z)\,\Gamma(1-z) = \pi/\sin \pi z.$ (5)

5·12. Evaluation of $\Gamma(z)$ when z is half an odd integer.
In (5) § 5·112 put $z = \frac{1}{2}$, and we get

$$[\Gamma(\tfrac{1}{2})]^2 = \pi,$$

so $$\Gamma(\tfrac{1}{2}) = \sqrt{\pi}, \tag{1}$$

since by (3) § 5·111 or (1) § 5·13, $\Gamma(\tfrac{1}{2}) > 0$. To evaluate, say, $\Gamma(\tfrac{9}{2})$, we use the factorial property of the gamma function. Thus

$$\Gamma(\tfrac{9}{2}) = \tfrac{7}{2}\Gamma(\tfrac{7}{2}) = \tfrac{7}{2}\cdot\tfrac{5}{2}\Gamma(\tfrac{5}{2}) = \tfrac{7}{2}\cdot\tfrac{5}{2}\cdot\tfrac{3}{2}\cdot\tfrac{1}{2}\Gamma(\tfrac{1}{2})$$

$$= \frac{1\,.\,3\,.\,5\,.\,7}{2^4}\sqrt{\pi}. \tag{2}$$

In like manner, if n is an even positive integer

$$\Gamma\left(\frac{n+1}{2}\right) = \left(\frac{n-1}{2}\right)\Gamma\left(\frac{n-1}{2}\right) = \left(\frac{n-1}{2}\right)\left(\frac{n-3}{2}\right)\Gamma\left(\frac{n-3}{2}\right)$$

$$= \frac{1\,.\,3\,.\,5\,\dots\,(n-1)}{2^{\frac{1}{2}n}}\sqrt{\pi}. \tag{3}$$

When z is half an odd negative integer, put $(z+m) = \frac{1}{2}$ or $m = \frac{1}{2} - z$, m being a positive integer. Then

$$\Gamma(z+m) = (z+m-1)(z+m-2)\dots(z+1)z\Gamma(z),$$

so $$\Gamma(z) = \Gamma(z+m)/[(z+m-1)(z+m-2)\dots z] \tag{4}$$

$$= \sqrt{\pi}/(-\tfrac{1}{2})(-\tfrac{3}{2})\dots(\tfrac{1}{2}-m)$$

$$= (-1)^m\,2^m\,\sqrt{\pi}/1\,.\,3\,.\,5\,\dots\,(2m-1). \tag{5}$$

As particular cases we have $\Gamma(-\tfrac{1}{2}) = -2\sqrt{\pi}$; $\Gamma(-\tfrac{3}{2}) = 4\sqrt{\pi}/1\,.\,3$; $\Gamma(-\tfrac{5}{2}) = -8\sqrt{\pi}/1\,.\,3\,.\,5$; $\Gamma(-\tfrac{7}{2}) = 16\sqrt{\pi}/1\,.\,3\,.\,5\,.\,7$, the signs being alternately positive and negative. The gamma function is plotted for real values of z in Fig. 30. It is never zero, and when $z > 0$, $\Gamma(z)$ is positive.

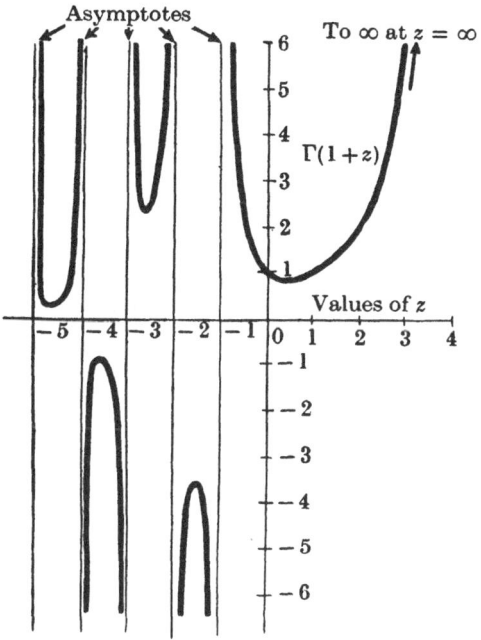

Fig. 30. The gamma function $\Gamma(1+z)$, z real.

5·13. $\Gamma(z)$ expressed as a real integral. Euler gave the definition

$$\Gamma(z) = \int_0^\infty e^{-x} x^{z-1} dx, \tag{1}$$

provided $R(z) > 0$. This restriction makes (1) less general than (1) § 5·111. Writing xt for x in (1), where t is real and positive, we have

$$\int_0^\infty e^{-xt} x^{z-1} dx = \frac{\Gamma(z)}{t^z}. \tag{2}$$

By aid of (1) we can evaluate

$$\int_0^\infty e^{-x^n} dx, \tag{3}$$

n being a positive integer > 0. Substituting $x^n = y$, we get

$$dx = \frac{dy}{nx^{n-1}} = y^{\frac{1}{n}-1}\frac{dy}{n},$$

so (3) becomes

$$\frac{1}{n}\int_0^\infty e^{-y}\, y^{\frac{1}{n}-1}\, dy = \frac{1}{n}\,\Gamma\left(\frac{1}{n}\right) = \Gamma(1+1/n), \tag{4}$$

by (1) and (5) § 5·111. When $n = 2$, we obtain the important result

$$\int_0^\infty e^{-x^2}\, dx = \Gamma(\tfrac{3}{2}) = \tfrac{1}{2}\sqrt{\pi}, \tag{5}$$

by § 5·12. More generally it can be shown by substituting $a^m x^n = y$ that, if $R(a) > 0$, m and n being non-zero positive integers,

$$\int_0^\infty e^{-a^m x^n}\, dx = \frac{\Gamma(1+1/n)}{a^{m/n}} = \Gamma(1+1/n)\,a^{-m/n}. \tag{6}$$

Another result of importance is

$$\Gamma(2z) = \frac{2^{2z-1}\Gamma(z)\,\Gamma(z+\tfrac{1}{2})}{\sqrt{\pi}}, \tag{7}$$

being the duplication formula for the gamma function.

5·141. Contour integral for $1/\Gamma(1+\nu)$, $R(\nu) < 0$. Consider the integral $\dfrac{1}{2\pi i}\displaystyle\int_{Br_1} \dfrac{e^z\, dz}{z^{\nu+1}}$ taken along the path shown in Fig. 31. On the small circle round O, put $z = r\,e^{i\theta}$, and we get*

$$I_1 = \frac{1}{2\pi}\int_{-\pi}^{\pi} e^{r(\cos\theta + i\sin\theta) - i\nu\theta}\, r^{-\nu}\, d\theta. \tag{1}$$

As $r \to 0$ the integral does likewise, provided $R(\nu) < 0$, so the value of the integral is zero in the limit. On the line below the barrier, we transfer to the real variable by writing $z = x\,e^{-i\pi}$, where x signifies $|x|$. Then

$$I_2 = \frac{e^{\nu\pi i}}{2\pi i}\int_\infty^0 \frac{e^{-x}\, dx}{x^{\nu+1}} = -\frac{e^{\nu\pi i}}{2\pi i}\int_0^\infty \frac{e^{-x}\, dx}{x^{\nu+1}}. \tag{2}$$

---

* See first footnote to § 4·25.

On the line above the barrier, we write $z = xe^{i\pi}$, and obtain

$$I_3 = \frac{e^{-\nu\pi i}}{2\pi i} \int_0^\infty \frac{e^{-x}\,dx}{x^{\nu+1}}. \tag{3}$$

Adding (2) and (3),

$$I_2 + I_3 = \frac{\sin(-\nu\pi)}{\pi} \int_0^\infty e^{-x} x^{-\nu-1}\,dx \tag{4}$$

$$= -\frac{\sin \nu\pi}{\pi} \Gamma(-\nu), \tag{5}$$

provided $R(\nu) < 0$, since the integral (4) is valid only on this condition.

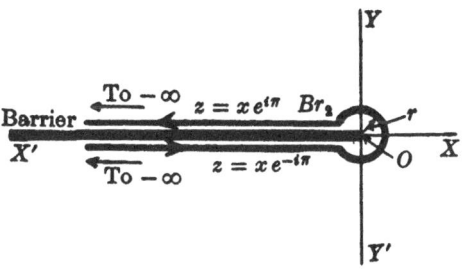

Fig. 31.

But by (5) § 5·112, with $-\nu$ for z,

$$-\frac{\sin \nu\pi}{\pi} = \frac{1}{\Gamma(-\nu)\,\Gamma(1+\nu)},$$

so (5) gives

$$\frac{1}{2\pi i} \int_{Br_2} \frac{e^z\,dz}{z^{\nu+1}} = \frac{1}{\Gamma(1+\nu)}, \tag{6}$$

$R(\nu) < 0$, since the contribution round the origin is zero.

5·142. Extension of range of ν in (6) §5·141. (1) §5·111 is analytic for all values of z, so that $1/\Gamma(1+\nu)$, as defined there, is analytic for all ν. But (6) § 5·141 holds for $R(\nu) < 0$, and as both sides are analytic for all ν, so by the principle of analytical continuation in § 2·72, (6) § 5·141 is valid for all ν.

5·15. By §4·31 the contours Br_1, Br_2 are equivalent for integral (6) § 5·141 when $R(\nu) > -1$, so we may write

$$\frac{1}{2\pi i}\int_{Br_1} \frac{e^z\,dz}{z^{\nu+1}} = \frac{1}{\Gamma(1+\nu)}, \tag{1}$$

provided $R(\nu) > -1$, and

$$\frac{1}{2\pi i}\int_{Br_2} \frac{e^z\,dz}{z^{\nu+1}} = \frac{1}{\Gamma(1+\nu)}, \tag{2}$$

for all values of ν. The contours are unaltered in type if z is replaced by zt, t being real and positive, so from (1), (2) we have

$$\frac{1}{2\pi i}\int_{Br_1} \frac{e^{zt}\,dz}{z^{\nu+1}} = \frac{t^\nu}{\Gamma(1+\nu)}, \tag{3}$$

provided $R(\nu) > -1$, and

$$\frac{1}{2\pi i}\int_{Br_2} \frac{e^{zt}\,dz}{z^{\nu+1}} = \frac{t^\nu}{\Gamma(1+\nu)}, \tag{4}$$

for all values of ν.

5·211. The error function. This function due to Gauss is defined by

$$Z(x) = e^{-\frac{1}{2}x^2}/\sqrt{(2\pi)}, \tag{1}$$

using the notation of reference [238]. It is illustrated by the curve of Fig. 32, which is symmetrical about the vertical axis. Statisticians call this a 'normal curve' or the Gaussian error function. The area between the curve and the x-axis from $-\infty$ to $+\infty$ is unity, which provides a convenient datum. From (1) we have

$$\int_{-\infty}^\infty Z(x)\,dx = \frac{1}{\sqrt{(2\pi)}}\int_{-\infty}^\infty e^{-\frac{1}{2}x^2}\,dx. \tag{2}$$

Writing $x^2 = 2y$, $dx = dy/\sqrt{(2y)}$, we get

$$\int_{-\infty}^\infty Z(x)\,dx = \frac{1}{\sqrt{\pi}}\int_0^\infty e^{-y}y^{-\frac{1}{2}}\,dy = \frac{\Gamma(\frac{1}{2})}{\sqrt{\pi}} = 1. \tag{3}$$

The area of the curve from $-\infty$ to x is represented by

$$\int_{-\infty}^x Z(x)\,dx = \frac{1}{\sqrt{(2\pi)}}\int_{-\infty}^x e^{-\frac{1}{2}x^2}\,dx, \tag{4}$$

and it is termed the error, normal, or Gaussian integral.

Comprehensive tables based upon formulae (1) and (4) are used extensively by statisticians (see [238]). The quantities tabulated are $\frac{1}{2}[1+\alpha(x)]$, this being integral (4), against x, x against $\frac{1}{2}[1+\alpha(x)]$ as independent variable and $Z(x)$ against x. $\alpha(x)$ is given by

$$\alpha(x) = \sqrt{\frac{2}{\pi}} \int_0^x e^{-\frac{1}{2}t^2}\, dt. \tag{5}$$

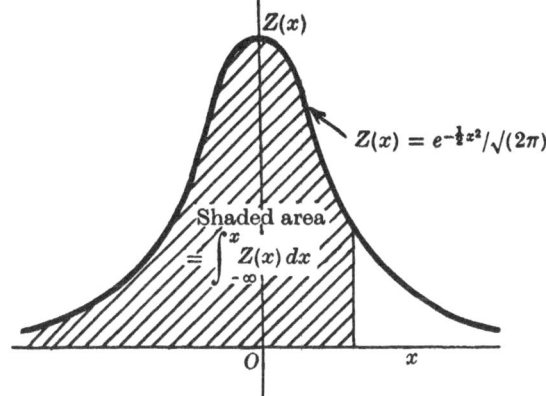

Fig. 32. The function $Z(x)$.

Since
$$\sqrt{\frac{2}{\pi}} \int_{-\infty}^0 e^{-\frac{1}{2}t^2}\, dt = 1, \tag{6}$$

from (2) and (3), we see that

$$\frac{\{1+\alpha(x)\}}{2} = \frac{1}{\sqrt{(2\pi)}} \int_{-\infty}^x e^{-\frac{1}{2}t^2}\, dt, \tag{7}$$

and when $x = \infty$, $\frac{1}{2}[1+\alpha(x)] = 1$, since $\alpha(x)$ varies from 0 to 1 as x varies from 0 to ∞.

5·212. Definition of erf x the error function integral.
In works on mathematical physics and technical mathematics, the following definition is used:

$$\text{erf}\, x = \frac{2}{\sqrt{\pi}} \int_0^x e^{-v^2}\, dy = \frac{1}{\sqrt{\pi}} \int_0^{x^2} e^{-v}\, v^{-\frac{1}{2}}\, dv. \tag{1}$$

If in (5) § 5·211 we write $x\sqrt{2}$ for x and $y\sqrt{2}$ for t, we obtain

$$\alpha(x\sqrt{2}) = \frac{2}{\sqrt{\pi}} \int_0^x e^{-y^2} dy = \mathrm{erf}\, x, \qquad (2)$$

so $\qquad \tfrac{1}{2}[1 + \alpha(x\sqrt{2})] = \tfrac{1}{2}(1 + \mathrm{erf}\, x) = G$ (after Gauss). $\qquad (3)$

Knowing the value of x, we calculate $x\sqrt{2}$, and then find $G = \tfrac{1}{2}[1 + \alpha(x\sqrt{2})]$ from statistical tables, which enables

$$\mathrm{erf}\, x = (2G - 1) \qquad (4)$$

to be found.

Expanding the exponential in (1), we have

$$\mathrm{erf}\, x = \frac{2}{\sqrt{\pi}} \int_0^x \left[1 - y^2 + \frac{y^4}{2!} - \frac{y^6}{3!} + \frac{y^8}{4!} - \ldots \right] dy \qquad (5)$$

$$= \frac{2}{\sqrt{\pi}} \left[x - \frac{x^3}{1!\,3} + \frac{x^5}{2!\,5} - \frac{x^7}{3!\,7} + \ldots \right]. \qquad (6)$$

term by term integration being permissible [215, § 176 B]. Series (6) is uniformly convergent, so $\mathrm{erf}\, x$ is a continuous function of x.

Writing $y^2 = v$ in (1) an integral identical with (3) § 5·211 is obtained, so

$$\frac{2}{\sqrt{\pi}} \int_0^\infty e^{-v^2} dy = 1, \qquad (7)$$

and from (1), $\qquad \dfrac{2}{\sqrt{\pi}} \displaystyle\int_x^\infty e^{-v^2} dy = 1 - \mathrm{erf}\, x = \mathrm{erfc}\, x. \qquad (8)$

This is defined to be the *complementary* error function integral. The graphical forms of $\mathrm{erf}\, u$ and $\mathrm{erfc}\, u$ are depicted in Fig. 33.

The reader will be able to verify the following:

$$\mathrm{erf}\,(-x) = -\mathrm{erf}\, x, \quad \mathrm{erfc}\,(-x) = 1 + \mathrm{erf}\, x = 2 - \mathrm{erfc}\, x,$$

$$\mathrm{erfc}\,(-x) + \mathrm{erfc}\, x = 2; \qquad (9)$$

$$\frac{d}{dx}(\mathrm{erf}\, ax) = \frac{2a}{\sqrt{\pi}} e^{-a^2 x^2}, \quad \frac{d}{dx}(\mathrm{erfc}\, ax) = -\frac{2a}{\sqrt{\pi}} e^{-a^2 x^2}; \qquad (10)$$

$$\int_0^t \mathrm{erf}\, ax\, dx = [x\,\mathrm{erf}\, ax]_0^t - \frac{2a}{\sqrt{\pi}} \int_0^t e^{-a^2 x^2} x\, dx \qquad (11)$$

$$= t\,\mathrm{erf}\, at + \frac{e^{-a^2 t^2}}{a\sqrt{\pi}}; \qquad (12)$$

$$\int_0^t \mathrm{erfc}\, ax\, dx = \int_0^t (1 - \mathrm{erf}\, ax)\, dx = t\,\mathrm{erfc}\, at - \frac{e^{-a^2 t^2}}{a\sqrt{\pi}}. \qquad (13)$$

5·221. Contour integral for erfc x. Writing $z = \frac{1}{2}s$ in (6) §5·212 yields

$$1 - \mathrm{erf}\,\tfrac{1}{2}s = 1 - \frac{2}{\sqrt{\pi}}\left[\frac{s}{2} - \frac{s^3}{1!\,2^3.3} + \frac{s^5}{2!\,2^5.5} - \frac{s^7}{3!\,2^7.7} + \dots\right], \quad (1)$$

so

$$\mathrm{erfc}\,\tfrac{1}{2}s = \sum_{r=0}^{\infty} \frac{(-1)^r\,s^r}{r!\,\Gamma(1 - \frac{1}{2}r)}. \quad (2)$$

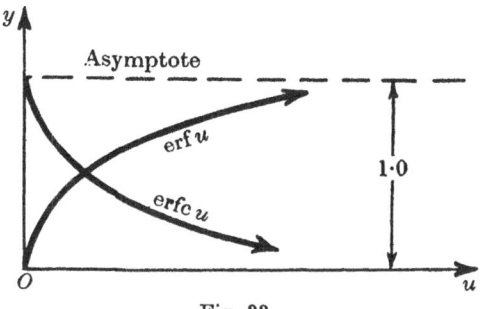

Fig. 33.

Expressing the gamma function as a contour integral using (2) §5·15—integral (1) §5·15 is divergent when $r \geqslant 2$—(2) becomes

$$\mathrm{erfc}\,\tfrac{1}{2}s = \sum_{r=0}^{\infty} \frac{(-1)^r}{2\pi i} \int_{Br_2} \frac{s^r}{r!} \frac{e^\zeta\,d\zeta}{\zeta^{1-\frac{1}{2}r}} \quad (3)$$

$$= \sum_{r=0}^{\infty} \frac{(-1)^r}{2\pi i} \int_{Br_2} \frac{e^\zeta}{\zeta} \frac{(s\sqrt{\zeta})^r}{r!}\,d\zeta \quad (4)$$

$$= \frac{1}{2\pi i} \int_{Br_1} e^{\zeta - s\sqrt{\zeta}} \frac{d\zeta}{\zeta}, \quad (5)$$

since it is permissible to take Σ inside the integral sign (see [215]). In (5) Br_1 is equivalent to Br_2, because the integrals along the arcs in the second and third quadrants $\to 0$ as the radius $\to \infty$, if s real > 0. The integrand in (5) has a branch point at the origin. Writing $\zeta = zt$ and $s = \sqrt{(a/t)}$ in (5), a real > 0, we get

$$\mathrm{erfc}\,\tfrac{1}{2}\sqrt{(a/t)} = \frac{1}{2\pi i} \int_{Br_1} e^{zt - \sqrt{(az)}} \frac{dz}{z} = \frac{2}{\sqrt{\pi}} \int_{\frac{1}{2}\sqrt{(a/t)}}^{\infty} e^{-v^2}\,dy, \quad (6)$$

from (8) §5·212, so

$$\mathrm{erf}\,\tfrac{1}{2}\sqrt{(a/t)} = 1 - \frac{1}{2\pi i} \int_{Br_1} e^{zt - \sqrt{(az)}} \frac{dz}{z}, \quad (7)$$

and $$\operatorname{erf}\tfrac{1}{2}(a/t) = 1 - \frac{1}{2\pi i}\int_{Br_1} e^{zt^2-a\sqrt{z}}\frac{dz}{z}. \tag{8}$$

Although (6) is convergent on Br_1, if $e^{-\sqrt{(az)}}$ be expanded, all integrals after the second diverge. Consequently, if (6) were to be evaluated term by term it would be necessary to integrate along the equivalent contour Br_2, on which all integrals converge. It should be observed that Br_1 and Br_2 are equivalent for the integral (6), but not for individual terms in the expansion, after the second (see §2, Appendix 7). This integral may be transformed to a real one, when integrating along Br_2. It then takes the form

$$\operatorname{erfc}\tfrac{1}{2}\sqrt{(a/t)} = 1 - \frac{1}{\pi}\int_0^\infty e^{-xt}\sin\sqrt{(ax)}\,\frac{dx}{x}, \tag{9}$$

which may be evaluated by expanding $\sin\sqrt{(ax)}$ and integrating term by term, since the requisite conditions are satisfied (see [215], §176B).

5·222. If both sides of (6) §5·221 are differentiated with respect to t,* using Br_2 we have, with a real > 0,

$$\frac{1}{2\pi i}\int_{Br_2} e^{zt-\sqrt{(az)}}\,dz = -\frac{d}{dt}\operatorname{erf}\tfrac{1}{2}\sqrt{(a/t)}. \tag{1}$$

From (6) §5·212 we find that when $x = \tfrac{1}{2}\sqrt{(a/t)}$,

$$-\frac{d}{dx}(\operatorname{erf} x)\frac{dx}{dt} = \tfrac{1}{2}\sqrt{(a/\pi t)}\,e^{-a/4t}/t, \tag{2}$$

so from (1) and (2)

$$\frac{1}{2\pi i}\int_{\substack{Br_1\\Br_2}} e^{zt-\sqrt{(az)}}\,dz = \tfrac{1}{2}\sqrt{(a/\pi t)}\,e^{-a/4t}/t. \tag{3}$$

In like manner by differentiating (6) §5·221 with respect to a,† we obtain $$\frac{1}{2\pi i}\int_{\substack{Br_1\\Br_2}} \frac{e^{zt-\sqrt{(az)}}\,dz}{\sqrt{z}} = \frac{e^{-a/4t}}{\sqrt{(\pi t)}}. \tag{4}$$

* For validity conditions see §7·11. The integral (1) is convergent on Br_1, if a real > 0.

† For the question of validity conditions see §7·11.

5·23. Asymptotic expansion of erfc x. It frequently happens that an asymptotic series is useful for computation for large values of x (see Appendix 3). From (8) §5·212, writing $y^2 = v$, and changing the limits of integration accordingly, we have

$$\operatorname{erfc} x = \frac{1}{\sqrt{\pi}} \int_{x^2}^{\infty} e^{-v} v^{-\frac{1}{2}} dv = \frac{1}{\sqrt{\pi}} \int_{\infty}^{x^2} v^{-\frac{1}{2}} d(e^{-v}) \tag{1}$$

$$= \frac{1}{\sqrt{\pi}} \left[e^{-v} v^{-\frac{1}{2}} \right]_{\infty}^{x^2} - \frac{1}{2\sqrt{\pi}} \int_{\infty}^{x^2} v^{-\frac{3}{2}} d(e^{-v}) \tag{2}$$

$$= \frac{1}{\sqrt{\pi}} \left[e^{-x^2} x^{-1} - \frac{e^{-x^2} x^{-3}}{2} + \frac{1.3}{2^2} \int_{\infty}^{x^2} v^{-\frac{5}{2}} d(e^{-v}) \right]. \tag{3}$$

Continuing the process of partial integration, we obtain

$$\operatorname{erfc} x = \frac{e^{-x^2}}{x\sqrt{\pi}} \left[1 - \frac{1}{2x^2} + \frac{1.3}{2^2 x^4} - \frac{1.3.5}{2^3 x^6} + \ldots \right.$$

$$\left. + (-1)^{n-1} \frac{1.3.5 \ldots (2n-3)}{2^{n-1} x^{2n-2}} \right]$$

$$+ \frac{(-1)^n}{\sqrt{\pi}} \frac{1.3.5 \ldots (2n-1)}{2^n} \int_{x^2}^{\infty} e^{-v} v^{-(n+\frac{1}{2})} dv. \tag{4}$$

The integral term represents R_n, the remainder after n terms of the series have been taken. To prove that the series is truly asymptotic in character we have to show that for every fixed value of n, $|x^{2n-1}R_n| \to 0$ as $x \to \infty$. Now the greatest value of e^{-v} in the range of integration is e^{-x^2}, so omitting the factor outside the integral in (4),

$$\int_{x^2}^{\infty} e^{-v} v^{-(n+\frac{1}{2})} dv < e^{-x^2} \int_{x^2}^{\infty} v^{-(n+\frac{1}{2})} dv = \frac{2e^{-x^2}}{(2n-1) x^{2n-1}}. \tag{5}$$

Thus

$$|x^{2n-1}R_n| < \frac{e^{-x^2}}{\sqrt{\pi}} \frac{1.3.5 \ldots (2n-3)}{2^{n-1}}, \tag{6}$$

and this tends to zero, for any $n > 1$, as $x \to \infty$, so series (4) is asymptotic (see Appendix 3). The modulus of the nth term of (4) is

$$\frac{e^{-x^2}}{\sqrt{\pi}} \frac{1.3.5 \ldots (2n-3)}{2^{n-1} x^{2n-1}}, \tag{7}$$

so the ratio of $|R_n|$ to this term is less than 1. Thus the error in stopping at the nth term is less than that term. The minimum term of (4) occurs when the nth and $(n+1)$th terms are approximately equal numerically, which gives $\dfrac{2n-1}{2x^2} \simeq 1$ or $n \sim x^2$ when x is large. It may be remarked that if $x = \frac{1}{2}\sqrt{(a/t)}$, the convergent expansion of erfc x is in inverse powers of t, whereas the asymptotic expansion is in ascending powers of t.

5·31. Bessel functions.* These functions may be defined in several ways, one of which depends upon the differential equation

$$\frac{d^2y}{dz^2} + \frac{1}{z}\frac{dy}{dz} + \left(1 - \frac{\nu^2}{z^2}\right)y = 0. \tag{1}$$

This linear equation of the second order has two independent solutions, which are denoted, respectively, by the symbols $J_\nu(z)$ and $Y_\nu(z)$. The complete solution of (1) with two arbitrary constants is

$$y = AJ_\nu(z) + BY_\nu(z), \tag{2}$$

where we define $J_\nu(z)$, $Y_\nu(z)$ to be Bessel functions of the first and second kinds, respectively, of order ν. Solution (2) is valid for all values of z and ν. When, however, ν is non-integral, the solution may also be written

$$y = A_1 J_\nu(z) + B_1 J_{-\nu}(z), \tag{3}$$

the relationship between $J_{-\nu}(z)$ and $Y_\nu(z)$ being

$$Y_\nu(z) = \frac{\cos \nu\pi J_\nu(z) - J_{-\nu}(z)}{\sin \nu\pi}. \tag{4}$$

When $\nu = n$ an integer, $Y_n(z)$ is the limit of (4) as $\nu \to n$. Substituting from (4) into (2), we get

$$y = J_\nu(z)[A + B \cot \nu\pi] - \frac{B}{\sin \nu\pi} J_{-\nu}(z), \tag{5}$$

so that in (3), if

$$A_1 = A + B \cot \nu\pi \quad \text{and} \quad B_1 = -B/\sin \nu\pi, \tag{6}$$

* The reader is expected to be familiar with, say, Chapters I, II, and the beginning of Chapter VII, reference [234].

(2) and (3) are identical. In technical applications, the constants are determined from the boundary conditions, so that (6) is satisfied automatically.

Since A and B in (2) are arbitrary, we may assign to them any real or complex values. Suppose we write

$$A = A_2 + B_2 \quad \text{and} \quad B = i(A_2 - B_2),$$

then (2) takes the form

$$y = A_2[J_\nu(z) + iY_\nu(z)] + B_2[J_\nu(z) - iY_\nu(z)] \qquad (7)$$

$$= A_2 H_\nu^{(1)}(z) + B_2 H_\nu^{(2)}(z), \qquad (8)$$

where $H_\nu^{(1)}(z)$ and $H_\nu^{(2)}(z)$ are defined to be Hankel functions or Bessel functions of the third kind of order ν. The solution (8) is valid for all values of ν and z, and is sometimes useful in problems involving wave propagation (see [234], p. 75).

From (7), (8) we obtain

$$J_\nu(z) = \tfrac{1}{2}[H_\nu^{(1)}(z) + H_\nu^{(2)}(z)], \qquad (9)$$

and

$$Y_\nu(z) = \frac{1}{2i}[H_\nu^{(1)}(z) - H_\nu^{(2)}(z)]. \qquad (10)$$

5·32. Contour integral for $J_\nu(z)$. One solution of (1) §5·31 is found by the method of Frobenius, namely, the substitution of a power series of the type

$$y = z^\mu \sum_{r=0}^{\infty} a_r z^r \qquad (1)$$

in the equation, the coefficients a_r being determined in the well-known way. The first solution is

$$J_\nu(z) = \sum_{r=0}^{\infty} (-1)^r \frac{(\tfrac{1}{2}z)^{\nu+2r}}{r!\,\Gamma(\nu+r+1)}. \qquad (2)$$

By (2) § 5·15, the right-hand side of (2) may be written

$$(\tfrac{1}{2}z)^\nu \sum_{r=0}^{\infty} (-1)^r \frac{1}{\Gamma(\nu+r+1)} \frac{(\tfrac{1}{4}z^2)^r}{r!}$$

$$= (\tfrac{1}{2}z)^\nu \sum_{r=0}^{\infty} (-1)^r \frac{1}{2\pi i} \int_{Br_1} e^t \frac{(\tfrac{1}{4}z^2)^r}{r!} \frac{dt}{t^{\mu+r+1}} \qquad (3)$$

$$= (\tfrac{1}{2}z)^\nu \sum_{r=0}^{\infty} (-1)^r \frac{1}{2\pi i} \int_{Br_1} \frac{(z^2/4t)^r}{r!} \frac{e^t dt}{t^{\nu+1}}. \qquad (4)$$

It is permissible to take Σ inside the integral sign (see [215]), so

$$J_\nu(z) = \frac{(\frac{1}{2}z)^\nu}{2\pi i} \int_{Br_2} \frac{e^{t-z^2/4t}\,dt}{t^{\nu+1}},$$ (5)

and $$J_\nu(az) = \frac{(\frac{1}{2}az)^\nu}{2\pi i} \int_{Br_2} \frac{e^{t-a^2z^2/4t}\,dt}{t^{\nu+1}},$$ (6)

t being the complex variable on Br_2 and a an arbitrary complex constant. Integral (6) is valid for all values of ν and z, the contour being Br_2. As shown in § 4·24, Br_1 and Br_2 are equivalent provided the contributions to the integrals from the arcs in the second and third quadrants vanish as the radius $\to \infty$. Arguing on the same lines as in § 4·31, it may be shown that this occurs when $R(\nu) > -1$. Consequently we may write

$$J_\nu(z) = \frac{(\frac{1}{2}z)^\nu}{2\pi i} \int_{Br_1} \frac{e^{t-z^2/4t}\,dt}{t^{\nu+1}} \qquad (R(\nu) > -1),$$ (7)

and $$J_\nu(az) = \frac{(\frac{1}{2}az)^\nu}{2\pi i} \int_{Br_1} \frac{e^{t-a^2z^2/4t}\,dt}{t^{\nu+1}} \qquad (R(\nu) > -1),$$ (8)

there being no restriction on a and z.

5·33. Modified Bessel functions of the first and second kinds. These functions, namely $I_\nu(z)$ and $K_\nu(z)$, are independent solutions of the equation

$$\frac{d^2y}{dz^2} + \frac{1}{z}\frac{dy}{dz} - \left(1 + \frac{\nu^2}{z^2}\right)y = 0.$$ (1)

The complete solution of (1) may be written

$$y = AI_\nu(z) + BK_\nu(z), \text{ for all values of } \nu,$$ (2)

and $$y = A_1 I_\nu(z) + B_1 I_{-\nu}(z), \text{ when } \nu \text{ is non-integral.}$$ (3)

$$I_\nu(z) = i^{-\nu}J_\nu(iz) = \sum_{r=0}^{\infty} \frac{(\frac{1}{2}z)^{\nu+2r}}{r!\,\Gamma(\nu+r+1)},$$ (4)

this being series (2) § 5·32 with all its terms made positive.

The modified function of the second kind is defined to be

$$K_\nu(z) = \frac{\frac{1}{2}\pi}{\sin \nu\pi}[I_{-\nu}(z) - I_\nu(z)].$$ (5)

When $\nu = n$ an integer, $K_n(z)$ is the limit of (5) as $\nu \to n$.

Since (5) § 5·32 is valid for all values of z, if we substitute zi for z and multiply by $i^{-\nu}$, we obtain

$$I_\nu(z) = \frac{(\tfrac{1}{2}z)^\nu}{2\pi i} \int_{Br_2} \frac{e^{t+z^2/4t}\,dt}{t^{\nu+1}}, \qquad (6)$$

which is valid for all values of ν and z.

From (7) § 5·32 we get

$$I_\nu(z) = \frac{(\tfrac{1}{2}z)^\nu}{2\pi i} \int_{Br_1} \frac{e^{t+z^2/4t}\,dt}{t^{\nu+1}} \qquad (R(\nu) > -1). \qquad (7)$$

5·34. Transformations. If in (6) §5·32 we write $\tfrac{1}{2}zt$ for t, the integral becomes

$$J_\nu(az) = \frac{a^\nu}{2\pi i} \int_{Br_2} \frac{e^{\tfrac{1}{2}z(t-a^2/t)}\,dt}{t^{\nu+1}}, \qquad (1)$$

for all values of ν, provided $R(z) > 0$. If Br_1 is used we must have $R(\nu) > -1$ for convergence, z being real and positive. Formula (6) § 5·33 gives

$$I_\nu(az) = \frac{a^\nu}{2\pi i} \int_{Br_2} \frac{e^{\tfrac{1}{2}z(t+a^2/t)}\,dt}{t^{\nu+1}}, \qquad (2)$$

for all ν, $R(z) > 0$. The contour Br_1 may be used provided $R(\nu) > -1$, z being real and positive. To effect another transformation of (6) § 5·32 we write (see p. 108)

$$\zeta z = t - a^2 z^2/4t \quad \text{or} \quad t = \tfrac{1}{2}z\{\zeta + \sqrt{(\zeta^2 + a^2)}\},$$

which gives

$$dt = \tfrac{1}{2}z\{1 + \zeta/\sqrt{(\zeta^2 + a^2)}\}\,d\zeta = t\,d\zeta/\sqrt{(\zeta^2 + a^2)}.$$

Hence (6) § 5·32 becomes

$$J_\nu(az) = \frac{a^\nu}{2\pi i} \int_{Br_2} \frac{e^{\zeta z}\,d\zeta}{\sqrt{(\zeta^2 + a^2)}\,\{\zeta + \sqrt{(\zeta^2 + a^2)}\}^\nu}, \qquad (3)$$

for all values of a and ν provided $R(z) > 0$. The preceding transformation applied to (8) § 5·32 gives

$$J_\nu(az) = \frac{a^\nu}{2\pi i} \int_{Br_1} \frac{e^{\zeta z}\,d\zeta}{\sqrt{(\zeta^2 + a^2)}\,\{\zeta + \sqrt{(\zeta^2 + a^2)}\}^\nu}, \qquad (4)$$

for all values of a, provided $R(\nu) > -1$, z real and positive. The method of ascertaining the various restrictions will be found in reference [125], and the transformations are treated in Appendix 5.

In like manner we obtain

$$I_\nu(az) = \frac{a^\nu}{2\pi i} \int_{Br_2} \frac{e^{\zeta z}\, d\zeta}{\sqrt{(\zeta^2 - a^2)}\,\{\zeta + \sqrt{(\zeta^2 - a^2)}\}^\nu}, \tag{5}$$

the restrictions being identical with those given in connection with (3) and (4) above. After transformation the singularities of the integrand may differ from the originals. For instance (3) and (4) have branch points at $\zeta = \pm ia$, whereas if ν is a positive integer, the integrand of (6) § 5·32 has an essential singularity at the origin. If ν is non-integral the singularity is a branch point.

5·351. Asymptotic expansion of $J_0(z)$. Writing $a = 1$ and $\nu = 0$ in (3) § 5·34, we get

$$J_0(z) = \frac{1}{2\pi i} \int_{Br_2} \frac{e^{\zeta z}\, d\zeta}{\sqrt{(1 + \zeta^2)}} = \frac{1}{2\pi i} \int_{Br_2} \frac{e^{\zeta z}\, d\zeta}{\sqrt{\{(\zeta - i)(\zeta + i)\}}}. \tag{1}$$

The integrand has two branch points, and the most suitable form of Br_2 is illustrated in Fig. 25 (b). Taking the upper part of the contour first, we move the origin to the point $\zeta = i$, by writing $(\zeta - i) = u$, where u is real and negative on the parallel lines above and below the barrier. Thus we obtain

$$I_1 = \frac{e^{iz}}{2\pi i} \int_{Br_2} \frac{e^{uz}\, du}{\sqrt{(2ui + u^2)}} = \frac{e^{iz}}{2\pi i} \int_{Br_2} \frac{e^{uz}\, du}{\sqrt{(2ui)}\sqrt{(1 - \tfrac{1}{2}ui)}}. \tag{2}$$

The denominator is now expanded, using the binomial theorem, on the assumption that, when $|z|$ is large enough and of appropriate phase, the bulk of the integral is contributed between $u = 0$ and $u = -2$, where $|\tfrac{1}{2}iu| = 1$ (see Appendix 3). Thus we obtain

$$I_1 \sim \frac{e^{i(z-\frac{1}{4}\pi)}}{\sqrt{2}.\,2\pi i} \int_{Br_2} \frac{e^{uz}\, du}{\sqrt{u}} \left[1 + \frac{iu}{2^2} - \frac{1.3u^2}{2^4.2!} - \frac{1.3.5iu^3}{2^6.3!} + \dots \right] \tag{3}$$

$$= \frac{e^{i(z-\frac{1}{4}\pi)}}{\sqrt{2}.\,2\pi i} \int_{Br_2} e^{uz}\, du \left[u^{-\frac{1}{2}} - \frac{1.3u^{\frac{3}{2}}}{2^4.2!} + \dots \right.$$
$$\left. + i\left\{\frac{\sqrt{u}}{2^2} - \frac{1.3.5u^{\frac{5}{2}}}{2^6.3!} + \dots\right\} \right] \tag{4}$$

$$= \frac{e^{i(z-\frac{1}{4}\pi)}}{\sqrt{(2\pi)}} \left[\frac{1}{\sqrt{z}} - \frac{1^2.3^2}{2^6.2!z^{\frac{5}{2}}} + \dots + i\left\{ -\frac{1}{2^3.z^{\frac{3}{2}}} + \frac{1^2.3^2.5^2}{2^9.3!z^{\frac{7}{2}}} - \dots\right\} \right], \tag{5}$$

by (4) § 5·15. Now the asymptotic expansion (5) is for the Hankel function of the first kind of order zero, and we have (see [234], Chapter IV)

$$\tfrac{1}{2}H_0^{(1)}(z) \sim \frac{e^{i(z-\frac{1}{4}\pi)}}{\sqrt{(2\pi z)}}\left[1 - \frac{1^2.3^2}{2!\,(8z)^2} + \frac{1^2.3^2.5^2.7^2}{4!\,(8z)^4} - \cdots \right.$$
$$\left. + i\left(-\frac{1}{1!\,(8z)} + \frac{1^2.3^2.5^2}{3!\,(8z)^3} - \cdots\right)\right]. \tag{6}$$

The integral round the branch point $\zeta = -i$ is found by writing $-i$ for i in (6), which gives the asymptotic expansion for the Hankel function of the second kind, namely,

$$\tfrac{1}{2}H_0^{(2)}(z) \sim \frac{e^{-i(z-\frac{1}{4}\pi)}}{\sqrt{(2\pi z)}}\left[1 - \frac{1^2.3^2}{2!\,(8z)^2} + \frac{1^2.3^2.5^2.7^2}{4!\,(8z)^4} - \cdots \right.$$
$$\left. + i\left(\frac{1}{1!\,(8z)} - \frac{1^2.3^2.5^2}{3!\,(8z)^3} + \cdots\right)\right]. \tag{7}$$

Applying (9) § 5·31, with $\nu = 0$, to (6) and (7), the required asymptotic expansion is

$$J_0(z) \sim \sqrt{\left(\frac{2}{\pi z}\right)}\left\{\cos\left(z - \tfrac{1}{4}\pi\right)\left[1 - \frac{1^2.3^2}{2!\,(8z)^2} + \frac{1^2.3^2.5^2.7^2}{4!\,(8z)^4} - \cdots\right]\right.$$
$$\left. + \sin\left(z - \tfrac{1}{4}\pi\right)\left[\frac{1}{1!\,(8z)} - \frac{1^2.3^2.5^2}{3!\,(8z)^3} + \cdots\right]\right\}. \tag{8}$$

The asymptotic expansion of $Y_0(z)$ may be obtained by applying (10) § 5·31 to (6), (7).

In (6), (7) it may be shown that

$$|R_n| < \frac{[\Gamma(n+\tfrac{1}{2})]^2}{\pi^{\frac{3}{2}}n!\,(2z)^{n+\frac{1}{2}}}. \tag{9}$$

When $|z| \to \infty$, n being fixed, $|z^{n-1}R_n| \to 0$, so by Appendix 3 the series for $H_0^{(1)}(z)$ and $H_0^{(2)}(z)$ are asymptotic. Hence (8) is the asymptotic expansion of $J_0(z)$. The modulus of the nth term is given by the right-hand side of (9). So the error incurred in stopping at the nth term is less than its modulus.

5·352. Angle range of z. (1°) For integral (1) § 5·351 to exist, $R(\zeta z)$ must be zero or negative when $|\zeta|$ is large, i.e. $R(\zeta z) \leqslant 0$, on those parts of Br_2 at the left-hand side of the imaginary axis. Now ζ is negatively infinite above and below the left end of the barrier in Fig. 25 (b), so $R(z)$ must be zero or positive, i.e. $R(z) \geqslant 0$. By § 2·72 integral (1) § 5·351 holds over a region of the z-plane for which both sides are analytic functions of z, and as just shown this condition obtains if $R(z) \geqslant 0$, or $-\frac{1}{2}\pi \leqslant \text{phase } z \leqslant \frac{1}{2}\pi$.

(2°) Integral (1) § 5·351 can be obtained from (5) § 5·32 by the substitutions $\zeta z = t - z^2/4t$, $\nu = 0$. When t is large, the first substitution is equivalent to $\zeta z \simeq t$, since $z^2/4t$ is negligible. The contour for t in (5) § 5·32 is Br_2 of Fig. 24 (b), and this is equivalent to any contour of the same type whose barrier lies between $\frac{3}{2}\pi$ and $\frac{1}{2}\pi$ (Fig. 24 (a), § 4·24). Hence the contour for ζ must lie within this range. Now phase $z =$ phase $t -$ phase ζ, and on the upper side of the barrier in Fig. 24 (b), phase $t = \pi$. Thus phase z lies between $(\pi - \frac{3}{2}\pi)$ and $(\pi - \frac{1}{2}\pi)$, i.e. $-\frac{1}{2}\pi$ and $\frac{1}{2}\pi$, so $R(z) \geqslant 0$. The angle range of z may be extended as shown in reference [234].

Note on § 5·34. The complete version of the second formula below (2) is

$$t = \tfrac{1}{2}z\{\zeta \pm \sqrt{(\zeta^2 + a^2)}\},$$

t being the complex variable on the original contour. The lower sign does not permit t to be infinite if z is real, finite, and positive, so it is inadmissible.

VI

EVALUATION OF $\dfrac{1}{2\pi i}\displaystyle\int_{Br_1} \dfrac{e^{zt}\phi(z)\,dz}{z}$ WHEN $\phi(z)$ HAS BRANCH POINTS

6·11. Example. To show the procedure adopted in evaluating integrals of the above type, we shall treat several cases which occur in technical applications. The first is associated with the determination of the input current to a uniform unloaded submarine cable of great length connected to a sine-wave generator. The current is obtained by evaluation of an integral of the type

$$I = \frac{1}{2\pi i}\int_{Br_1} \frac{e^{zt}\sqrt{z}\,dz}{z^2+\omega^2},\tag{1}$$

ω being real and positive. The integrand has three singularities, namely, a branch point at $z = 0$ and simple poles at $z = \pm i\omega$, as shown in Fig. 34 (c), where the Br_2 contour equivalent to Br_1 is depicted. The contour comprising Br_1, Br_2 and the arcs in the second and third quadrants encloses no singularity of the integrand. There is no contribution to the integral from the arcs as the radius $\to\infty$, so Br_1 and Br_2 described positively are equivalent.

6·12. I is the sum of the integrals round the poles, round the branch point at O and along the lines above and below the barrier. The integrals along the lines parallel to the imaginary axis neutralize each other, since they are equal but opposite when the line separation $\to 0$. This is due to the function being analytic and single-valued, thereby returning to its original value after each pole is rounded. The integral (1) § 6·11 may be written

$$I = \frac{1}{2\pi i}\int_{Br_2} \frac{e^{zt}\sqrt{z}\,dz}{(z+i\omega)(z-i\omega)}.\tag{1}$$

The residue at the pole $z = i\omega$ is $e^{i\omega t}\sqrt{(i\omega)}/2i\omega$, and that at $z = -i\omega$ is found by writing $-i$ for i in the foregoing. The sum of these residues is

$$\frac{1}{\sqrt{\omega}}\sin(\omega t + \tfrac{1}{4}\pi).\tag{2}$$

On the small circle δ of radius ϵ at the origin $|z| \ll \omega$, so we have $\dfrac{1}{2\pi i} \displaystyle\int_\delta \dfrac{e^{zt}\sqrt{z}\,dz}{\omega^2}$, and when $|z| \to 0$ this vanishes. On the line below the barrier on the negative side of the real axis $z = x\,e^{-i\pi}$, so I yields

$$\frac{i}{2\pi i}\int_\infty^0 \frac{e^{-xt}\sqrt{x}\,dx}{x^2+\omega^2}\,.$$

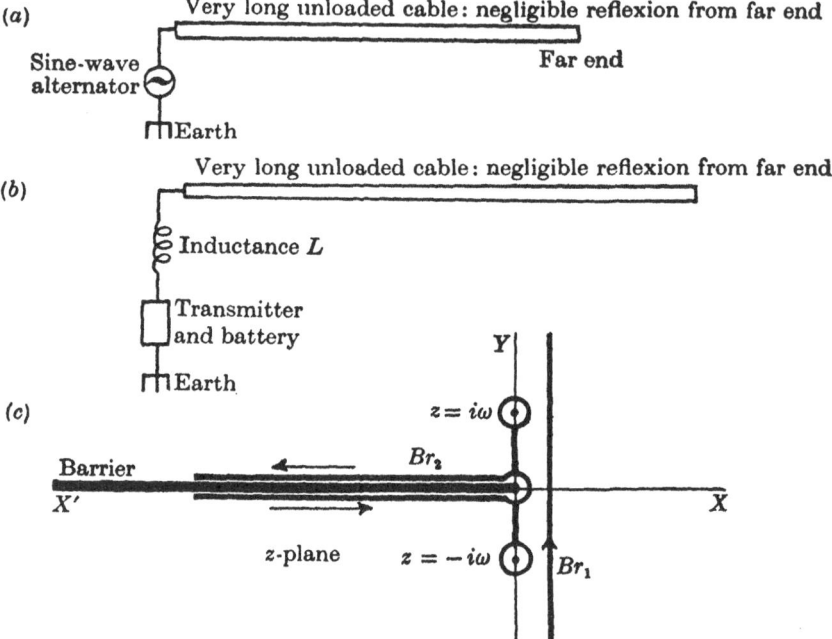

(a)   Very long unloaded cable: negligible reflexion from far end

Sine-wave alternator   Far end

Earth

(b)   Very long unloaded cable: negligible reflexion from far end

Inductance L

Transmitter and battery

Earth

(c)

Barrier

X'

$z = i\omega$

Br_2

X

z-plane $z = -i\omega$ Br_1

Y

Fig. 34 (a), (b), (c). Diagrams illustrating submarine cable problems. In text x signifies $|x|$.

On the upper line $z = x\,e^{i\pi}$, yielding

$$-\frac{i}{2\pi i}\int_0^\infty \frac{e^{-xt}\sqrt{x}\,dx}{x^2+\omega^2}\,.$$

The sum of these two integrals is

$$-\frac{1}{\pi}\int_0^\infty \frac{e^{-xt}\sqrt{x}\,dx}{x^2+\omega^2}\,. \tag{3}$$

By virtue of the factor e^{-xt}, when t is large, the major part of integral (3) will be obtained for values of x in the neighbourhood of the origin. If the denominator is expanded in ascending powers of x with a remainder term, and the expansion integrated term by term, we shall obtain a series which will probably turn out to be asymptotic (see Appendix 3). Thus we get

$$-\frac{(1/\omega^2)}{\pi}\int_0^\infty e^{-xt}\sqrt{x}\left[1-\frac{x^2}{\omega^2}+\frac{x^4}{\omega^4}-\cdots\right.$$

$$\left.+(-1)^{n-1}\left(\frac{x}{\omega}\right)^{2n-2}+\frac{(-1)^n(x/\omega)^{2n}}{1+x^2/\omega^2}\right]dx \quad (4)$$

$$=-\frac{(1/\omega^2)}{\pi}\left[\frac{\Gamma(\frac{3}{2})}{t^{\frac{3}{2}}}-\frac{\Gamma(\frac{7}{2})}{\omega^2 t^{\frac{7}{2}}}+\frac{\Gamma(\frac{11}{2})}{\omega^4 t^{\frac{11}{2}}}-\cdots+\frac{(-1)^{n-1}\Gamma(2n-\frac{1}{2})}{\omega^{2n-2}t^{2n-\frac{1}{2}}}\right.$$

$$\left.+\frac{(-1)^n}{\omega^{2n-2}}\int_0^\infty\frac{e^{-xt}x^{2n+\frac{1}{2}}\,dx}{x^2+\omega^2}\right]. \quad (5)$$

The modulus of the remainder after n terms in [ ] is

$$|R_n|=\left|\frac{(-1)^n}{\omega^{2n-2}}\int_0^\infty\frac{e^{-xt}x^{2n+\frac{1}{2}}\,dx}{x^2+\omega^2}\right|<\frac{1}{\omega^{2n-2}}\int_0^\infty e^{-xt}x^{2n-\frac{3}{2}}\,dx, \quad (6)$$

so
$$|R_n|<\frac{1}{\omega^{2n-2}}\frac{\Gamma(2n-\frac{1}{2})}{t^{2n-\frac{1}{2}}}. \quad (7)$$

Formula (7) is the modulus of the nth term, so that if n terms are taken to compute the value of the series, the error in stopping (5) at the nth term is less than that term. Also as t tends to infinity, $|t^{2n-2}R_n|\to 0$, so series (5) is asymptotic. It can also be shown that $|R_n|<$ the numerical value of the $(n+1)$th term.

6·13. We infer from Appendix 3, that (5) §6·12 has a minimum term and its number may be found approximately by equating the formulae for the nth and $(n+1)$th terms, which gives

$$\frac{1}{\omega^{2n}}\frac{\Gamma(2n+\frac{3}{2})}{t^{2n+\frac{3}{2}}}=\frac{1}{\omega^{2n-2}}\frac{\Gamma(2n-\frac{1}{2})}{t^{2n-\frac{1}{2}}}, \quad (1)$$

or
$$(2n+\tfrac{1}{2})(2n-\tfrac{1}{2})/\omega^2=t^2, \quad (2)$$

so
$$n=\tfrac{1}{2}\sqrt{(\omega^2 t^2+\tfrac{1}{4})}\sim\tfrac{1}{2}\omega t, \quad (3)$$

for practical values of ω and t. The nearest integer would be chosen, and with $\omega = 120$, $t = 0 \cdot 25$ sec., $n = 15$. However, by (7) §6·12 the error in taking the first term alone is less than the second term, and since the ratio of the two is 240/1, this approximation would be adequate in practice with the given data.

6·14. The complete solution of the problem is the sum of (2) and (5) §6·12; so taking the first term only from (5), we have

$$\frac{1}{2\pi i}\int_{Br_1} \frac{e^{zt}\sqrt{z}\,dz}{z^2 + \omega^2} \sim \frac{1}{\sqrt{\omega}}\sin{(\omega t + \tfrac{1}{4}\pi)} - \frac{1}{2\sqrt{\pi}\,\omega^2 t^{\frac{3}{2}}}. \tag{1}$$

Owing to the character of the last term, formula (1) is applicable only when t is appropriately large. To obtain the value of the integral for *any* value of $t > 0$, it is essential to determine the solution as a convergent expansion in rising powers of t. Thus

$$\frac{1}{2\pi i}\int_{Br_1} \frac{e^{zt}\sqrt{z}\,dz}{z^2 + \omega^2} = \frac{1}{2\pi i}\int_{Br_1} \frac{e^{zt}\sqrt{z}\,dz}{z^2\left(1 + \dfrac{\omega^2}{z^2}\right)} \tag{2}$$

$$= \frac{1}{2\pi i}\int_{Br_1} \frac{e^{zt}\,dz}{z^{\frac{3}{2}}}\left[1 - \frac{\omega^2}{z^2} + \frac{\omega^4}{z^4} - \frac{\omega^6}{z^6} + \ldots\right] \tag{3}$$

$$= \frac{t^{\frac{1}{2}}}{\Gamma(\frac{3}{2})} - \frac{\omega^2 t^{\frac{5}{2}}}{\Gamma(\frac{7}{2})} + \frac{\omega^4 t^{\frac{9}{2}}}{\Gamma(\frac{11}{2})} - \ldots \tag{4}$$

$$= 2\sqrt{\frac{t}{\pi}}\left\{1 - \frac{(2\omega t)^2}{1 \cdot 3 \cdot 5} + \frac{(2\omega t)^4}{1 \cdot 3 \cdot 5 \cdot 7 \cdot 9} - \ldots\right\}. \tag{5}$$

Since c in Fig. 22 (a) may be made as large as we please (finite), in (2) $|\,\omega/z\,| < 1$, so the expansion in (3) is convergent. Term by term integration is permissible (see [215]). When t is in the neighbourhood of the origin of time

$$I \simeq 2\sqrt{\frac{t}{\pi}}, \tag{6}$$

provided ω is not too large.

As a rule a convergent series is useless for computation when t exceeds a certain value, owing to the slowness of convergence. Under this condition an asymptotic series is used, provided it can be found.

6·21. Example. Evaluate

$$I = \frac{1}{2\pi i} \int_{Br_1} \frac{e^{zt}\,dz}{z[(az)^{\frac{1}{2}} + 1]} \quad (a \text{ real and} > 0). \quad (1)$$

This refers to the problem of finding the input p.d. when an inductance L is connected between the sending battery and a uniform unloaded cable of great length (see Fig. 34 (b)). It also refers to the mechanical problem stated in Appendix 3, p. 339.

Resolving the integrand of (1) into partial fractions, we have

$$I = \frac{1}{2\pi i} \int_{Br_1} e^{zt} \left[\frac{1}{z} - \frac{a^{\frac{1}{2}} z^{\frac{1}{2}}}{(az)^{\frac{1}{2}} + 1} \right] dz. \quad (2)$$

The value of the first integral in (2) is unity, so we pass on to evaluate the second, namely,

$$I_2 = -\frac{a^{\frac{1}{2}}}{2\pi i} \int_{Br_1} \frac{e^{zt} \sqrt{z}\,dz}{(az)^{\frac{1}{2}} + 1}. \quad (3)$$

The factors of the denominator are

$$[\sqrt{(az)} + 1] \quad \text{and} \quad [az - \sqrt{(az)} + 1].$$

By §§ 1·64, 2·64 the relevant singularity due to the first factor is a branch point at the origin.* The second factor vanishes when $az - \sqrt{(az)} + 1 = 0$, or $z = \frac{1}{2a}(-1 \pm i\sqrt{3})$. Each of these values of z corresponds to a pole of the integrand on the first branch.

To evaluate (3) we choose a suitable contour which is equivalent to Br_1, as shown in Fig. 35. The integrals along the broken arcs vanish as the radius $\to \infty$, as can be seen by considering (3) or by § 4·31, when $|z| \to \infty$. To evaluate (3) round the small circle of radius r at the origin, put $z = r e^{i\theta}$, and we get

$$-\frac{ia^{\frac{1}{2}}}{2\pi} \int_{-\pi}^{\pi} e^{rt(\cos\theta + i\sin\theta) + \frac{3}{2}\theta i} \frac{r^{\frac{3}{2}}\,d\theta}{1 + (ar)^{\frac{1}{2}} e^{\frac{1}{2}\theta i}} \to 0, \quad (4)$$

as $r \to 0$.

---

* A simple pole occurs on the second branch of the function, but it is inadmissible in the present problem (see § 2·64).

The residue at the pole $z = \dfrac{1}{2a}(-1 + i\sqrt{3}) = b$ is by § 3·231

$$\text{Res.}_1 = -\left[\frac{a^{\frac{3}{2}} e^{zt} z^{\frac{1}{2}}}{\dfrac{d}{dz}[1 + (az)^{\frac{3}{2}}]}\right]_{z=b} = -\tfrac{2}{3}[e^{zt}]_{z=b};$$

so $$\text{Res.}_1 = -\tfrac{2}{3}e^{\frac{1}{2}w(-1+i\sqrt{3})}, \tag{5}$$

where $w = t/a$.

The residue at the other pole is obtained by writing $-i$ for i in (5), so

$$\text{Res.}_2 = -\tfrac{2}{3}e^{-\frac{1}{2}w(1+i\sqrt{3})}. \tag{6}$$

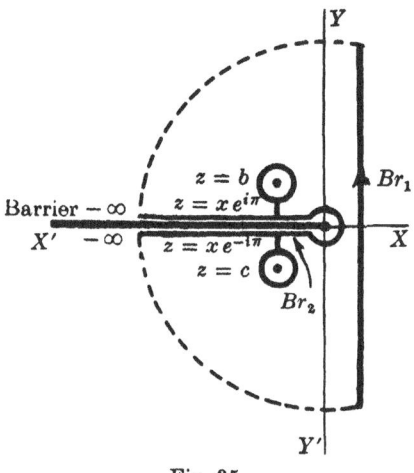

Fig. 35.

The contribution to (3) from the poles is the sum of (5) and (6), namely,

$$\Sigma\,\text{Res.} = -\tfrac{2}{3}e^{-\frac{1}{2}w}[e^{\frac{1}{2}\sqrt{3}wi} + e^{-\frac{1}{2}\sqrt{3}wi}]$$
$$= -\tfrac{4}{3}e^{-\frac{1}{2}w}\cos \tfrac{1}{2}\sqrt{3}w. \tag{7}$$

6·22. We have now to evaluate the integral (3) §6·21 along the two lines on each side of the negative part of the real axis. On the lower line $z = x\,e^{-i\pi}$, which gives

$$-\frac{a^{\frac{3}{2}}}{2\pi}\int_\infty^0 \frac{e^{-xt}\sqrt{x}\,dx}{1 + i(ax)^{\frac{3}{2}}}, \tag{1}$$

and on the upper line $z = x\,e^{i\pi}$, which gives

$$\frac{a^{\frac{3}{2}}}{2\pi}\int_0^\infty \frac{e^{-xt}\sqrt{x}\,dx}{1-i(ax)^{\frac{3}{2}}}. \tag{2}$$

The sum of (1) and (2) is

$$\frac{a^{\frac{3}{2}}}{\pi}\int_0^\infty \frac{e^{-xt}\sqrt{x}\,dx}{1+a^3x^3}. \tag{3}$$

This integral without $a^{\frac{3}{2}}/\pi$ is treated in Appendix 3 (see (8)). Using the numerical data there, we find that the integral is given with adequate accuracy by the first term of (10). Hence

$$\frac{a^{\frac{3}{2}}}{\pi}\int_0^\infty \frac{e^{-xt}\sqrt{x}\,dx}{1+a^3x^3} \sim \frac{a^{\frac{3}{2}}}{\pi}\frac{\Gamma(\frac{3}{2})}{t^{\frac{3}{2}}} = 1/2\sqrt{\pi\,w^{\frac{3}{2}}}. \tag{4}$$

6·23. The complete solution of the problem is unity (from the first integral in (2) § 6·21) plus (7) § 6·21 and (4) § 6·22. Whence

$$\frac{1}{2\pi i}\int_{Br_1} \frac{e^{zt}\,dz}{z[1+(az)^{\frac{3}{2}}]} \sim 1 - \tfrac{4}{3}e^{-\frac{1}{2}w}\cos\frac{\sqrt{3}\,w}{2} + \frac{1}{2\sqrt{\pi\,w^{\frac{3}{2}}}}. \tag{1}$$

As in (1) § 6·14 this solution applies only when t exceeds a certain value. For any value of $t > 0$, the procedure is to obtain a convergent series as at (4) § 6·14. Thus we find that

$$I = \frac{4\pi}{3}\left(\frac{w}{\pi}\right)^{\frac{3}{2}}\left[1 + \frac{(2w)^3}{5.7.9} + \frac{(2w)^6}{5.7\ldots15} + \ldots\right]$$
$$- \left[\frac{w^3}{3!} + \frac{w^6}{6!} + \frac{w^9}{9!} + \ldots\right], \tag{2}$$

where $w = t/a$. When t is small enough

$$I \simeq \frac{4\pi}{3}\left(\frac{t}{\pi a}\right)^{\frac{3}{2}}. \tag{3}$$

The physical interpretation of (1) is that, although the force applied to the mass m (see pp. 339, 340) is constant, the interaction of the mass with the spring stiffness and resistance causes a damped oscillation of m whose frequency $n = \sqrt{3}/4\pi a$, where $a = (m^2/rs)^{\frac{1}{3}}$, s and r being the stiffness and resistance per unit length, respectively. Formula (3) shows that the force-time curve has a finite slope at the origin.

In the electrical case the current into the cable oscillates at a frequency given by the above formula, where $a = (CL^2/R)^{\frac{1}{2}}$, C and R being the capacitance and resistance per unit length, respectively.

6·24. Application of product theorem, Appendix 8.* Evaluate

$$I = \frac{1}{2\pi i} \int_{Br_1} \frac{e^{zt} dz}{\sqrt{(z^4 - 1)}}. \tag{1}$$

The integrand has four branch points at $z = \pm 1$, $\pm i$, as illustrated in Fig. 36, where the 'cross' contour is equivalent to Br_1. The integral may be expressed in the form

$$I = \frac{1}{2\pi i} \int_{Br_1} e^{zt} \left[\frac{z}{\sqrt{(z^2+1)}} \frac{z}{\sqrt{(z^2-1)z}} \frac{1}{z} \right] \frac{dz}{z}. \tag{2}$$

Using § 1 Appendix 8, (5) § 8·21, (1), (2) § 8·422, we have

$$\phi_1 = p/\sqrt{(p^2+1)} \subset J_0(t), \quad \phi_2 = p/\sqrt{(p^2-1)} \subset I_0(t), \quad \phi = \phi_1 \phi_2/p \subset f(t). \tag{3}$$

Then $\quad I = f(t) = \int_0^t J_0(t-\lambda) I_0(\lambda) d\lambda = \int_0^t J_0(\lambda) I_0(t-\lambda) d\lambda. \tag{4}$

(2) may be evaluated also by expansion and term by term integration. The part in [ ] is equal to $1/z \sqrt{(1 - 1/z^4)}$, and since c in Fig. 36 may be as large as we please, $|z^{-4}|$ may be made less than unity, so a binomial expansion is valid. Thus

$$I = \frac{1}{2\pi i} \int_{Br_1} e^{zt} \frac{dz}{z} \left[\frac{1}{z} + \frac{1}{2.1!z^5} + \frac{1.3}{2^2.2!z^9} + \frac{1.3.5}{2^3.3!z^{13}} + \dots \right] \tag{5}$$

$$= t + \frac{t^5}{2.1!5!} + \frac{1.3t^9}{2^2.2!9!} + \frac{1.3.5t^{13}}{2^3.3!13!} + \dots \tag{6}$$

$$= \sum_{r=0}^{\infty} \frac{t^{4r+1}}{2^r r!(4r+1)!}. \tag{7}$$

Term by term integration is permissible, since the conditions in [215], § 176 are satisfied. (7) is absolutely and uniformly convergent in every closed interval of t, so the function it represents is continuous.

6·31. Approximation to $\dfrac{1}{2\pi i} \displaystyle\int_{Br_1} \dfrac{e^{zt}\phi(z)\,dz}{z}$, **when $\phi(z)$ has a branch point and t is small.** Let

$$\frac{\phi(z)}{z} = \frac{a_n z^\mu + a_{n-1} z^{\mu-1} + \dots}{b_m z^\nu + b_{m-1} z^{\nu-1} + \dots}, \tag{1}$$

___
* This section may be read after studying Chapter VIII.

the number of terms in numerator and denominator being finite, while $(\nu - \mu) > 0$, μ and ν being such that a branch point occurs.

On writing $\zeta = zt$, (1) becomes

$$\frac{\phi(z)}{z} = t^{\nu-\mu}\left[\frac{a_n\zeta^\mu + a_{n-1}\zeta^{\mu-1}t + \dots}{b_m\zeta^\nu + b_{m-1}\zeta^{\nu-1}t + \dots}\right]. \tag{2}$$

When t is small enough (2) may be written

$$\frac{\phi(z)}{z} \sim \frac{t^{\nu-\mu}a_n}{b_m\zeta^{\nu-\mu}}, \tag{3}$$

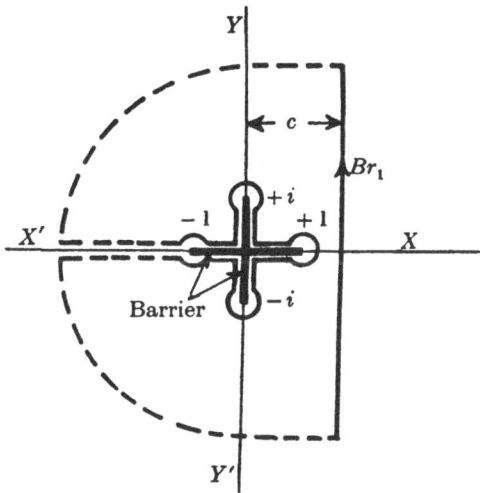

Fig. 36.

following the argument given in §§ 4·61, 4·62. Thus an integral of the above type gives

$$I \simeq \frac{(a_n t^{\nu-\mu-1}/b_m)}{2\pi i}\int_{Br_1}\frac{e^\zeta d\zeta}{\zeta^{\nu-\mu}} = \frac{(a_n/b_m)}{2\pi i}\int_{Br_1}\frac{e^{zt}dz}{z^{\nu-\mu}} = \frac{a_n}{b_m}\frac{t^{\nu-\mu-1}}{\Gamma(\nu-\mu)}, \tag{4}$$

which is identical with the result found from (1) if all the terms in the numerator and denominator are neglected except the first, i.e. if $|z|$ is assumed to be very large. Such procedure is identical with that in § 4·61, where the singularities of $\phi(z)$ are poles only.

Examples. 1°. Applying the above to (1) § 6·21 yields

$$I \sim \frac{1}{2\pi i}\int_{Br_1}\frac{e^{zt}dz}{a^{\frac{3}{2}}z^{\frac{5}{2}}} = t^{\frac{3}{2}}/a^{\frac{3}{2}}\Gamma(\tfrac{5}{2})$$

$$= 4t^{\frac{3}{2}}/3\sqrt{\pi}\,a^{\frac{3}{2}}, \tag{5}$$

when t is small enough in the neighbourhood of the origin of time, the result being identical with (3) § 6·23, which was obtained from a formal expansion and integration term by term.

2°. We now consider a case where the numerator and denominator of the integrand may each be represented by infinite series, e.g.

$$\frac{1}{2\pi i}\int_{Br_1}\frac{e^{zt}J_1(b\sqrt{z})}{z^{\frac{3}{2}}J_0(b\sqrt{z})}\,dz. \tag{6}$$

Writing $\zeta = zt$ or $z = \zeta/t$, (6) becomes

$$\frac{\sqrt{t}}{2\pi i}\int_{Br_1}\frac{e^{\zeta}J_1[b\sqrt{(\zeta/t)}]}{\zeta^{\frac{3}{2}}J_0[b\sqrt{(\zeta/t)}]}\,d\zeta. \tag{7}$$

Taking $c \pm i\infty$ for Br_1 in (7), if $t > 0$ we may choose c so that the smallest value of $|\zeta| = ct$ (on the positive real axis) although finite, is as large as we please. Thus with t small, $b\sqrt{(\zeta/t)}$ will always be large enough to permit the use of asymptotic formulae for the Bessel functions in (7), as well as in (6). The integral is evaluated in § 14·2.

6·32. Example. What is the form of $\dfrac{1}{2\pi i}\displaystyle\int_{Br_1}\dfrac{e^{zt}\tanh\sqrt{z}\,dz}{z^{\frac{3}{2}}}$, when t is small?

Since $\tanh\sqrt{z} = 1 - \dfrac{2e^{-2\sqrt{z}}}{1+e^{-2\sqrt{z}}}$, the integral may be written

$$I = \frac{1}{2\pi i}\int_{Br_1}e^{zt}\left[\frac{1}{z^{\frac{3}{2}}} - \frac{2e^{-2\sqrt{z}}}{(1+e^{-2\sqrt{z}})z^{\frac{3}{2}}}\right]dz. \tag{1}$$

By (3) § 5·15 the value of the first integral is $2\sqrt{(t/\pi)}$.

The singularities of the second part of the integral are (i) a branch point at $z = 0$, (ii) simple poles on the negative real axis at

$$z = -\tfrac{1}{4}(2n+1)^2\pi^2,$$

n a positive integer, and a limit point of poles at infinity. Thus Br_1 may be taken as a semicircle of radius $1/t$ on the right of the imaginary axis and those parts of the axis above and below the semicircle. Thereon $z = e^{i\theta}/t$, $2\sqrt{z} = 2e^{\frac{1}{2}i\theta}/\sqrt{t}$, so

$$e^{zt-2\sqrt{z}} = e^{(\cos\theta - 2t^{-\frac{1}{2}}\cos\frac{1}{2}\theta)+i(\sin\theta - 2t^{-\frac{1}{2}}\sin\frac{1}{2}\theta)}. \tag{2}$$

Thus as $t \to 0$, the modulus of the expression (2) is exponentially small and $\to 0$ in $-\frac{1}{2}\pi \leqslant \theta \leqslant \frac{1}{2}\pi$. This applies also to $e^{-2\sqrt{z}}$ in the denominator of (1). On the imaginary axis, $z = \pm iy$, $y \geqslant 1/t$, so $2\sqrt{z} = \sqrt{2}(1\pm i)y$, and, therefore, $e^{zt-2\sqrt{z}} = e^{\pm iy(t-\sqrt{2})-\sqrt{2}y}$. As $t \to 0$, $y \to \infty$, so the modulus of (2) is exponentially small and $\to 0$. Hence when t is small enough, the second part of the integral (1) may be neglected, and the form of (1) is $2\sqrt{(t/\pi)}$. If we take $\sqrt{\tanh z}$ large enough $z \simeq 1$, and the preceding result is obtained immediately. It is used in § 15·43 to obtain the value of (12) § 15·42 when $\beta = 0$.

6·41. Value of $\dfrac{1}{2\pi i}\displaystyle\int_{Br_1}\dfrac{e^{zt}\phi(z)\,dz}{z}$, **when** $\dfrac{\phi(z)}{z}$ **has branch points and**
t **is large.** In practical applications it is often expedient to approximate
to the value of an integral when t is large, by making $|z|$ small in com-
parison with other quantities in the integrand, and evaluating the
integral so reduced. For instance, when t is large, make $|z| \ll |a|$, and then

$$\frac{1}{2\pi i}\int_{Br_1}\frac{e^{zt}dz}{z^{\frac{1}{2}}(z+a)} \simeq \frac{1}{2\pi i}\int_{Br_1}\frac{e^{zt}dz}{az^{\frac{1}{2}}} = \frac{2}{a}\sqrt{\left(\frac{t}{\pi}\right)}, \tag{1}$$

provided $R(a) > 0$, a point treated in § 6·43. We shall now consider the
validity of this procedure in typical cases.

6·42. When the singularities of $\phi(z)/z$ are branch points only, suppose
that the contour can be drawn in a form akin to that shown in Fig. 25 (b).
If a branch point occurs at $z = -a$, then moving the origin to $z = -a$ by
writing $z = \zeta - a$, the integral can be evaluated along the lines $\zeta = xe^{\pm i\pi}$.
The real integral contains the factor e^{-xt}, obtained by writing $\zeta = xe^{\pm i\pi}$
in $e^{\zeta t}$. Consequently, in cases associated with our work herein, when t is
large, the major contribution to the integral from the region of the branch
point in question is obtained for values of ζ in the neighbourhood of
$\zeta = 0$ (see Appendix 3). By making $|\zeta|$ small compared with other
quantities in the integrand, as in (1) § 6·41, an approximation to the
evaluation at the branch point is found. Moving the origin to the other
branch points in turn the procedure is repeated, and the sum of the
contributions gives the desired result. In § 5·351 the integrand has
branch points at $z = \pm i$. By following the above procedure, the first
term of the asymptotic expansion of $J_0(t)$, namely,

$$\sqrt{\frac{2}{\pi t}}\cos(t - \tfrac{1}{4}\pi),$$

is obtained. As an example of a branch point at the origin we have

$$\frac{1}{2\pi i}\int_{Br_1}\frac{e^{zt}\,dz}{\sqrt{z}\,\sqrt{(z^2+1)}} \simeq \frac{1}{2\pi i}\int_{Br_1}\frac{e^{zt}dz}{\sqrt{z}} = 1/\sqrt{(\pi t)}, \tag{1}$$

this being the approximation for the branch point at $z = 0$.

6·43. When the singularities of $\phi(z)/z$ are poles and branch points,
the contributions from both must be considered. In (1) § 6·41 there is
a pole at $z = -a \neq 0$. Moving the origin to this point, we get

$$\frac{e^{-at}}{2\pi i}\int_{Br_1}\frac{e^{\zeta t}d\zeta}{(\zeta-a)^{\frac{1}{2}}\zeta} \simeq \frac{e^{-at}}{2\pi i}\int_{Br_1}\frac{e^{\zeta t}d\zeta}{(-a)^{\frac{1}{2}}\zeta} = \frac{e^{-at-\frac{1}{2}\pi i}}{a^{\frac{1}{2}}}, \tag{1}$$

this being the contribution for the pole (exact since $\zeta = 0$).

When $R(a) > 0$, (1) $\to 0$ as $t \to \infty$, and can be disregarded in comparison
with the approximation in (1) § 6·41.

Now consider the integral

$$\frac{1}{2\pi i}\int_{Br_1}\frac{e^{zt}\,dz}{z^\nu(z+a)^n},\tag{2}$$

ν real but non-integral, n a positive integer >0, $a\neq 0$. For the integral to be convergent $\nu+n>0$ (see Appendix 7). Evaluating at the branch point $z=0$, we have, approximately,

$$\frac{1}{2\pi i}\int_{Br_1}\frac{e^{zt}\,dz}{z^\nu a^n}=\frac{t^{\nu-1}}{a^n\Gamma(\nu)}.\tag{3}$$

To evaluate (2) at the nth order pole, move the origin to $z=-a$, thereby obtaining

$$\frac{e^{-at}}{2\pi i}\int_{Br_1}\frac{e^{\zeta t}\,d\zeta}{(\zeta-a)^\nu\,\zeta^n}.\tag{4}$$

Writing $-a$ for $\zeta-a$, by (1) § 4·51, we have approximately

$$e^{-at-\nu\pi i}t^{n-1}/a^\nu(n-1)!.\tag{5}$$

Whether (3) or (5) is the more important contribution depends upon the values of ν, n and a. If $R(a)=0$, but $n>\nu$, (5) will be the more important. But if $R(a)>0$, (3) takes precedence.

6·44. Consider the integral

$$\frac{1}{2\pi i}\int_{Br_1}\frac{e^{zt}\,dz}{z^n(z+a)^\nu},\tag{1}$$

a, n and ν being restricted as at (2) § 6·43. The nth order pole introduces a contribution of the type

$$a_{n-1}t^{n-1}.\tag{2}$$

For the branch point we move the origin to $z=-a$, and get

$$\frac{e^{-at}}{2\pi i}\int_{Br_1}\frac{e^{\zeta t}\,d\zeta}{(\zeta-a)^n\,\zeta^\nu}\simeq\frac{(-1)^n\,e^{-at}\,t^{\nu-1}}{a^n\Gamma(\nu)}.\tag{3}$$

The relative importance of (2) and (3) depends upon the values of ν, n and a. The sum of (2) and (3) is the complete approximation.

(1) § 6·14 provides an example of the type (2) § 6·43. When t is large enough the second term can be neglected. (1) § 6·23 is another instance, but here $1+(az)^{\frac{3}{2}}$ introduces two poles in addition to a branch point. When t is large enough the integral is substantially unity.

VII

DIFFERENTIATION AND INTEGRATION UNDER THE INTEGRAL SIGN

7·11. Introducing the subject in a concrete way, suppose we know that

$$I = \frac{1}{2\pi i} \int_{Br_1} \frac{e^{zt}\,dz}{z^{n+1}} = \frac{t^n}{n!}, \tag{1}$$

t real > 0, n a positive integer, and that we want to evaluate

$$I_1 = \frac{1}{2\pi i} \int_{Br_1} \frac{e^{zt}\,dz}{z^n}. \tag{2}$$

Differentiating (1) under the integral sign, with respect to t, yields

$$\frac{dI}{dt} = \frac{1}{2\pi i} \int_{Br_1} \frac{z\,e^{zt}\,dz}{z^{n+1}} = \frac{1}{2\pi i} \int_{Br_1} \frac{e^{zt}\,dz}{z^n}. \tag{3}$$

By differentiating $t^n/n!$ we obtain $t^{n-1}/(n-1)!$, and this is the value of (3) provided the above process is legitimate, which happens to be the case, for the reason given below.

Rigorous proofs of the conditions to be satisfied for differentiating and integrating under the integral sign,* in the various cases treated in this book, would take us beyond the scope of a text intended for technologists. Unfortunately, therefore, we shall have to be content with the statement that the procedure in cases of the type encountered herein usually leads to correct results, provided the integrals so obtained are themselves convergent.

It may be well to cite a divergent case. Referring to (4) § 8·23, suppose we write

$$\frac{d(\cos t)}{dt} = \frac{1}{2\pi i} \int_{Br_1} \frac{e^{zt}\,z^2\,dz}{1+z^2}$$

$$= \frac{1}{2\pi i} \int_{Br_1} e^{zt}\left(1 - \frac{1}{1+z^2}\right)dz. \tag{4}$$

---

* See references [215, 220, 226, 237, 240, 243].

By (3) § 8·23 the second part of the second integral is $-\sin t$, but by Appendix 7 the first part is divergent. Hence differentiation of (4) § 8·23 under the integral sign is not permissible. In both of the integrals (3), (4) § 8·23, however, Br_2 is equivalent to Br_1 $(t > 0)$. By Appendix 6,

$$\frac{1}{2\pi i} \int_{Br_2} e^{zt} z^n dz = 0,$$

n a positive integer, and differentiation of the above integrals under the integral sign is valid any number of times, when Br_2 replaces Br_1.

7·12. As another example consider the derivatives of

$$f(a) = \frac{1}{2\pi i} \int_C \frac{f(z)\,dz}{(z-a)}, \tag{1}$$

with respect to a. Then differentiating (1) under the integral sign,

$$f'(a) = \frac{1}{2\pi i} \int_C \frac{f(z)\,dz}{(z-a)^2}, \tag{2}$$

$$f''(a) = \frac{2!}{2\pi i} \int_C \frac{f(z)\,dz}{(z-a)^3}, \tag{3}$$

and so on, as in § 2·41, C being a contour surrounding the point $z = a$, on and within which $f(z)$ is analytic. In this instance all the resulting integrals are convergent, provided (1) is convergent.

7·13. As an example of integration under the integral sign, if $n \geqslant 1$, we have

$$\int_0^t dt \int_{Br_1} \frac{e^{zt}\,dz}{z^n} = \int_{Br_1} \left[\frac{e^{zt}\,dz}{z^{n+1}}\right]_0^t = \int_{Br_1} \frac{(e^{zt}-1)\,dz}{z^{n+1}} \tag{1}$$

$$= \int_{Br_1} \frac{e^{zt}\,dz}{z^{n+1}}, \tag{2}$$

since the second integral on the right-hand side of (1) is zero. Accordingly (1) § 7·11 is reproduced.

7·21. Changing the order of integration. In evaluating double integrals it is often expedient to change the order of integration. For instance suppose we want to evaluate

$$\int_0^\infty \int_{Br_1} \frac{e^{z-at/z}\, dz\, dt}{z^2}. \tag{1}$$

Changing the order of integration we have

$$I = \int_{Br_1} \frac{e^z\, dz}{z^2} \int_0^\infty e^{-at/z}\, dt \tag{2}$$

$$= \frac{1}{a} \int_{Br_1} \frac{e^z\, dz}{z} = \frac{2\pi i}{a}, \tag{3}$$

for by §4·22 Br_1 is equivalent to a circle round the origin, the residue there being unity. A discussion of the conditions to be satisfied for inverting the order of integration is beyond the scope of the text, and the reader is referred to [215, 220, 226, 237, 240, 243]. Usually the procedure is valid if convergence ensues.

7·22. Example. Verify that

$$\frac{1}{2\pi i} \int_{Br_1} \frac{e^{zt} K_0(az)\, dz}{z} = \cosh^{-1}(t/a), \qquad (t > a)$$

given that the modified Bessel function

$$K_0(az) = \int_0^\infty e^{-az \cosh \theta}\, d\theta.$$

Substituting in the original integral for $K_0(az)$, we obtain

$$I = \frac{1}{2\pi i} \int_{Br_1} e^{zt} \left[\int_0^\infty e^{-az \cosh \theta}\, d\theta \right] \frac{dz}{z} \tag{1}$$

$$= \frac{1}{2\pi i} \int_0^{\theta_1} d\theta \int_{Br_1} e^{z(t - a \cosh \theta)} \frac{dz}{z}, \tag{2}$$

the changed order of integration being permissible. By §4·41 the contour integral is zero when $t < a \cosh \theta$, so that we must have $t > a \cosh \theta$, or $\theta < \cosh^{-1} t/a$. Accordingly, the upper limit in the second integral is $\theta_1 = \cosh^{-1} t/a$. Evaluating (2), the contour integral yields unity, so we get

$$I = \int_0^{\cosh^{-1} t/a} d\theta = \cosh^{-1}(t/a). \tag{3}$$

It is permissible to differentiate the original integral under the sign with respect to t, and we find that

$$\frac{1}{2\pi i}\int_{Br_1} e^{zt} K_0(az)\, dz = \frac{1}{\sqrt{(t^2 - a^2)}} \qquad (t > a). \tag{4}$$

MISCELLANEOUS EXAMPLES

7·31. Example. Evaluate

$$I = \frac{1}{2\pi i}\int_{Br_1} \frac{e^{zt}\, dz}{\sqrt{(1+z)}}.$$

Writing $(1+z) = \zeta$, $dz = d\zeta$, this transformation being treated in Appendix 5, we get

$$I = \frac{1}{2\pi i}\int_{Br_1} \frac{e^{(\zeta-1)t}\, d\zeta}{\sqrt{\zeta}} = \frac{e^{-t}}{2\pi i}\int_{Br_1} \frac{e^{\zeta t}\, d\zeta}{\sqrt{\zeta}} \tag{1}$$

$$= e^{-t}/\sqrt{(\pi t)}. \tag{2}$$

7·32. Example. Evaluate

$$I = \frac{1}{2\pi i}\int_{Br_1} \frac{e^{zt}\, dz}{z\sqrt{(1+z)}}.$$

This integral may be differentiated under the sign, so

$$\frac{dI}{dt} = \frac{1}{2\pi i}\int_{Br_1} \frac{e^{zt}\, dz}{\sqrt{(1+z)}} = \frac{e^{-t}}{\sqrt{(\pi t)}}, \tag{1}$$

from (2) § 7·31. Hence

$$I = \frac{1}{\sqrt{\pi}}\int_0^t \frac{e^{-\tau}\, d\tau}{\sqrt{\tau}} = \operatorname{erf}\sqrt{t}, \tag{2}$$

from (1) § 5·212, since $I = 0$ when $t = 0$, a point the reader should confirm (see § 4·71).

7·33. Example. Show that $\dfrac{1}{2\pi i}\displaystyle\int_{Br_1}\dfrac{e^{zt}dz}{\sqrt{z(z-1)}}=e^t\operatorname{erf}\sqrt{t}.$ (1)

Put $z=\zeta+1$ (see §2, Appendix 5), and we get

$$\frac{e^t}{2\pi i}\int_{Br_1}\frac{e^{\zeta t}d\zeta}{\zeta\sqrt{(\zeta+1)}}=e^t\operatorname{erf}\sqrt{t}\qquad(2)$$

by §7·32.

7·34. Example. Evaluate

$$I=\frac{1}{2\pi i}\int_{Br_2}\frac{z\,e^{zt}dz}{\sqrt{(1+z)}},$$

by differentiating under the integral sign.
Now

$$\frac{d}{dt}\left[\frac{1}{2\pi i}\int_{Br_2}\frac{e^{zt}dz}{\sqrt{(1+z)}}\right]=I,\qquad(1)$$

so from (2) §7·31,

$$\frac{d}{dt}\left(\frac{e^{-t}}{\sqrt{(\pi t)}}\right)=I=-\frac{e^{-t}}{\sqrt{(\pi t)}}\left(1+\frac{1}{2t}\right).\qquad(2)$$

This result may be obtained also by writing $(1+z)=\zeta$.

7·35. Example. Evaluate

$$I=\frac{1}{2\pi i}\int_{Br_2}\frac{e^{zt}\sqrt{(z^2+2az)}\,dz}{z}.$$

Then $\qquad I=\dfrac{1}{2\pi i}\displaystyle\int_{Br_2}\dfrac{e^{zt}\sqrt{\{(z+a)^2-a^2\}}\,dz}{z}.$ (1)

Put $(z+a)=\zeta$, $dz=d\zeta$, and (1) becomes (see Appendix 5)

$$I=\frac{e^{-at}}{2\pi i}\int_{Br_2}\frac{e^{\zeta t}(\zeta+a)\,d\zeta}{\sqrt{(\zeta^2-a^2)}}\qquad(2)$$

$$=\frac{a\,e^{-at}}{2\pi i}\int_{Br_2}\frac{e^{\zeta t}d\zeta}{\sqrt{(\zeta^2-a^2)}}+\frac{e^{-at}}{2\pi i}\frac{d}{dt}\int_{Br_2}\frac{e^{\zeta t}d\zeta}{\sqrt{(\zeta^2-a^2)}}\qquad(3)$$

$$=a\,e^{-at}[I_0(at)+I_1(at)],\qquad(4)$$

from (5) §5·34, and the recurrence formula $I_1(v)=I_0'(v)$.

7·36. Example. Evaluate the following integral which occurs in a problem on loaded telegraph cables:

$$I=\int_0^\infty\cos mx\,\frac{\sin ut}{u}\,dm,\qquad(1)$$

where $\qquad u=\sqrt{(m^2v^2-a^2)}.$

From [234], p. 64

$$\frac{\sin ut}{u} = \sqrt{\frac{\pi t}{2u}} \, J_{\frac{1}{2}}(ut)$$

$$= \frac{t\sqrt{\pi}}{2} \frac{1}{2\pi i} \int_{Br_1} e^{z - u^2 t^2/4z} \frac{dz}{z^{\frac{3}{2}}}, \tag{2}$$

by (8) § 5·32. Substituting from (2) into (1), we obtain

$$I = \frac{t\sqrt{\pi}}{2} \frac{1}{2\pi i} \int_{Br_1} e^{z + a^2 t^2/4z} \frac{dz}{z^{\frac{3}{2}}} \int_0^\infty \cos mx \, e^{-m^2 v^2 t^2/4z} \, dm. \tag{3}$$

Now

$$\int_0^\infty \cos mx \, e^{-b^2 m^2} \, dm = \frac{\sqrt{\pi}}{2b} \, e^{-x^2/4b^2},$$

so (3) becomes

$$I = \frac{\pi}{2v} \frac{1}{2\pi i} \int_{Br_1} e^{z(1 - x^2/v^2 t^2) + a^2 t^2/4z} \frac{dz}{z}. \tag{4}$$

Substituting $\zeta = z(1 - x^2/v^2 t^2)$ (which is valid by Appendix 5 since x, v, t are all real and positive), and writing z for ζ thereafter, we have

$$I = \frac{\pi}{2v} \frac{1}{2\pi i} \int_{Br_1} e^{z + a^2(t^2 - x^2/v^2)/4z} \frac{dz}{z} \tag{5}$$

$$= \frac{\pi}{2v} I_0\{a \sqrt{(t^2 - x^2/v^2)}\}, \tag{6}$$

by (7) § 5·33, when $t > x/v$.

This is an excellent example of the evaluation of a real integral by a contour method, the analysis being brief and concise. Where intricate real integrals are concerned, the contour method is sometimes very powerful.

It is important to notice that the Bessel function in (2) is expressed as a contour integral on Br_1 but not Br_2. This is essential because $R(z) > 0$ to make the infinite integral in (3) convergent.

7·37. Example. Evaluate

$$I = \frac{1}{2\pi i} \int_{Br_1} \frac{e^{zt} \, dz}{z + b\sqrt{z}}. \tag{1}$$

Writing $z = \zeta^2$, $dz = 2\zeta d\zeta$, we transfer to the Br_3 contour in Fig. 107 (c), Appendix 5. Thus

$$I = \frac{1}{\pi i} \int_{Br_3} \frac{e^{\zeta^2 t} \, d\zeta}{\zeta + b}. \tag{2}$$

Put $v = \zeta + b$, and (2) becomes

$$I = \frac{e^{b^2 t}}{\pi i} \int_{Br_2} \frac{e^{v^2 t - 2bvt} \, dv}{v}. \tag{3}$$

Now transfer to a new Br_1 contour by writing $v^2 = u$, $du/2u = dv/v$, then

$$I = \frac{e^{b^2 t}}{2\pi i} \int_{Br_1} \frac{e^{ut - (2bt)\sqrt{u}} \, du}{u} \tag{4}$$

$$= e^{b^2 t} \operatorname{erfc}(b\sqrt{t}), \tag{5}$$

by (7) § 5·221.

7·38. Example. Evaluate

$$I = \frac{1}{2\pi i} \int_{Br_1} \frac{e^{zt - a\sqrt{z}} \, dz}{z(\sqrt{z} + b)}.$$

Now

$$\frac{1}{z(\sqrt{z} + b)} = \frac{(1/b)}{z} - \frac{(1/b)}{\sqrt{z}(\sqrt{z} + b)}, \tag{1}$$

and the integral corresponding to the second member of (1) is

$$I_1 = \frac{(1/b)}{2\pi i} \int_{Br_1} \frac{e^{zt - a\sqrt{z}} \, dz}{z} = \frac{1}{b} \operatorname{erfc} \frac{a}{2\sqrt{t}}, \tag{2}$$

by (7) § 5·221.

Writing $z = \zeta^2$, $dz = 2\zeta d\zeta$, $dz/\sqrt{z} = 2d\zeta$, we transfer to the contour Br_3, Appendix 5, Fig. 107 (c), so the integral corresponding to the third member of (1) is

$$I_2 = -\frac{(2/b)}{2\pi i} \int_{Br_3} \frac{e^{\zeta^2 t - a\zeta} \, d\zeta}{\zeta + b}. \tag{3}$$

Put $\zeta + b = v$ and (3) becomes

$$I_2 = -\frac{(2/b)}{2\pi i} \int_{Br_3} \frac{e^{(v-b)^2 t - a(v-b)} \, dv}{v} = -\frac{(2/b)}{2\pi i} e^{b^2 t + ab} \int_{Br_3} \frac{e^{v^2 t - v(2bt + a)} \, dv}{v} \tag{4}$$

$$= -(1/b) e^{b^2 t + ab} \operatorname{erfc}[(2bt + a)/2\sqrt{t}], \tag{5}$$

by (3), (5) § 7·37. Hence on adding (2) and (5), we obtain

$$I = (1/b) \{\operatorname{erfc}(a/2\sqrt{t}) - e^{b^2 t + ab} \operatorname{erfc}[(2bt + a)/2\sqrt{t}]\}. \tag{6}$$

7·39. Example. Evaluate

$$I = \frac{1}{2\pi i} \int_{Br_1} \frac{e^{zt - b\sqrt{(z + a^2)}} \, dz}{z}.$$

Write $z + a^2 = v^2$, $dz = 2v\,dv$, $dz/z = 2v\,dv/(v^2 - a^2)$, thereby transferring to the contour Br_3 in Fig. 107 (c). Then we get

$$I = \frac{e^{-a^2 t}}{\pi i} \int_{Br_3} \frac{e^{v^2 t - bv}\, v\, dv}{v^2 - a^2} \tag{1}$$

$$= \frac{e^{-a^2 t}}{2\pi i} \int_{Br_3} e^{v^2 t - bv} \left[\frac{1}{v - a} + \frac{1}{v + a} \right] dv \tag{2}$$

$$= (e^{-a^2 t}/2)\, \{ e^{a^2 t - ab}\, \mathrm{erfc}\,[(b - 2at)/2\sqrt{t}] + e^{a^2 t + ab}\, \mathrm{erfc}\,[(b + 2at)/2\sqrt{t}] \}, \tag{3}$$

by (3) and (5) § 7·38. Clearing the exponentials in t leads to the result

$$I = (1/2)\{ e^{ab}\, \mathrm{erfc}\,[(b + 2at)/2\sqrt{t}] + e^{-ab}\, \mathrm{erfc}\,[(b - 2at)/2\sqrt{t}] \}. \tag{4}$$

7·41. Integration by parts.

It can be proved that the procedure used for the real variable, namely,

$$\int w\, dv = wv - \int v\frac{dw}{dz}\, dz, \tag{1}$$

where w and v are functions of z, is applicable to the complex variable (see [237, 243]).

7·42. Example.

Show that

$$\frac{1}{2\pi i} \int_{Br_1} \frac{e^{zt}\, dz}{z^{n+1}} = \frac{t^n}{n!},$$

by aid of partial integration.

$$I = -\frac{1/n}{2\pi i} \int_{Br_1} e^{zt}\, d\left(\frac{1}{z^n}\right) = -\left[\frac{(1/n)}{2\pi i} \frac{e^{zt}}{z^n} \right] + \frac{1}{2\pi i} \frac{t}{n} \int_{Br_1} \frac{e^{zt}\, dz}{z^n}. \tag{1}$$

The question of the limits for the first expression on the right-hand side of (1) now arises. Since there is a pole of order $n + 1$ at the origin, Br_1 is equivalent to a circle about O. If z is taken round this once, it regains its original value, so that both limits are identical and the bracketed quantity vanishes. Hence

$$I = \frac{t/n}{2\pi i} \int_C \frac{e^{zt}\, dz}{z^n} \tag{2}$$

$$= -\frac{t/n(n-1)}{2\pi i} \int_C e^{zt}\, d\left(\frac{1}{z^{n-1}}\right) = \frac{1}{2\pi i} \frac{t^2}{n(n-1)} \int_C \frac{e^{zt}\, dz}{z^{n-1}}, \tag{3}$$

since the term corresponding to wv vanishes as before. Proceeding in this way, we get ultimately

$$I = \frac{t^n/n!}{2\pi i} \int_C \frac{e^{zt}\,dz}{z} = \frac{t^n}{n!}. \tag{4}$$

7·43. Example. Evaluate

$$\frac{1}{2\pi i} \int_{Br_1} \frac{e^{zt-\sqrt{(az)}}\,dz}{z^{\frac{3}{2}}}.$$

Integrating by parts,

$$I = -\frac{2}{2\pi i} \int_{Br_1} e^{zt-\sqrt{(az)}}\,d(z^{-\frac{1}{2}})$$

$$= -\left[\frac{2}{2\pi i}\frac{e^{zt-\sqrt{(az)}}}{\sqrt{z}}\right] + \frac{2}{2\pi i}\int_{Br_1}[t-\tfrac{1}{2}\sqrt{(a/z)}]\frac{e^{zt-\sqrt{(az)}}\,dz}{\sqrt{z}}. \tag{1}$$

Now in the original integral, Br_2 is equivalent to Br_1, so the limits for the first term in (1) are $x\,e^{-i\pi}$ and $x\,e^{i\pi}$ when $x\to\infty$. Consequently the first term vanishes, leaving

$$I = \frac{2t}{2\pi i}\int_{Br_1}\frac{e^{zt-\sqrt{(az)}}\,dz}{\sqrt{z}} - \frac{\sqrt{a}}{2\pi i}\int_{Br_1}\frac{e^{zt-\sqrt{(az)}}\,dz}{z} \tag{2}$$

$$= 2\sqrt{\left(\frac{t}{\pi}\right)}e^{-a/4t} - \sqrt{a}\,\text{erfc}\frac{1}{2}\sqrt{\frac{a}{t}}, \tag{3}$$

from (4) §5·222 and (7) §5·221.

In using partial integration it happens frequently that the term wv in §7·41 vanishes at the limits of integration, but each case must be checked individually (see example §7·45).

7·44. Example. Evaluate

$$I = \frac{1}{2\pi i}\int_{Br_1}\frac{e^{zt-a\sqrt{z}}\,dz}{z^2}.$$

Then

$$I = -\frac{1}{2\pi i}\int_{Br_2}e^{zt-a\sqrt{z}}\,d(z^{-1}), \tag{1}$$

Br_1 and Br_2 being equivalent. Taking (1) by parts, we obtain

$$I = \lim_{x\to\infty}\left\{-\frac{1}{2\pi i}\left[\frac{e^{zt-a\sqrt{z}}}{z}\right]_{z=xe^{-i\pi}}^{z=xe^{i\pi}}\right\} + \frac{1}{2\pi i}\int_{Br_1}\frac{(t-a)}{2\sqrt{z}}\frac{e^{zt-a\sqrt{z}}\,dz}{z} \tag{2}$$

$$= 0 + t\,\text{erfc}\,(a/2\sqrt{t}) - (a/2)\{2\sqrt{(t/\pi)}\,e^{-a^2/4t} - a\,\text{erfc}\,(a/2\sqrt{t})\}, \tag{3}$$

by (6) § 5·221 and (3) § 7·43. Thus

$$I = (t + a^2/2)\, \text{erfc}\,(a/2\sqrt t) - a\sqrt{(t/\pi)}\,e^{-a^2/4t}, \tag{4}$$

so by (7) § 15·23, with $\tfrac12 a$ for a, and (4) above, we have

$$\frac{1}{2\pi i}\int_{Br_1}\frac{e^{zt-a\sqrt z}\,dz}{z^2} = \int_0^t \text{erfc}\,(a/2\sqrt t)\,dt. \tag{5}$$

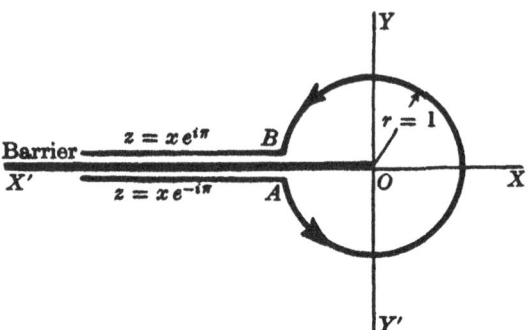

Fig. 37. At A, $z = e^{-i\pi}$, at B, $z = e^{i\pi}$.

7·45. Example. Evaluate $\dfrac{1}{2\pi i}\displaystyle\int_{Br_2} e^{zt}\log z\,dz$, where Br_2 consists of a circle of unit radius round the origin and the lines $x\,e^{i\pi}$ and $x\,e^{-i\pi}$ from $x = 1$ to ∞, as illustrated in Fig. 37.

On the circle in Fig. 37,

$$\frac{1}{2\pi i}\int_C e^{zt}\log z\,dz = \frac{1/t}{2\pi i}\int_C \log z\, de^{zt}$$

$$= \frac{1/t}{2\pi i}\left[e^{zt}\log z\right]_{z=e^{-i\pi}}^{z=e^{i\pi}} - \frac{1/t}{2\pi i}\int_C \frac{e^{zt}\,dz}{z} \tag{1}$$

$$= \frac{1/t}{2\pi i}[e^{-t}\{\log e^{i\pi} - \log e^{-i\pi}\}] - \frac{1}{t}$$

$$= \frac{(e^{-t}/t)}{2\pi i}\log e^{2\pi i} - \frac{1}{t}$$

$$= \frac{1}{t}\,(e^{-t} - 1). \tag{2}$$

On the parallel lines we get

$$-\frac{1}{2\pi i}\int_1^\infty e^{-xt}[\log x\,e^{i\pi} - \log x\,e^{-i\pi}]\,dx = -\int_1^\infty e^{-xt}\,dx = -\frac{e^{-t}}{t}. \tag{3}$$

Adding (2) and (3), we obtain

$$\frac{1}{2\pi i}\int_{Br_s} e^{zt}\log z\,dz = -\frac{1}{t}. \tag{4}$$

This example has been worked out in detail to illustrate the non-vanishing of the term in (1), corresponding to vw in (1) § 7·41, after partial integration. The evaluation may be effected also as follows:

$$\frac{1}{2\pi i}\int_{Br_s} e^{zt}\log z\,dz = \lim_{x\to\infty}\left[\frac{1/t}{2\pi i}e^{zt}\log z\right]_{z=x\,e^{-i\pi}}^{z=x\,e^{i\pi}} - \frac{1/t}{2\pi i}\int_{Br_s}\frac{e^{zt}\,dz}{z} \tag{5}$$

$$= \lim_{x\to\infty}(1/t)\,[e^{-xt}] - 1/t = -1/t. \tag{6}$$

Repeated differentiation of the first integral in (5) with regard to t is valid, since all integrals converge uniformly. Thus we obtain

$$\frac{1}{2\pi i}\int_{Br_s} e^{zt} z^n \log z\,dz = \frac{(-1)^{n+1}n!}{t^{n+1}}, \tag{7}$$

n finite $\geqslant 0$, t real > 0.

Notes. (i) The integral in § 7·38 may be evaluated also by applying the theorem in Problem 55, p. 329, taking $\phi(p^{\frac{1}{2}}) = e^{-ap^{\frac{1}{2}}}/(p^{\frac{1}{2}} + b)$.

(ii) (5) § 7·44 may be obtained by integrating (6) § 5·221 under the integral sign with respect to t.

PART II
THEORY OF TRANSFORM CALCULUS

VIII

MELLIN INVERSION THEOREM: TRANSFORM THEORY

8·11. Mellin theorem. The transform theory herein is based upon a particular case of this theorem treated in § 6, Appendix 4, which may be stated as follows:

If

$$\phi(p) = p \int_0^\infty e^{-pt} f(t)\, dt \quad (R(p) > 0), \tag{1}*$$

then will

$$f(t) = \frac{1}{2\pi i} \int_{Br_1} \frac{e^{pt}\, \phi(p)\, dp}{p} = \frac{1}{2\pi i} \int_{Br_1} \frac{e^{zt}\, \phi(z)\, dz}{z}, \tag{2}$$

and vice versa. The conditions for validity are given in § 6, Appendix 4. By § 4·41, integral (2), subject to condition 1° (c), is zero when $t < 0$. Although in many cases $f(t)$ exists when $t < 0$, in our work we consider only that part of the function where $t > 0$. $\phi(p)$ is known as the p-multiplied Laplace transform of $f(t)$,† while (2) is the Bromwich-Wagner integral (see [130]).

8·12. Suppose
$$f(t) = t^\nu, \tag{1}$$

then by (1) § 8·11 and (2) § 5·13,

$$\phi(p) = p \int_0^\infty e^{-pt} t^\nu\, dt = \frac{\Gamma(1 + \nu)}{p^\nu}, \tag{2}$$

provided $R(\nu) > -1$. When $R(\nu) \leqslant -1$, the integral in (2) is divergent, and $\phi(p)$ does not exist. Inserting the result from (2) into (2) § 8·11, and writing z for p, we have

$$f(t) = t^\nu = \frac{\Gamma(1 + \nu)}{2\pi i} \int_{Br_1} \frac{e^{zt}\, dz}{z^{\nu+1}} \tag{3}$$

$$= \Gamma(1 + \nu)\, t^\nu / \Gamma(1 + \nu) = t^\nu, \tag{4}$$

---

* Transforms obtained from (1) do not always satisfy (2). Inversion of a transform of this description is accomplished by solving (1) as an integral equation for $f(t)$ using a list of transforms (see [128]).

† $\phi(p)/p$ is the Laplace transform (see [236]).

by (3) § 5·15, which illustrates the theorem when $R(\nu) > -1$. From (2), we have $\phi(p)/p = \Gamma(1+\nu)/p^{\nu+1}$, and the modulus of this tends to zero uniformly as $|p| \to \infty$, $-\frac{1}{2}\pi \leqslant$ phase $p \leqslant \frac{1}{2}\pi$.

We call $\Gamma(1+\nu)/p^{\nu}$ the transform of t^{ν}, *referred to the Mellin inversion theorem*, and shall write

$$t^{\nu} \Longrightarrow \Gamma(1+\nu)/p^{\nu}, \qquad (5)$$

or $\qquad\qquad t^{\nu}/\Gamma(1+\nu) \Longrightarrow 1/p^{\nu}. \qquad (6)$

$t^{\nu}/\Gamma(1+\nu)$ is the inverse transform of $p^{-\nu}$ in terms of the variable t. In general $\phi(p)$ defined in (1) or in (2) § 8·11 is taken as the transform of $f(t)$, while integral (2) § 8·11 gives the inverse transform of $\phi(p)$.* Since (2) § 8·11 is zero when $t < 0$—under restriction (c) Appendix 4—the transform derived from either (1) or (2) § 8·11 refers to $f(t)$ when $t > 0$.

The transforms are referred to the contour Br_1, upon which the above case of the Mellin theorem is based, but they comply also with the Laplace integral (1) § 8·11. If the contours Br_1 and Br_2 are equivalent (see § 4·31 for conditions), the transforms may then be referred to the latter contour also. As shown in Appendix 6, it is possible to have transforms referred to Br_2 but not to Br_1. These transforms are not associated with either part of the Mellin theorem. Additional information will be found in Appendix 6 and reference [128].

8·21. Example. Find the transform of $f(t) = e^{at}$, a being finite but either real or complex.

(1°) From (1) § 8·11 when $R(p) > R(a)$,

$$\phi(p) = p \int_0^{\infty} e^{-pt} e^{at}\, dt = p \int_0^{\infty} e^{(-p+a)t}\, dt = \frac{p}{p-a}, \qquad (1)$$

or $\qquad\qquad e^{at} \Longrightarrow p/(p-a). \qquad (2)$

$\phi(p)/p = 1/(p-a)$ and the modulus $\to 0$ uniformly as $|p| \to \infty$, $-\frac{1}{2}\pi \leqslant$ phase $z \leqslant \frac{1}{2}\pi$. Hence the transform (2) satisfies both parts of the Mellin theorem.

---

* These statements are valid only if both parts of the Mellin theorem, § 8·11, are satisfied. It is possible to find a $\phi(p)$ from (1) which makes (2) divergent (see [128]).

(2°) From (2) § 8·11,

$$f(t) = \frac{1}{2\pi i}\int_{Br_1}\frac{e^{(z+a)t}\,dz}{z} = \frac{1}{2\pi i}\int_{Br_1}e^{\zeta t}\left(\frac{\zeta}{\zeta-a}\right)\frac{d\zeta}{\zeta}, \tag{3}$$

after substituting $\zeta = (z+a)$, this transformation being valid as shown in Appendix 5. Hence from (2) § 8·11 and the second integral in (3), it follows that

$$\phi(p) = p/(p-a). \tag{4}$$

From above we see that when $f(t) = e^{at}$, $\phi(p)$ can be obtained from either formula (1) or (2) § 8·11. Generally it is preferable to use that formula which yields $\phi(p)$ the more readily, e.g. (1) § 8·11 in this instance.

8·22. The addition theorem. If

$$f_1(t) \rightleftharpoons \phi_1(p), \quad f_2(t) \rightleftharpoons \phi_2(p), \quad \ldots, \quad f_n(t) \rightleftharpoons \phi_n(p);$$

then
$$\sum_{m=1}^{n} f_m(t) \rightleftharpoons \sum_{m=1}^{n} \phi_m(p). \tag{1}$$

For a finite number of functions the proof is immediate. As an example, write i for a in (2) of § 8·21 and we obtain

$$e^{it} = \cos t + i\sin t \rightleftharpoons p/(p-i) = p(p+i)/(p^2+1). \tag{2}$$

Separating the real and imaginary parts (assuming p to be real), we get
$$\sin t \rightleftharpoons p/(p^2+1), \tag{3}$$

and
$$\cos t \rightleftharpoons p^2/(p^2+1). \tag{4}$$

If certain conditions given in [236] p. 28 are satisfied, n may be infinite. To illustrate this we shall determine the transform of $\sin t$ by expansion. Then

$$\sum_{m=1}^{\infty} \phi_m(p) = \phi(p) = p\int_0^{\infty} e^{-pt}\left[t - \frac{t^3}{3!} + \frac{t^5}{5!} - \ldots\right]dt \tag{5}$$

$$= \frac{1}{p} - \frac{1}{p^3} + \frac{1}{p^5} - \ldots, \tag{6}$$

by (2) § 8·12.

By virtue of Theorem 15 in [236] p. 175, term-by-term integration in (5) is valid, and (6) is absolutely convergent if $p > 1$, this being permissible.

Series (6) is the expansion of $1/p(1 + 1/p^2)$, so

$$\phi(p) = p/(p^2 + 1), \tag{7}$$

or

$$\sin t \rightleftharpoons p/(p^2 + 1). \tag{8}$$

8·23. Contour integral corresponding to transform. If $\phi(p)$, the p-multiplied Laplace transform of $f(t)$,[*] is known, then provided that $|\phi(p)/p| \to 0$ uniformly[†] as $|p| \to \infty$, $-\tfrac{1}{2}\pi \leqslant \theta \leqslant \tfrac{1}{2}\pi$, a contour integral for $f(t)$ is

$$f(t) = \frac{1}{2\pi i} \int_{Br_1} \frac{e^{zt}\,\phi(z)\,dz}{z}, \tag{1}$$

$t > 0$, by § 8·11. As a simple illustration, $e^{at} \rightleftharpoons p/(p - a)$, so

$$e^{at} = \frac{1}{2\pi i} \int_{Br_1} \frac{e^{zt}\,dz}{z - a}, \tag{2}$$

since e^{at} is the residue of $e^{zt}/(z - a)$ at the pole $z = a$.

Also $\lim_{|p| \to \infty} |\phi(p)/p| \to 0$ in both (3) and (4) § 8·22, so

$$\sin t = \frac{1}{2\pi i} \int_{Br_1} \frac{e^{zt}\,dz}{1 + z^2}, \tag{3}$$

and

$$\cos t = \frac{d}{dt}\sin t = \frac{1}{2\pi i} \int_{Br_1} \frac{e^{zt}\,z\,dz}{1 + z^2}. \tag{4}$$

8·3. Change of origin. In applications to electrical wave transmission and to acoustics, the argument of the function may assume the form $a\sqrt{(t^2 - b^2)}$. The corresponding versions of (1), (2) § 8·11 are, respectively,

$$\phi(p) = p \int_b^\infty e^{-pt} f[a\sqrt{(t^2 - b^2)}]\,dt, \tag{1}$$

and

$$f[a\sqrt{(t^2 - b^2)}] = \frac{1}{2\pi i} \int_{Br_1} \frac{e^{zt}\,\phi(z)\,dz}{z}. \qquad (t > b) \tag{2}$$

An example will be found at (7), (8) § 8·7.

---

[*] Unless stated to the contrary, the transforms used in the text are the p-multiplied type.

[†] See footnote on p. 346.

8·411. Exponential multiplier. Multiplying both sides of (2) § 8·11 by e^{at}, we have

$$e^{at}f(t) = \frac{1}{2\pi i}\int_{Br_1} e^{(z+a)t}\phi(z)\frac{dz}{z}. \tag{1}$$

Writing $(z+a) = \zeta$, or $z = (\zeta - a)$, with $dz = d\zeta$, gives

$$e^{at}f(t) = \frac{1}{2\pi i}\int_{Br_1} e^{\zeta t}\phi(\zeta - a)\frac{d\zeta}{(\zeta - a)} = \frac{1}{2\pi i}\int_{Br_1} e^{\zeta t}\left[\frac{\zeta\phi(\zeta - a)}{(\zeta - a)}\right]\frac{d\zeta}{\zeta}. \tag{2}$$

Hence
$$e^{at}f(t) \Rightarrow p\phi(p-a)/(p-a). \tag{3}$$

8·412. Example. Find the transforms of $e^{at}\sin t$ and $e^{at}\cos t$. From (3), (4) § 8·22 and (3) § 8·411,

$$e^{at}\sin t \Rightarrow p/[(p-a)^2+1], \tag{1}$$

$$e^{at}\cos t \Rightarrow p(p-a)/[(p-a)^2+1]. \tag{2}$$

8·421. In (2) § 8·11 write at for t, a being real and positive, then

$$f(at) = \frac{1}{2\pi i}\int_{Br_1} e^{zat}\phi(z)\frac{dz}{z}. \tag{1}$$

Putting $\zeta = az$, thereby transferring to an equivalent Br_1 path (see Appendix 5), and writing z for ζ thereafter, we get

$$f(at) = \frac{1}{2\pi i}\int_{Br_1} e^{zt}\phi(z/a)\frac{dz}{z}. \tag{2}$$

Hence
$$f(at) \Rightarrow \phi(p/a), \tag{3}$$

and
$$f(t/a) \Rightarrow \phi(pa), \tag{4}$$

on replacing a in (3) by $1/a$.

8·422. In § 8·421 when a is complex, a new contour is involved. Let Br_1 be a straight line parallel to $Y'OY$ as shown in Fig. 38 (a), and distant c from it, then $z = c + iy = re^{i\theta}$ is any point thereon. If $a = e + if = r_0 e^{i\theta_0}$, then $\zeta = za = rr_0 e^{i(\theta+\theta_0)}$. In Fig. 38 (a), $OP = |z| = r$, the line NQ is parallel to Br_1 and distant cr_0 from it, so that $OQ = |za| = rr_0$. Thus if ON and NQ are rotated counter-clockwise bodily about O through an angle θ_0, the appropriate contour is Br_1' in Fig. 38 (b). If it can be shown that Br_1' is equivalent to Br_1, formulae (3), (4) § 8·421—*mutatis mutandis* —are valid.

As a case in point, consider the Bessel function $J_0(it) = I_0(t)$.

$$J_0(t) \rightleftharpoons p/\sqrt{(p^2+1)},$$

so using (3) § 8·421 with $a = i = e^{\frac{1}{2}\pi i}$, we obtain

$$J_0(it) = I_0(t) \rightleftharpoons (p/i)/\sqrt{\{(p/i)^2+1\}} = p/\sqrt{(p^2-1)}. \tag{1}$$

Now by § 8·11 and the preceding transform

$$J_0(t) = \frac{1}{2\pi i} \int_{Br_1} \frac{e^{zt}\,dz}{\sqrt{(z^2+1)}}, \tag{2}$$

so the integrand has branch points at $z = \pm i$, the contour being the dumb-bell type of Fig. 25 (a). The new contour for the above transformation is obtained by revolving that of Fig. 25 (a) counter-clockwise until it lies along the real axis. It may be shown that this contour is equivalent to Br_1 for the integral $\dfrac{1}{2\pi i} \displaystyle\int \dfrac{e^{zt}\,dz}{\sqrt{(z^2-1)}}$, so (1) is the transform of $I_0(t)$.

8·423. In (3) § 8·411 write bt for t, b real > 0, and c real > 0 for ab, then

$$e^{ct}f(bt) \rightleftharpoons p\phi[(p-c)/b]/(p-c). \tag{1}$$

8·424. Example. Find the transforms of $\sin at$, $\cos at$, $e^{bt}\sin \omega t$, and $e^{bt}\cos \omega t$.

Writing at for t and p/a for p in (3), (4) § 8·22, we get

$$\sin at \rightleftharpoons \frac{pa}{p^2+a^2}, \quad \text{or} \quad \sin at = \frac{a}{2\pi i} \int_{Br_1} \frac{e^{zt}\,dz}{z^2+a^2}, \tag{1}$$

and $\quad \cos at \rightleftharpoons \dfrac{p^2}{p^2+a^2}, \quad$ or $\quad \cos at = \dfrac{1}{2\pi i} \displaystyle\int_{Br_1} \dfrac{e^{zt}\,z\,dz}{z^2+a^2}. \tag{2}$

In both of these cases as in § 8·422, it can be proved valid for a to be complex.

For the next two examples we take

$$f_1(t) = \sin t \rightleftharpoons p/(p^2+1), \tag{3}$$

$$f_2(t) = \cos t \rightleftharpoons p^2/(p^2+1). \tag{4}$$

Applying (1) § 8·423 to these, writing b for c and ω for b, we obtain

$$e^{bt}\sin \omega t \rightleftharpoons \frac{p(p-b)/\omega}{(p-b)\{[(p-b)/\omega]^2+1\}} = \frac{p\omega}{(p-b)^2+\omega^2}, \tag{5}$$

and $\quad e^{bt}\cos \omega t \rightleftharpoons \dfrac{p[(p-b)/\omega]^2}{(p-b)\{[(p-b)/\omega]^2+1\}} = \dfrac{p(p-b)}{(p-b)^2+\omega^2}. \tag{6}$

8·43. The differentiation theorem. By (1) §8·11

$$p\phi(p) = p^2 \int_0^\infty e^{-pt} f(t)\, dt = -p \int_0^\infty f(t)\, d\, e^{-pt}. \tag{1}$$

Integrating by parts, we have

$$p\phi(p) = [pf(t)\, e^{-pt}]_{t=\infty}^{t=0} + p \int_0^\infty e^{-pt} f'(t)\, dt \tag{2}$$

$$= pf(0) + p \int_0^\infty e^{-pt} f'(t)\, dt, \tag{3}$$

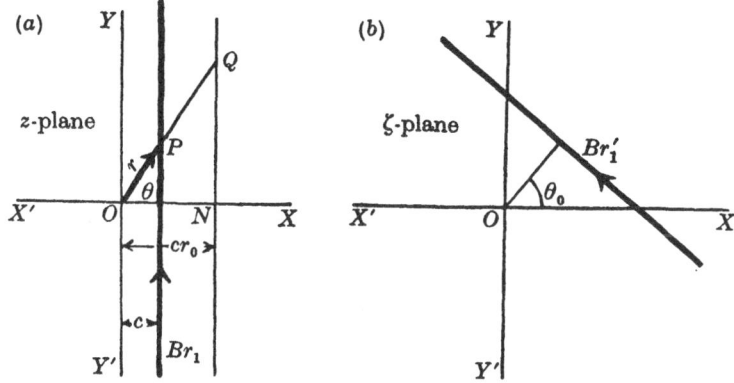

Fig. 38.

provided that $e^{-pt} f(t) \to 0$ as $t \to +\infty$ and that the integral converges (see [236] pp. 8, 9, 21 for conditions). Thus

$$f'(t) \Rightarrow p\phi(p) - pf(0), \tag{4}$$

and $$f'(t) \Rightarrow p\phi(p) \tag{5}$$

if $f(0) = 0$.

In general, if all the integrals converge

$$f^{(n)}(t) \Rightarrow p^n \phi(p) - \sum_{r=0}^{n-1} p^{n-r} f^{(r)}(0), \tag{6}$$

provided $e^{-pt} f^{(r)}(t) \to 0$ as $t \to +\infty$.

If $f(0) = f'(0) = \ldots = f^{(n-1)}(0) = 0$, the 'sigma' term in (6) is zero and differentiation of the function n times corresponds to multiplication of its transform by p^n.

If
$$f[a\sqrt{(t^2-b^2)}] \Rrightarrow \phi(p), \quad (t>b) \tag{7}$$
then
$$f^{(n)}[a\sqrt{(t^2-b^2)}] \Rrightarrow p^n\phi(p) - e^{-bp}\sum_{r=0}^{n-1}p^{n-r}\frac{d^r}{dt^r}\{f[(a\sqrt{(t^2-b^2)})]\}_{t=b}. \tag{8}$$

The proof is left as an exercise for the reader. See [131].

8·44. Example. Find the transform of $\cos t$ from that of $\sin t$. From (3) § 8·22, $\sin t \Rrightarrow p/(p^2+1)$, while $f(0) = 0$. Hence by (5) § 8·43

$$\cos t = \frac{d}{dt}\sin t \Rrightarrow \frac{p^2}{p^2+1}. \tag{1}$$

By (4) § 8·43,

$$\sin t = -\frac{d}{dt}\cos t \Rrightarrow -p[p^2/(p^2+1) - \cos(0)] \tag{2}$$

$$= -p^3/(p^2+1) + p = p/(p^2+1), \tag{3}$$

so
$$\sin t \Rrightarrow p/(p^2+1). \tag{4}$$

8·45. The integration theorem. By (1) § 8·11 if $\int_0^t f(\tau)\,d\tau \Rrightarrow \phi_1(p)$, then

$$\phi_1(p) = p\int_0^\infty e^{-pt}\left[\int_0^t f(\tau)\,d\tau\right]dt = -\int_0^\infty \left[\int_0^t f(\tau)\,d\tau\right]d\,e^{-pt} \tag{1}$$

$$= \left[e^{-pt}\int_0^t f(\tau)\,d\tau\right]_{t=\infty}^{t=0} + \int_0^\infty e^{-pt}f(t)\,dt. \tag{2}$$

If $e^{-pt}\int_0^t f(\tau)\,d\tau \to 0$ as $t \to \infty$ the first term of (2) vanishes at both limits. Hence by (1) and (2)

$$\int_0^t f(\tau)\,d\tau \Rrightarrow \phi_1(p) = p^{-1}\phi(p), \tag{3}$$

so that integration of the function corresponds to multiplication of its transform by p^{-1}. In general we find that (see [236])

$$\left[\int_0^t d\tau\right]^n f(\tau) \Rrightarrow p^{-n}\phi(p). \tag{4}$$

The above analysis is, of course, based upon the hypothesis that $f(t)$ has a transform obtainable from § 8·11.

Corresponding to (7) § 8·43, we get

$$\int_b^t f[a\sqrt{(\tau^2 - b^2)}]\,d\tau \rightleftharpoons p^{-1}\phi(p). \tag{5}$$

8·46. Example. Given that

$$J_1(t) \rightleftharpoons p/[\sqrt{(p^2+1)}\{p + \sqrt{(p^2+1)}\}] = p[1 - p/\sqrt{(p^2+1)}],$$

find the transform of $J_0(t)$.

Since

$$J_1(t) = -J_0'(t) = -\frac{d}{dt}J_0(t), \tag{1}$$

$$\int_0^t J_1(t)\,dt = -[J_0(t)]_0^t = 1 - J_0(t), \tag{2}$$

or

$$J_0(t) = 1 - \int_0^t J_1(t)\,dt. \tag{3}$$

When $t \to \infty$, by Prob. 32, p. 323, the value of the integral in (3) is unity, so $e^{-pt}\int_0^t J_1(t)\,dt \to 0$ as $t \to \infty$. Hence by § 8·45

$$J_0(t) \rightleftharpoons 1 - 1/[\sqrt{(p^2+1)}\{p + \sqrt{(p^2+1)}\}] = p/\sqrt{(p^2+1)}. \tag{4}$$

8·51. The shift theorem. Consider the integral relationship

$$\chi(t-h) = \frac{1}{2\pi i}\int_{Br_1} \frac{e^{z(t-h)}\phi(z)\,dz}{z}, \tag{1}$$

where $\chi(t)$ is continuous or piecewise continuous for $t > h$, its transform being obtainable from (1) § 8·11. By § 4·41 the value of the integral (1) is zero when $t < h$, so the origin of $\chi(t)$ is moved to the point $t = h$ real > 0. (1) may be written

$$\chi(t-h) = \frac{1}{2\pi i}\int_{Br_1} \frac{e^{zt}\,e^{-zh}\phi(z)\,dz}{z}, \tag{2}$$

so by § 8·11

$$\chi(t-h) \rightleftharpoons e^{-ph}\phi(p). \tag{3}$$

Let $\chi_1(t) = \sin\omega t$, then by (1) § 8·424,

$$\sin\omega t \rightleftharpoons \omega p/(p^2 + \omega^2), \tag{4}$$

and $\chi_1(t-h) = \sin \omega(t-h)$, so by (3), (4),

$$\sin \omega(t-h) \Rightarrow e^{-ph} \omega p/(p^2+\omega^2) \quad (t>h). \tag{5}$$

Formula (5) is illustrated in Fig. 39 (a), which might represent a disturbance applied to a dynamical system at $t = h$. This transform must not be confused with that for $\sin \omega(t-h)$, for $t > 0$, when expanded, i.e. $(\omega \cos \omega h - p \sin \omega h) p/(p^2 + \omega^2)$.

(a)

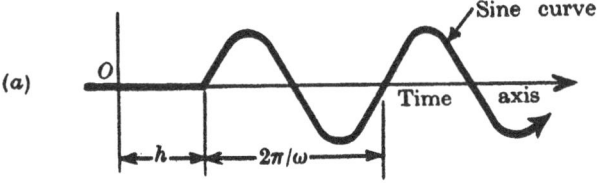

(b)

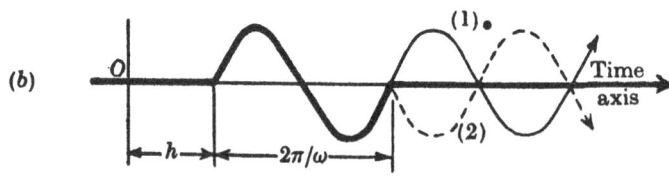

Fig. 39. The shift function e^{-ph} applied to $\sin \omega t$.

8·52. Suppose it is desired to suppress the sine wave of Fig. 39 (a) after one oscillation. To do so, another sine wave is introduced at time $t = (h + 2\pi/\omega)$ in opposite phase, which neutralizes the first wave. For the neutralizing wave

$$\chi_2(t-h-2\pi/\omega) = -\sin \omega(t-h-2\pi/\omega) \quad (t>h+2\pi/\omega),$$

$$\Rightarrow -e^{-p(h+2\pi/\omega)} \omega p/(p^2+\omega^2), \tag{1}$$

this being shown in Fig. 39 (b) by curve 2. The transform of the heavy curve is the sum of (5) § 8·51 and (1), namely,

$$\left.\begin{array}{l} \xi(t) = \sin \omega t \bigm| \quad (h \leqslant t \leqslant h + 2\pi/\omega) \\ \quad = 0 \quad \bigm| \quad \text{(for all other } t). \\ \Rightarrow e^{-ph}(1 - e^{-2\pi p/\omega}) \omega p/(p^2+\omega^2). \end{array}\right\} \tag{2}$$

When this is the transform of the right-hand side of a differential equation, the solution in terms of t applies only after a time interval $(h + 2\pi/\omega)$. This follows from the fact that the integral $\dfrac{1}{2\pi i}\displaystyle\int_{Br_1} \dfrac{e^{z(t-h-2\pi/\omega)}\,\phi(z)\,dz}{z}$ is zero when $t < (h + 2\pi/\omega)$, as shown in § 4·41. A number of cases illustrating the use of the shift function e^{-ph} is given in Figs. 48, 49.

8·53.

The problem in §§ 8·51, 8·52 may be treated in another way. In Fig. 39 (b), with $h = 0$, the transform of $\sin \omega t$ for one cycle is*

$$\psi(p) = p\int_0^{2\pi/\omega} e^{-pt}\sin \omega t\,dt = \frac{\omega p}{p^2 + \omega^2}(1 - e^{-2\pi p/\omega}), \qquad (1)$$

which agrees with (2) § 8·52 when $h = 0$.

As another example, the transform of a Morse dot (see Fig. 66 (b)) is

$$\psi_1(p) = E_0 p\int_0^{\tau} e^{-pt}\,dt = E_0(1 - e^{-p\tau}). \qquad (2)$$

The transform of the second dot is, (Fig. 66 (c))

$$\psi_2(p) = E_0 p\int_{2\tau}^{3\tau} e^{-pt}\,dt = E_0(e^{-2p\tau} - e^{-3p\tau}), \qquad (3)$$

and so on. For a semi-infinite sequence of dots,

$$\phi(p) = \sum_{n=1}^{\infty}\psi_n(p) = E_0(1 - e^{-p\tau} + e^{-2p\tau} - e^{-3p\tau} + \ldots)$$
$$= E_0/(1 + e^{-p\tau}), \qquad (4)$$

$R(p\tau) > 0$, so $|e^{-p\tau}| < 1$.

8·54.

For Fig. 39 (b) with $h > 0$, we have

$$\psi(p) = p\int_h^{h+2\pi/\omega} e^{-pt}\sin \omega(t-h)\,dt \qquad (1)$$

$$= e^{-ph}(1 - e^{-2\pi p/\omega})\,\omega p/(p^2 + \omega^2), \qquad (2)$$

as at (2) § 8·52.

---

* In general the upper limit is ∞ as in § 8·11, this covering the range of the function from $t = 0$ to $t = \infty$. For certain purposes, where a finite range of the function is required, the procedure given here may be useful. See also [236].

8·55. Application of §8·51 to Laplace's integral. If $f(t) \rightleftharpoons \phi(p)$ and $h > 0$, the transform of $f(t-h)$ is

$$\phi(p) = p \int_h^\infty e^{-pt} f(t-h)\, dt, \qquad (1)$$

where $t > h$. Substituting $\tau = (t-h)$ in (1) and writing t for τ thereafter, we have

$$\phi(p) = pe^{-ph} \int_0^\infty e^{-pt} f(t)\, dt, \qquad (2)$$

so $\qquad\qquad f(t-h) \rightleftharpoons e^{-ph}\phi(p). \qquad (3)$

8·61. Fourier expansion by transform procedure [127]. Let (i) $f(t)$ be a function of period $2h$ commencing at $t = 0$, being single-valued, except at its discontinuities where it is finite; (ii) the discontinuities, maxima and minima in the interval be limited in number. Then the transform of $f(t)$ for the range $t = 0$ to ∞ is

$$\phi(p) = \phi_0(p)/(1 - e^{-2ph}), \qquad (1)$$

where $\phi_0(p)$ is the transform of the zeroth period (see Fig. 40).

The transform of the first period is $\phi_0 e^{-2ph}$, since it commences at $t = 2h$, and in general the transform of the mth period is $\phi_0 e^{-2mph}$. Hence

$$\phi = \phi_0 \sum_{m=0}^\infty e^{-2mph} \qquad (2)$$

$$= \phi_0/(1 - e^{-2ph}), \qquad (3)$$

since, by Appendix 4, $R(p)$ can be as large as we please, so $|e^{-2ph}| < 1$.

If $|\phi(p)/p| \to 0$ uniformly as $|p| \to \infty$, $-\tfrac{1}{2}\pi \leqslant \text{phase}\, p \leqslant \tfrac{1}{2}\pi$, then

$$f(t) = \frac{1}{2\pi i} \int_{Br_1} \frac{e^{zt}\,\phi_0(z)\,dz}{z(1 - e^{-2zh})} \qquad (t > 0). \qquad (4)$$

The evaluation of this integral will yield a Fourier expansion. In this respect it should be observed that by §4·41 integral (4) is zero when $t < 0$. Since the period of the function is $2h$, so also is that of the series. Moreover *the series* represents the function behind the origin, except at points of discontinuity.

8·62. Example [127]. Find the transform of and the Fourier expansion corresponding to Fig. 40 if each period is a half sine wave, i.e. the function is $|\sin t|$.

The transform of $\sin t$ is $p/(p^2+1)$. If at $t = \pi$—the period of $|\sin t|$—a second sine wave starts, it will neutralize the first, except for the zeroth period. The transform of this wave train is $p\,e^{-\pi p}/(p^2+1)$. The transform of the zeroth period is the sum of these two transforms, namely,*

$$\phi_0(p) = \frac{p}{(p^2+1)}(1+e^{-\pi p}), \qquad (1)$$

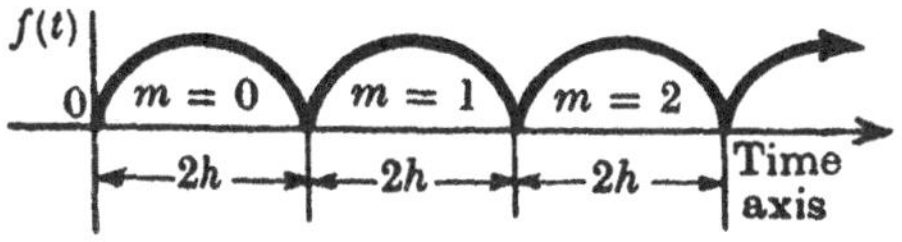

Fig. 40. Iterated function commencing at $t = 0$.

so by (1) §8·61 the transform of $|\sin t|$ is

$$\phi(p) = \frac{p}{(p^2+1)}\left(\frac{1+e^{-\pi p}}{1-e^{-\pi p}}\right). \qquad (2)$$

By (4) §8·61 the Fourier expansion of $|\sin t|$ is represented by

$$\frac{1}{2\pi i}\int_{Br_1} \frac{e^{zt}(1+e^{-\pi z})\,dz}{(z^2+1)(1-e^{-\pi z})}. \qquad (3)$$

Owing to the factor $(1+e^{-\pi z})$ the integrand has no poles at $z = \pm i$, so there is no component of fundamental period 2π in the expansion. The zeros of $(1-e^{-\pi z})$ occur when $z = 2ni$, $n = -\infty$ to ∞. For the pole at $z = 0$, the contribution is

$$\left[\frac{e^{zt}(1+e^{-z\pi})}{(z^2+1)\dfrac{d}{dz}(1-e^{-z\pi})}\right]_{z=0} = \frac{1}{\pi}\left[\frac{e^{zt}(1+e^{-z\pi})}{(z^2+1)\,e^{-z\pi}}\right]_{z=0} = \frac{2}{\pi}. \qquad (4)$$

The contribution from the remaining singularities (poles and a

---

* Alternatively $\phi_0(p) = p\displaystyle\int_0^{\pi} e^{-pt}\sin t\,dt.$

limit point at ∞) is obtained by putting $z = 2ni$ in the second member of (4), which gives (see § 4·54)

$$\frac{2}{\pi} \Sigma \frac{e^{2nti}}{(1 - 4n^2)} = -\frac{4}{\pi} \sum_{n=1}^{\infty} \frac{\cos 2nt}{(4n^2 - 1)}. \tag{5}$$

Hence by (4) and (5) the expansion of $|\sin t|$ is

$$\frac{2}{\pi} \left[1 - 2 \sum_{n=1}^{\infty} \frac{\cos 2nt}{4n^2 - 1} \right], \tag{6}$$

so the components have periods which are even sub-multiples of that of the original $\sin t$. $2/\pi$ represents the mean value of $|\sin t|$.

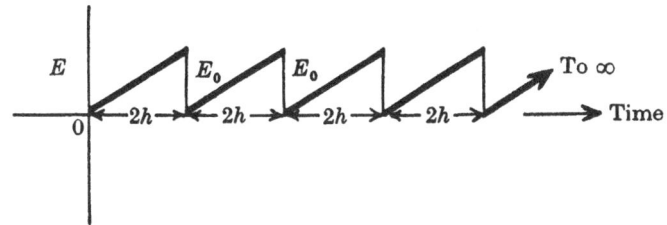

Fig. 41. Saw-tooth voltage wave form, which is piecewise continuous, with finite discontinuities at $t = 2nh$.

8·63. Example [127]. A voltage E of saw-tooth wave form, shown in Fig. 41, is applied to an electrical circuit of the type in Fig. 42, initially quiescent. What is the current in resistance R when $t > 0$?

From the circuital diagram, if

$$I_0 \Rrightarrow \phi_0, \quad I_1 \Rrightarrow \phi_1, \quad I_2 \Rrightarrow \phi_2(p),^*$$

then
$$\phi_0 = \phi_1 + \phi_2. \tag{1}$$

The p.d.'s across C and R are equal, so with $E(t) \Rrightarrow \varphi(p)$

$$\dot{\phi}_2 R = \phi_1/Cp = \varphi - \phi_0 R_0. \tag{2}$$

By (1) and (2),
$$\phi_2 = \phi_0 - \phi_1$$
$$= \{(\varphi - \phi_2 R)/R_0\} - CRp\phi_2,$$

or
$$\phi_2 = (\varphi/CRR_0)/(p + a), \tag{3}$$

where
$$a = (R + R_0)/CRR_0.$$

* These are the transform versions of I_0, I_1, I_2 as functions of t.

We have now to find the transform of the zeroth period of the saw-tooth wave form. The transform of the straight line from $t = 0$ with slope $\tan \alpha$ is $E_0/2hp$, provided it continues indefinitely. It stops at $t = 2h$, and this can be effected by introducing a line on the negative side of the time axis, starting from the point $t = 2h$, $E = -E_0$ and having a slope $-\tan \alpha$, as shown in Fig. 43. The transform of this line is $-E_0 e^{-2ph} - \dfrac{E_0 e^{-2ph}}{2ph}$. Adding these two transforms, we get that of the zeroth period, namely,

$$\varphi_0(p) = \frac{E_0}{2ph}(1 - e^{-2ph}) - E_0 e^{-2ph}. \tag{4}$$

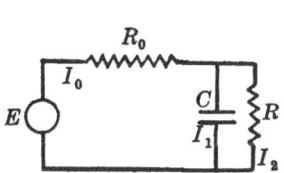

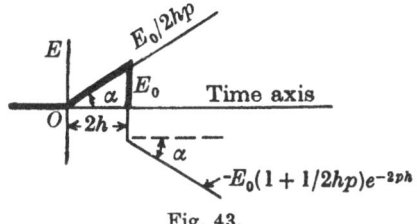

Fig. 42. Circuit to which saw-tooth voltage is applied.

Fig. 43.

Alternatively we can use the method of § 8·53. The equation of the straight line from 0 to $E = E_0$ in Fig. 41 is $f(t) = E_0 t/2h$, so its transform is

$$\varphi_0(p) = \frac{E_0}{2h} p \int_0^{2h} e^{-pt} t\,dt, \tag{4a}$$

which is identical with (4). By (1) § 8·61 the transform of the saw-tooth wave continued indefinitely is

$$\varphi(p) = E_0 \left[\frac{1}{2hp} - \frac{e^{-2ph}}{1 - e^{-2ph}} \right]. \tag{5}$$

Substituting this value for φ in (3), we obtain

$$I_2 \Rightarrow \phi_2 = \frac{E_0}{CRR_0} \left[\frac{1}{2ph} - \frac{e^{-2ph}}{1 - e^{-2ph}} \right] \Big/ (p + a). \tag{6}$$

Using the Mellin theorem,

$$I_2 = \frac{E_0}{CRR_0} \frac{1}{2\pi i} \int_{Br_1} \left[\frac{1}{2z^2 h} - \frac{e^{-2zh}}{z(1 - e^{-2zh})} \right] \frac{e^{zt}\,dz}{(z + a)}. \tag{7}$$

Neglecting the external factor, the first integral yields on evaluation at the simple pole $z = -a$, and the double pole $z = 0$,

$$\frac{1}{2ha^2}(e^{-at} + at - 1). \tag{8}$$

For the pole at $z = -a$, the second integral contributes

$$-\frac{e^{-at+2ah}}{a(e^{2ah} - 1)}. \tag{9}$$

Near the double pole $z = 0$, we expand both numerator and denominator of the integrand, which gives

$$-\frac{1 + z(t - 2h) + \dots}{az(1 + z/a)\, 2zh(1 - zh + \dots)}$$

$$= -\frac{[1 + z(t - 2h) + \dots][1 - z/a + \dots][1 + zh - \dots]}{2ahz^2}. \tag{10}$$

The contribution to the integral from the double pole at $z = 0$ is the coefficient of $1/z$, namely,

$$(1 + ah - at)/2a^2h. \tag{11}$$

The contribution from the singularities at $z = \pi n i/h$, $n = -\infty$ to ∞, excepting $n = 0$, is by §4·54

$$-\sum_{n=-\infty}^{\infty}{}' \left[\frac{e^{z(t-2h)}}{z(z+a)\dfrac{d}{dz}(1 - e^{-2zh})}\right]_{z=\pi ni/h} = -\frac{1}{2}\sum_{n=-\infty}^{\infty}{}' \frac{e^{(t-2h)\pi ni/h}}{\pi ni\left(\dfrac{\pi ni}{h} + a\right)}$$

$$= -\frac{h}{2\pi i}\sum_{n=1}^{\infty}\left[\frac{e^{\pi nti/h}}{n(ah + n\pi i)} - \frac{e^{-\pi nti/h}}{n(ah - n\pi i)}\right]$$

$$= -\frac{h}{\pi}\sum_{n=1}^{\infty}\left[\frac{ah\sin\dfrac{n\pi t}{h} - n\pi\cos\dfrac{n\pi t}{h}}{n(a^2h^2 + n^2\pi^2)}\right]$$

$$= -\frac{h}{\pi}\sum_{n=1}^{\infty}\left[\frac{\sin\left(\dfrac{n\pi t}{h} - \tan^{-1}\dfrac{n\pi}{ah}\right)}{n\sqrt{(a^2h^2 + n^2\pi^2)}}\right]. \tag{12}$$

Adding (8), (9), (11), (12) and introducing the external factor from (7), we obtain

$$I_2 = \frac{E_0}{CRR_0}\left\{ \frac{1}{2a} + \frac{e^{-at}}{a}\left[\frac{1}{2ha} - \frac{e^{2ha}}{(e^{2ha}-1)}\right]\right.$$
$$\left. - \frac{h}{\pi}\sum_{n=1}^{\infty}\frac{\sin\left(\frac{n\pi t}{h} - \tan^{-1}\frac{n\pi}{ah}\right)}{n\sqrt{(a^2h^2+n^2\pi^2)}}\right\}. \tag{13}$$

The unidirectional current is $E_0/2(R+R_0)$, and the fundamental period of the alternating current is that of the saw-tooth wave, namely, $2h$. The second term gives the transient, while the first and third give the Fourier expansion of the current in R when the steady state obtains (in a practical sense).

8·7. Example. Find the transform of $I_0[a\sqrt{(t^2-b^2)}]$, $(t>b)$.
By (7) § 5·33,
$$I_0[a\sqrt{(t^2-b^2)}] = \frac{1}{2\pi i}\int_{Br_1}\frac{e^{z+a^2(t^2-b^2)/4z}\,dz}{z}. \tag{1}$$

We have now to transform the integrand, so that it assumes the form at (2) § 8·11, i.e. the exponential index has the form $zt+f(z)$. Substituting $z = \frac{1}{2}(t-b)[\zeta+\sqrt{(\zeta^2-a^2)}]$, the index becomes

$$v = \frac{1}{2}(t-b)[\zeta+\sqrt{(\zeta^2-a^2)}]+a^2(t^2-b^2)/2(t-b)[\zeta+\sqrt{(\zeta^2-a^2)}] \tag{2}$$

$$= \frac{1}{2}(t-b)[\zeta+\sqrt{(\zeta^2-a^2)}]+\frac{1}{2}(t+b)[\zeta-\sqrt{(\zeta^2-a^2)}], \tag{3}$$

after multiplying the second term in (2) above and below by $[\zeta-\sqrt{(\zeta^2-a^2)}]$. Thus from (3),
$$v = \zeta t - b\sqrt{(\zeta^2-a^2)}. \tag{4}$$

This transformation is justifiable by § 10, Appendix 5.
Now
$$\frac{dz}{d\zeta} = \frac{1}{2}(t-b)\left[1+\frac{\zeta}{\sqrt{(\zeta^2-a^2)}}\right] = \frac{z}{\sqrt{(\zeta^2-a^2)}}, \tag{5}$$

so
$$\frac{dz}{z} = \frac{d\zeta}{\sqrt{(\zeta^2-a^2)}}. \tag{6}$$

Substituting from (4) and (6) into (1), we obtain
$$I_0[a\sqrt{(t^2-b^2)}] = \frac{1}{2\pi i}\int_{Br_1}e^{\zeta t - b\sqrt{(\zeta^2-a^2)}}\frac{d\zeta}{\sqrt{(\zeta^2-a^2)}}, \tag{7}$$

and, therefore, by (1), (2) § 8·3,

$$I_0[a\,\sqrt{(t^2-b^2)}] \Rrightarrow \frac{p\,e^{-b\sqrt{(p^2-a^2)}}}{\sqrt{(p^2-a^2)}} = p\int_b^\infty e^{-pt}I_0[a\,\sqrt{(t^2-b^2)}]\,dt. \qquad (8)$$

Substituting β for a, and x/v for b in (8), we get

$$I_0[\beta\,\sqrt{(t^2-x^2/v^2)}] \Rrightarrow \frac{p\,e^{-(x/v)\sqrt{(p^2-\beta^2)}}}{\sqrt{(p^2-\beta^2)}}. \qquad (9)$$

Applying (3) § 8·411 to (9), with $-\alpha$ for a, leads to

$$e^{-\alpha t}I_0[\beta\,\sqrt{(t^2-x^2/v^2)}] \Rrightarrow \frac{p\,e^{-(x/v)\sqrt{\{(p+\alpha)^2-\beta^2\}}}}{\sqrt{\{(p+\alpha)^2-\beta^2\}}} \qquad (10)$$

$$= \frac{p\,e^{-(x/v)\sqrt{\{(p+2a)(p+2b)\}}}}{\sqrt{\{(p+2a)(p+2b)\}}}, \qquad (11)$$

where $\alpha = (a+b)$ and $\beta = (a-b)$.

Differentiation of (8) under the integral sign, with respect to b, is permissible, since the resulting integral is uniformly convergent (see [236], p. 120), so we obtain

$$\frac{I_1\{a\,\sqrt{(t^2-b^2)}\}}{\sqrt{(t^2-b^2)}} \Rrightarrow p[e^{-b\sqrt{(p^2-a^2)}} - e^{-bp}]/ab \qquad (12)$$

$$= p\int_b^\infty \frac{e^{-pt}I_1\{a\,\sqrt{(t^2-b^2)}\}}{\sqrt{(t^2-b^2)}}\,dt. \qquad (13)$$

Applying (3) § 8·411 to (12) yields

$$e^{-\alpha t}\frac{I_1\{a\,\sqrt{(t^2-b^2)}\}}{\sqrt{(t^2-b^2)}} \Rrightarrow \frac{p}{ab}\,[e^{-b\sqrt{\{(p+\alpha)^2-a^2\}}} - e^{-(p+\alpha)b}]. \qquad (14)$$

Referring to (13) we have $-\displaystyle\int_t^\infty = \int_b^t - \int_b^\infty$, so writing α for p, β for a, and x/v for b,

$$-\int_t^\infty \frac{e^{-\alpha t}I_1\{\beta\,\sqrt{(t^2-x^2/v^2)}\}}{\sqrt{(t^2-x^2/v^2)}}\,dt$$

$$= \int_{x/v}^t \frac{e^{-\alpha t}I_1\{\beta\,\sqrt{(t^2-x^2/v^2)}\}}{\sqrt{(t^2-x^2/v^2)}}\,dt - \frac{e^{-x\sqrt{(\alpha^2-\beta^2)}/v} - e^{-\alpha x/v}}{\beta x/v}, \qquad (15)$$

a result used in § 13·34.

8·8. Remarks on convergence.

The analysis above, and in the next chapter, depends in part upon convergence of infinite integrals. In certain cases this may be automatic, e.g. if $f(t) = e^{at}$, then (1) § 8·11 converges provided it is implied tacitly that $R(p) > R(a)$. If $f(t) = e^{at^2}$, $R(a) > 0$, the integral diverges, so e^{at^2} has no transform referred to the Mellin theorem. Again, if $f(t) = t^\nu$, ν a real fraction > -1,

$$f^{(m)}(t) = \nu(\nu - 1) \dots (\nu - m + 1) t^{\nu - m},$$

and when $(\nu - m) < -1$, the integral $p \int_0^\infty e^{-pt} f^{(m)}(t)\, dt$ diverges. For example, if $\nu = -\frac{1}{2}$, $f'(t) = -\frac{1}{2}t^{-\frac{3}{2}}$, and the integral in (3) § 8·43 diverges, so $t^{-\frac{1}{2}}$ has no transform referred to the Mellin theorem.

IX

SOLUTION OF ORDINARY LINEAR DIFFERENTIAL EQUATIONS WITH CONSTANT COEFFICIENTS

9·11. Example illustrating method. Suppose that the equation to be solved is

$$\frac{d^2y}{dt^2} + a^2y = c\,e^{-bt}, \tag{1}$$

subject to the initial conditions $y = 0$, $dy/dt = y' = 0$, at $t = 0$. Multiply both sides of (1) by $p\,e^{-pt}$ and integrate with respect to t from 0 to ∞ [6, 160, 236]. Then (1) gives

$$p\int_0^\infty e^{-pt}\frac{d^2y}{dt^2}\,dt + a^2p\int_0^\infty e^{-pt}y\,dt = cp\int_0^\infty e^{-(p+b)t}\,dt, \tag{2}$$

or
$$p\int_0^\infty e^{-pt}\,dy' + a^2p\int_0^\infty e^{-pt}y\,dt = \frac{cp}{(p+b)}, \tag{3}$$

provided $R(p+b) > 0$ to ensure convergence of the third integral in (2). Taking the first integral separately by parts, we have

$$p\int_0^\infty e^{-pt}\,dy' = p\left[e^{-pt}y'\right]_0^\infty + p^2\int_0^\infty e^{-pt}y'\,dt$$

$$= 0 + p^2\int_0^\infty e^{-pt}\,dy, \tag{4}$$

since $y'(0) = 0$, and if $e^{-pt}y' \to 0$ as $t \to \infty$. Continuing,

$$p^2\int_0^\infty e^{-pt}\,dy = p^2\left[e^{-pt}y\right]_0^\infty + p^3\int_0^\infty e^{-pt}y\,dt, \tag{5}$$

so
$$p\int_0^\infty e^{-pt}\frac{d^2y}{dt^2}\,dt = 0 + p^3\int_0^\infty e^{-pt}y\,dt, \tag{6}$$

since $y(0) = 0$, and if $e^{-pt}y(t) \to 0$ as $t \to \infty$.

Substituting from (6) into (2), we find that

$$(p^2+a^2)p \int_0^\infty e^{-pt}y\,dt = \frac{cp}{(p+b)}, \tag{7}$$

the right-hand side being the transform of $c\,e^{-bt}$. Thus

$$p \int_0^\infty e^{-pt}y\,dt = \frac{cp}{(p+b)(p^2+a^2)}, \tag{8}$$

and on applying the Mellin inversion theorem § 8·11 to (8), we obtain

$$y = \frac{c}{2\pi i}\int_{Br_1} \frac{e^{zt}\,dz}{(z+b)(z^2+a^2)}. \tag{9}$$

Evaluating (9) at the simple poles $z = -b,\ -ia,\ +ia$, yields

$$y = c\left[\frac{e^{-bt}}{(a^2+b^2)} + \frac{e^{iat}}{(ia+b)\,2ia} + \frac{e^{-iat}}{(ia-b)\,2ia}\right] \tag{10}$$

$$= \frac{c}{(a^2+b^2)}\left[e^{-bt} - \cos at + \frac{b}{a}\sin at\right]. \tag{11}$$

An alternative method of solution is given in [236], chapter 3.

9·12. Equation (1) § 9·11 can be interpreted in a practical way. Referring to the mechanical system shown schematically in Fig. 44 (a), m and s are, respectively, a mass and a massless helical spring having stiffness s. One end of the spring is fixed, and the mass is acted upon by a driving force

$$f = cm\,e^{-bt} \quad (t>0).$$

The differential equation for the system is

$$m\frac{d^2y}{dt^2} + sy = cm\,e^{-bt}, \tag{1}$$

or
$$\frac{d^2y}{dt^2} + \frac{s}{m}y = c\,e^{-bt}. \tag{2}$$

If in (1) § 9·11 we write s/m for a^2, (2) above is reproduced, and its solution is given by (11) § 9·11.

9·13. System not quiescent at $t = 0$. In §9·11 let the initial conditions be $y = y_0$, $y' = y_1$ at $t = 0$. Then in (4) and (6) § 9·11, instead of zero we get $-py_1$ and $-p^2 y_0$, respectively, for the evaluations at the lower limit. Thus we have

$$(p^2 + a^2) p \int_0^\infty e^{-pt} y \, dt - py_1 - p^2 y_0 = \frac{cp}{(p+b)}, \tag{1}$$

or $$p \int_0^\infty e^{-pt} y \, dt = \frac{cp}{(p+b)(p^2+a^2)} + \frac{y_1 p}{(p^2+a^2)} + \frac{y_0 p^2}{(p^2+a^2)}. \tag{2}$$

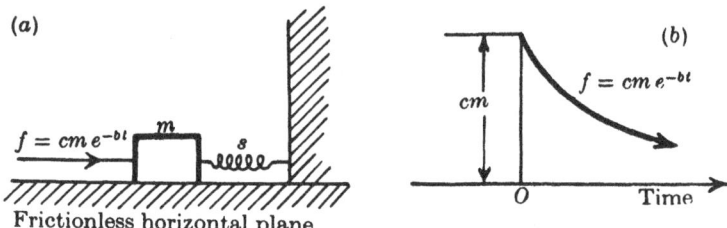

Fig. 44.

The first term on the right-hand side of (2) is that part of the transform solution corresponding to the action of the impressed force, while the sum of the second and third terms is that part arising from the initial conditions. The inverse of the first part of (2) is given at (11) § 9·11, and by (1), (2) § 8·424 that of the remainder is

$$y_1 \frac{\sin at}{a} + y_0 \cos at. \tag{3}$$

This is the solution of $d^2 y/dt^2 + a^2 y = 0$, under the prescribed initial conditions, i.e. it is the complementary function in the solution of (1) § 9·11. Moreover, any second order differential equation (having constant coefficients), whose right-hand side is zero, can be solved in the above manner provided the transforms of the initial conditions are inserted.

The complete solution of (1) under the prescribed initial conditions is the sum of (11) § 9·11 and (3), so

$$y = \frac{c e^{-bt}}{(a^2+b^2)} + \left(y_0 - \frac{c}{(a^2+b^2)} \right) \cos at + \left(y_1 + \frac{cb}{(a^2+b^2)} \right) \frac{\sin at}{a}. \tag{4}$$

9·21. Ordinary linear differential equation of the nth order. In technical applications herein, the equation will as a general rule be for a dynamical, electrical, or other system to which a disturbance expressed analytically by $\xi(t)$ is applied at $t = 0$. The method of solution is identical with that in § 9·11. The equation is

$$a_0\frac{d^ny}{dt^n} + a_1\frac{d^{n-1}y}{dt^{n-1}} + a_2\frac{d^{n-2}y}{dt^{n-2}} + \ldots + a_n y = \xi(t). \tag{1}$$

To solve (1) multiply both sides by $p\,e^{-pt}$, p being a parameter whose real part > 0, and integrate with respect to t from 0 to ∞ [6, 160, 236], then

$$p\int_0^\infty e^{-pt}(a_0 y^{(n)} + a_1 y^{(n-1)} + \ldots + a_n y)\,dt = p\int_0^\infty e^{-pt}\xi(t)\,dt. \tag{2}$$

Taking a typical term on the left-hand side of (2) and omitting the external multiplier $a_{n-r}p$, we get

$$\int_0^\infty e^{-pt} y^{(r)}\,dt = \int_0^\infty e^{-pt}\,dy^{(r-1)} = \left[e^{-pt}y^{(r-1)}\right]_0^\infty + p\int_0^\infty e^{-pt}y^{(r-1)}dt$$

$$= -y^{(r-1)}(0) + p\int_0^\infty e^{-pt}\,dy^{(r-2)}$$

$$= -y^{(r-1)}(0) - py^{(r-2)}(0) + p^2\int_0^\infty e^{-pt}y^{(r-3)}dt$$

$$= -\left[y_{r-1} + py_{r-2} + p^2 y_{r-3} + \ldots - p^r\int_0^\infty e^{-pt}y\,dt\right], \tag{3}$$

provided that $e^{-pt}y^{(m)} \to 0$ as $t \to +\infty$, $m = 0, 1, 2, \ldots, r-1$, where $y_{r-1} = y^{(r-1)}(0)$ is the value of $y^{(r-1)}$ at $t = 0$.

The left-hand side of (2) is, therefore,

$$-\{a_0[py_{n-1} + p^2 y_{n-2} + p^3 y_{n-3} + \ldots \qquad\qquad + p^n y_0]$$
$$+ a_1[py_{n-2} + p^2 y_{n-3} + \ldots \qquad\qquad + p^{n-1}y_0]$$
$$+ a_2[py_{n-3} + p^2 y_{n-4} + \ldots + p^{n-2}y_0]$$
$$\ldots\ldots\ldots\ldots\ldots\ldots\ldots\ldots\ldots\ldots\ldots\ldots$$
$$+ a_{n-1}py_0\}$$
$$+ [a_0 p^n + a_1 p^{n-1} + \ldots + a_n]p\int_0^\infty e^{-pt}y\,dt, \tag{4}$$

and this may be expressed concisely in the form

$$-\phi_3(p) + \phi_2(p)\, p \int_0^\infty e^{-pt}\, y\, dt. \tag{5}$$

$\phi_3(p)$ is the expression in $\{\ \}$ and $\phi_2(p)$ that in $[\ ]$ preceding the integral in (4). The former is the transform of the contribution arising from the initial conditions, i.e. pertaining to the state of the system as $t \to -0$.

By (1) § 8·11 the right-hand side of (2) is $\phi_1(p)$, the transform of $\xi(t)$. Thus, equating $\phi_1(p)$ to (5), we obtain

$$p \int_0^\infty e^{-pt}\, y\, dt = \frac{\phi_1(p)}{\phi_2(p)} + \frac{\phi_3(p)}{\phi_2(p)}. \tag{6}$$

Hence by the Mellin theorem the solution of (1) in terms of t is

$$y = f(t) = \frac{1}{2\pi i} \int_{Br_1} \frac{e^{zt}\,\phi_1(z)\,dz}{\phi_2(z)}\,\frac{dz}{z} + \frac{1}{2\pi i} \int_{Br_1} \frac{e^{zt}\,\phi_3(z)\,dz}{\phi_2(z)}\,\frac{dz}{z}. \tag{7}$$

The first contour integral in (7), when evaluated, gives the solution provided the system is quiescent initially, since $\phi_3(p)$ is then zero. The second integral gives that part of the solution arising from the initial conditions, i.e. from the state of the system as $t \to -0$. It is the solution of (1) in the absence of an applied disturbance, i.e. when $\xi(t) = 0$. Since the system contemplated is a linear one, the two solutions can be superimposed (added) to give the complete solution.

9·22. To facilitate the solution of differential equations, $\phi_3(p)$ may be written in the form of an array thus:

$$\phi_3(p) = \begin{array}{c} \\ y_0 \\ y_1 \\ y_2 \\ \\ y_{n-1} \end{array} \begin{array}{ccccccc} p^n & p^{n-1} & p^{n-2} & & & & p \\ \hline a_0 & a_1 & a_2 & \cdots & \cdots & a_{n-1} \\ 0 & a_0 & a_1 & \cdots & \cdots & a_{n-2} \\ 0 & 0 & a_0 & a_1 & \cdots & a_{n-3} \\ \multicolumn{6}{c}{\cdots\cdots\cdots\cdots\cdots\cdots} \\ 0 & 0 & 0 & 0 & \cdots & a_0 \end{array}. \tag{1}$$

Multiply each a by the corresponding y on its left, and then by the p immediately above. $\phi_3(p)$ is the sum of all terms so formed.

Thus for an equation of the second order, with $dy/dt = y_1$ and $y = y_0$ at $t = 0$,

$$\phi_3(p) = \frac{\begin{array}{cc} p^2 & p \end{array}}{y_1} \, y_0 \begin{vmatrix} a_0 & a_1 \\ 0 & a_0 \end{vmatrix} = y_0 a_0 p^2 + (y_0 a_1 + y_1 a_0)\, p. \qquad (2)$$

Formula (2) should be memorized.

In solving an equation like (1) § 9·21 when the system is not quiescent initially, the appropriate procedure is as follows: (i) obtain $\phi_1(p)$, the transform of $\xi(t)$, either by integration or from Appendix 10; (ii) write p for d/dt on the left-hand side thereby getting $\phi_2(p)$; (iii) form $\phi_3(p)$, the initial conditions transform, then the solution is obtained by (6) and (7) § 9·21.

9·23. Example. Solve

$$\frac{d^2y}{dt^2} + 4\frac{dy}{dt} + 4y = f\sin \omega t, \qquad (1)$$

given that $y_1 = y_0 = 0$ at $t = 0$.

This is the equation for a damped mechanical, electrical or acoustical system having one degree of freedom, driven by a sinusoidal force $f\sin \omega t$, which is applied at $t = 0$, the system being quiescent initially.

From (1) § 8·424, the transform of $f\sin \omega t$ is $\phi_1 = f\omega p/(p^2 + \omega^2)$. Also $\phi_2 = p^2 + 4p + 4$, while $\phi_3 = 0$. Whence by (6) § 9·21

$$\phi(p) = \frac{\phi_1}{\phi_2} = \frac{f\omega p}{(p^2 + \omega^2)(p + 2)^2}, \qquad (2)$$

this being the transform solution. Using (7) § 9·21, the solution in terms of t is

$$y = \frac{f\omega}{2\pi i}\int_{Br_1} \frac{e^{zt}\,dz}{(z - i\omega)(z + i\omega)(z + 2)^2}. \qquad (3)$$

The integrand of (3) has two simple poles at $z = \pm i\omega$, and a double pole at $z = -2$. The residues at the simple poles are

$$\frac{e^{i\omega t}}{2i\omega(2 + i\omega)^2} \quad \text{and} \quad -\frac{e^{-i\omega t}}{2i\omega(2 - i\omega)^2}. \qquad (4)$$

The residue at $z = -2$ is, by § 3·241,

$$\frac{d}{dz}\left(\frac{e^{zt}}{z^2+\omega^2}\right)_{z=-2} = \left[e^{zt}\left(\frac{t}{z^2+\omega^2} - \frac{2z}{(z^2+\omega^2)^2}\right)\right]_{z=-2}$$

$$= \frac{e^{-2t}}{(\omega^2+4)}\left[t + \frac{4}{(\omega^2+4)}\right]. \tag{5}$$

Adding the three residues and multiplying by $f\omega$, the displacement from the equilibrium position is

$$y = \frac{f}{2i}\left[\frac{e^{i\omega t}}{(2+i\omega)^2} - \frac{e^{-i\omega t}}{(2-i\omega)^2}\right] + \frac{f\omega e^{-2t}}{(\omega^2+4)^2}[4 + t(\omega^2+4)] \tag{6}$$

$$= \frac{f}{(\omega^2+4)^2}[(4-\omega^2)\sin\omega t - 4\omega\cos\omega t]$$

$$+ \frac{f\omega e^{-2t}}{(\omega^2+4)^2}[4 + t(\omega^2+4)] \tag{7}$$

$$= \frac{f}{(\omega^2+4)}\left[\sin(\omega t - \alpha) + \frac{\omega e^{-2t}}{(\omega^2+4)}\{4 + (\omega^2+4)t\}\right], \tag{8}$$

where the angle $\alpha = \tan^{-1}\{4\omega/(4-\omega^2)\}$.

The solution (8) has two parts, (i) a steady oscillation out of phase with the driving force by an angle α (this being dependent upon the inertia, stiffness, and resistance), (ii) a transient displacement which decays asymptotically to zero with increase in time. The occurrence of a double pole signifies that the system is critically damped, i.e. it is just aperiodic.

9·24. Example. Solve the differential equation in §9·23 if the initial conditions are $y' = v$ and $y = y_0$ when $t = 0$.

By (2) § 9·22,

$$\phi_3(p) = y_0(p^2+4p) + vp = y_0[p^2 + (4 + v/y_0)p]. \tag{1}$$

By (6) § 9·21 the transform solution is

$$\phi(p) = \frac{\phi_1}{\phi_2} + \frac{\phi_3}{\phi_2}, \tag{2}$$

of which the inverse of the first term on the right-hand side is given at (8) § 9·23. Inverting the second term, we obtain

$$f_2(t) = \frac{y_0}{2\pi i} \int_{Br_1} \frac{e^{zt}(z+a)\,dz}{(z+2)^2}, \tag{3}$$

where $a = (4 + v/y_0)$. Evaluating (3) at the double pole, by § 3·241

$$\frac{d}{dz}\left[e^{zt}(z+a)\right]_{z=-2} = e^{zt}[t(z+a)+1]_{z=-2}, \tag{4}$$

so $$f_2(t) = e^{-2t}[(2y_0+v)\,t+y_0], \tag{5}$$

and this is the contribution due to the initial velocity v and displacement y_0. The complete solution is the sum of (8) § 9·23 and (5).

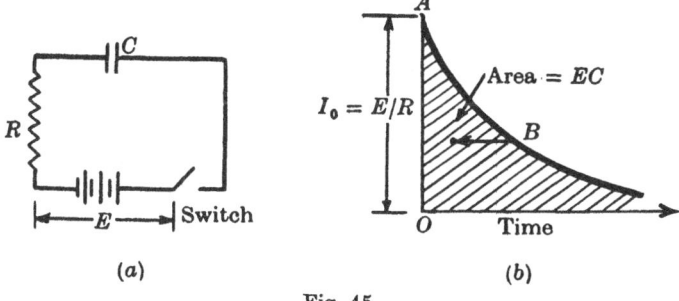

(a)   (b)

Fig. 45.

For an equation of the nth order the contribution arising from any of the initial conditions $y_0, y_1, \ldots, y_{n-1}$ can be found as follows: Using the array (1) § 9·22, obtain the corresponding polynomial in p; divide this by $\phi_2(p)$ and invert by the Mellin theorem, or by using partial fractions and Appendix 10.

9·25. **Example.** At time $t = 0$ a constant p.d. E is applied by a battery to a resistance R in series with an uncharged capacitance C, as shown schematically in Fig. 45 (a). Find the current at any time $t > 0$.

The differential equation for Fig. 45 (a) after closing the switch is

$$RI + \frac{1}{C}\int_0^t I\,dt = E. \tag{1}$$

Taking $I \Rightarrow \phi(p)$, by (3) § 8·45

$$\phi(R + 1/pC) = E,$$

or $$\phi = (E/R)p/(p + 1/CR). \tag{2}$$

Inverting (2) by the Mellin theorem, or by (10), p. 361,

$$I = \frac{(E/R)}{2\pi i} \int_{Br_1} \frac{e^{zt}\,dz}{z + 1/CR} = E\,e^{-t/CR}/R, \tag{3}$$

this being the well-known formula for the current into a capacitance C charged through a resistance R. The relationship between current and time is shown in Fig. 45 (b). At $t = 0$, I rises precipitately to the value E/R, since the capacitance is uncharged and acts momentarily as a short circuit. The current then decays exponentially with increase in time. If R is decreased gradually, the charging process being repeated at each stage, point A rises steadily while B approaches the vertical axis. When R is small enough the current at $t = 0$ is correspondingly large, and the current to the capacitance can be regarded as an impulse. For instance, if $E = 100$ volts, $R = 10^{-3}$ ohm, the initial current is 10^5 amperes and, as shown below, a capacitance of 1 microfarad is fully charged in a practical sense in a small fraction of a microsecond.

The quantity of electricity in the capacitance at any time t is

$$Q = \int_0^t I\,dt = \frac{E}{R}\int_0^t e^{-t/CR}\,dt = EC(1 - e^{-t/CR}). \tag{4}$$

When t is large enough, Q is independent of R and has the value EC. This is represented in Fig. 45 (b) by the shaded area, which is constant for all positive values of R. If $C = 1$ microfarad and $R = 10^{-3}$ ohm, $1/CR = 10^9$ sec.$^{-1}$, so in one-hundredth of a microsecond the exponential term in (4) would be

$$e^{-10} \simeq 4·6 \times 10^{-5},$$

i.e. for all practical purposes the capacitance would be fully charged. From a practical viewpoint, therefore, a charging time of 10^{-8} sec. is just as effective as an infinite time.

X

DISCONTINUOUS FUNCTIONS: IMPULSES: FREQUENCY SPECTRA

10·1. The unit function. The function represented in Fig. 46 (a) is zero when $t < 0$ and unity when $t > 0$, being discontinuous and undefined at $t = 0$. This is Heaviside's unit function designated by the symbol $H(t)$. Suppose the disturbance acting upon a dynamical system takes the form $\xi(t) = \sin \omega t$, and we wish to select the part from $t = 0$ to ∞. $\sin \omega t$ covers the range $t = -\infty$ to ∞, so if we write $\xi(t) H(t) = \sin \omega t \cdot H(t)$, the portion from $t = -\infty$ to 0 is obliterated. By aid of the unit function we can ascertain the effect of applying to a system a disturbance represented by an arbitrary function of t, and make it commence at $t = 0$. By moving the origin of the unit function to the point $t = h$, that portion of the applied disturbance (expressed as a function of t) from $t = h$ to ∞ is obtained, the part from $t = -\infty$ to h being omitted.

Symbolically we write

$$
\left.
\begin{aligned}
H(t) &= 1 \quad\left.\begin{aligned} & \quad 0 < t < \infty \\ \end{aligned}\right\} \\
&= 0 \quad -\infty < t < 0, \\
\end{aligned}
\right\}
$$

and
$$H(t) \Rrightarrow 1; \tag{1}$$

$$
\left.
\begin{aligned}
\sin t\, H(t) &= \sin t \quad\left.\begin{aligned} & \quad 0 \leqslant t < \infty \\ \end{aligned}\right\} \\
&= 0 \quad\quad -\infty < t < 0, \\
\end{aligned}
\right\}
$$

and
$$\sin t\, H(t) \Rrightarrow p/(p^2 + 1); \tag{2}$$

$$
\left.
\begin{aligned}
H(t - h) &= 1 \quad\left.\begin{aligned} & \quad h < t < \infty \\ \end{aligned}\right\} \\
&= 0 \quad -\infty < t < h, \\
\end{aligned}
\right\}
$$

and
$$H(t - h) \Rrightarrow e^{-ph}; \tag{3}$$

$$
\left.
\begin{aligned}
\sin(t - h)\, H(t - h) &= \sin(t - h) \quad\left.\begin{aligned} & \quad h \leqslant t < \infty \\ \end{aligned}\right\} \\
&= 0 \quad\quad\quad\quad\quad -\infty < t < h, \\
\end{aligned}
\right\}
$$

and
$$\sin(t - h)\, H(t - h) \Rrightarrow e^{-ph}\, p/(p^2 + 1). \tag{4}$$

The functions in (3), (4) are depicted graphically in Fig. 47 (a), (b). In (4) the sine function commences at $t = h$, being zero in $t < h$. If we write $\sin t$ for $\sin (t - h)$ in (4), then at $t = h$ the function has the value $\sin h$ as shown in Fig. 47 (c). Since $H(t)$ is undefined at $t = 0$, it cannot be differentiated there.

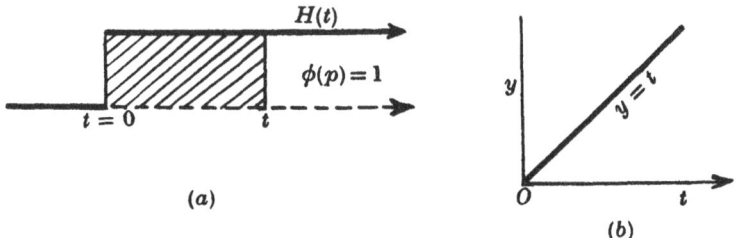

(a)

(b)

Fig. 46. (a) Heaviside's unit funtion $H(t)$. (b) $\int_0^t H(t)\, dt = t,\; t > 0$.

The unit function may be expressed as follows:

$$H(t) = \frac{1}{2} + \frac{1}{2\pi} \int_{-\infty}^{\infty} \frac{\sin yt\, dy}{y} \tag{5}$$

$$= \frac{1}{2} + \frac{1}{\pi} \int_0^{\infty} \frac{\sin yt\, dy}{y} \tag{6}*$$

$$= \frac{1}{2\pi i} \int_{Br_1} \frac{e^{zt}\, dz}{z}, \tag{7}$$

all for $t > 0$. Integrals (5), (6) may be evaluated as shown in $3° \S 3.3$. If $t > 0$, each has the value $\frac{1}{2}$, but they vanish for $t = 0$. The value of (7) is then $\frac{1}{2}$.

In (7) the contour may be either the imaginary axis indented on the right at the origin, or a path wholly to the right of this axis, e.g. $c - i\infty$ to $c + i\infty$ as in Fig. 22 (a).

10·21. Impulses. An impulse due to a force $f(t)$ (electrical or mechanical) either constant or variable, which acts upon a system for a time t, is defined to be $\int_0^t f(t)\, dt$. The force-time

* The integral represents the summation of an infinite spectrum of sine waves of angular frequency y and amplitude $dy/\pi y$ (see $\S 10\cdot514$).

diagrams corresponding to two simple cases are illustrated in Fig. 50 (a), (b). The transform of the impulse of Fig. 50 (a) is

$$\phi(p) = \frac{f_0}{ph}(1 - e^{-ph})^2. \tag{1}$$

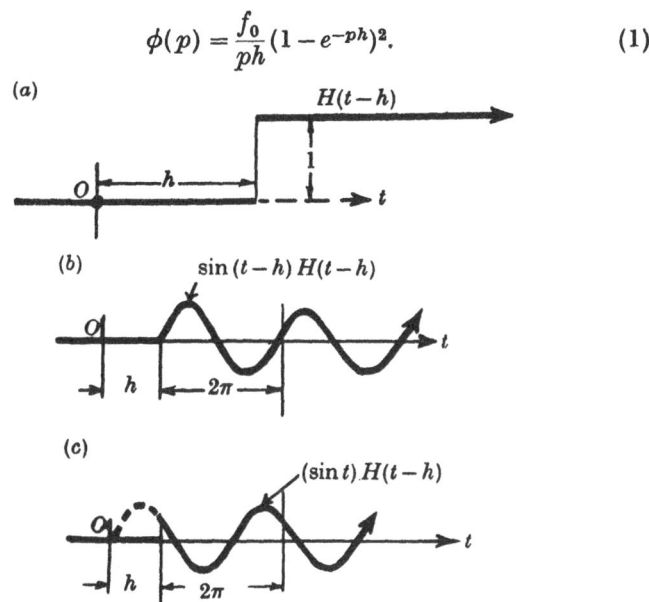

Fig. 47 (a). Illustrating the unit function with origin at $t = h > 0$, where it is finitely discontinuous and has no definite value; (b) $\sin t$, commencing at $t = h$; (c) $(\sin t)\,H(t-h)$, thereby suppressing the sine function up to $t = h$.

If the strength of the impulse is unity, then

$$\int_0^{2h} f(t)\,dt = 1, \tag{2}$$

so $f_0 h = 1$ or $f_0 = 1/h$. Substituting this value of f_0 in (1) and expanding, we obtain

$$\phi(p) = \frac{1}{ph^2}\left(ph - \frac{p^2h^2}{2!} + \frac{p^3h^3}{3!} - \ldots\right)^2 \tag{3}$$

$$= p - p^2h + \text{higher orders involving } h. \tag{4}$$

When $h \to 0$, $\phi(p) \to p$,   (5)

this being the transform of the impulse of infinite amplitude but unit strength defined above, *referred to the Laplace integral*

(1) §8·11. This transform does not conform with the Mellin theorem* (see Appendix 7 and reference [128]). In practice h could never be zero, but it might be very small.

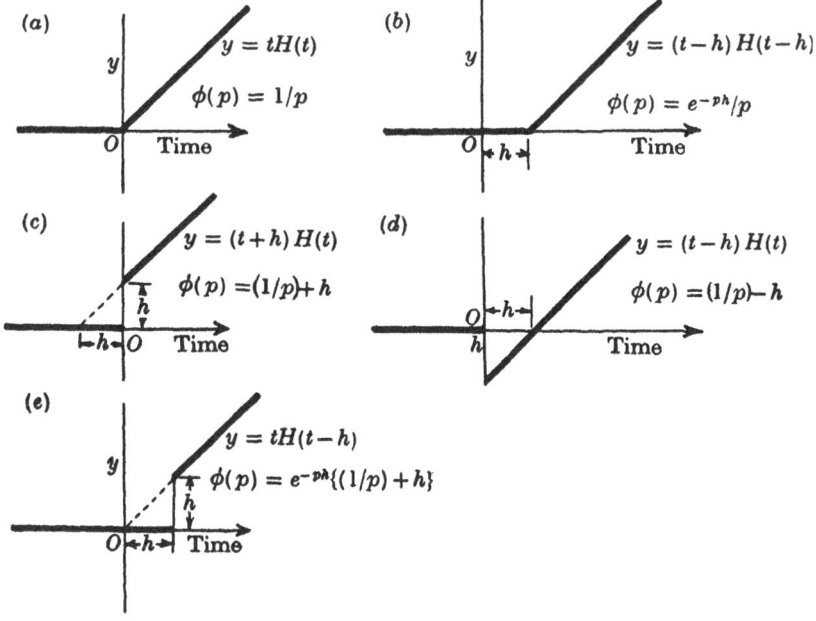

Fig. 48. Diagram illustrating various functions. (a), (b) are continuous in $-\infty < t < \infty$. (c), (d) are discontinuous at $t = 0$. (e) is discontinuous at $t = h$. All have discontinuous first derivatives.

If the strength of the impulse of Fig. 50 (b) is $\int_0^h f(t)\,dt = 1$, we have $f_0 h = 1$ or $f_0 = 1/h$. The transform is

$$\phi(p) = f_0(1 - e^{-ph}) \tag{6}$$

$$= \frac{1}{h}\left(ph - \frac{p^2 h^2}{2!} + \frac{p^3 h^3}{3!} - \dots\right) \tag{7}$$

$$= p - \frac{p^2 h}{2!} + \text{higher orders involving } h. \tag{8}$$

* When the transform is defined by (1) §8·11 only, it may have terms of the type p^n, n an integer $\geqslant 1$. As an example we have

$$e^{-t^2} \rightleftharpoons \tfrac{1}{2}\sqrt{\pi p}\, e^{\frac{1}{4}p^2} \operatorname{erfc} \tfrac{1}{2}p,$$

the expansion of this transform yielding an infinite series in positive integral powers of p. Other examples will be found in references [56, 111, 160, 162, 200 (no. 2), 201, 202].

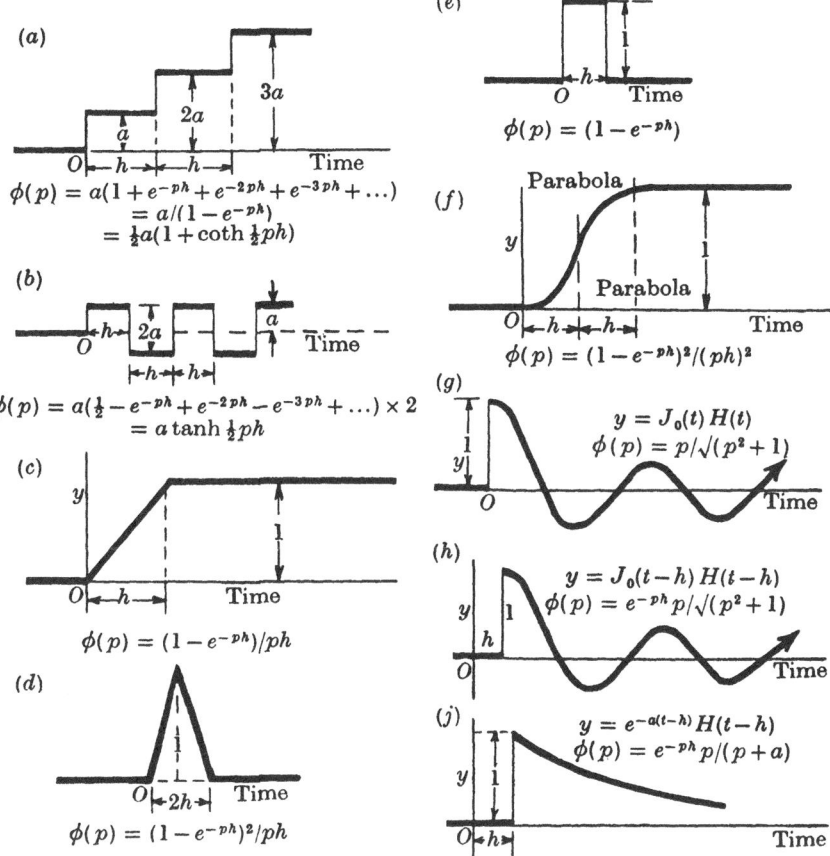

Fig. 49. (c), (d), (f) are continuous in $-\infty < t < \infty$. The remainder have discontinuities. Apart from the discontinuity at $t = 0$ (or $t = h$). (g), (h), (j) are continuous in $t > 0$ (or $t > h$).

When $h \to 0$, $\qquad\qquad \phi(p) \to p,$ $\qquad\qquad\qquad$ (9)

this being the transform of the impulse of infinite amplitude but unit strength defined above, and referred to (1) § 8·11. A more general type of impulse is treated in §§ 10·22, 10·31.

10·22. Example. A large force $f(t)$ such that

$$\int_0^\epsilon f(t)\,dt = A_0 \quad (0 < t < \epsilon)$$

is applied by the 'monkey' of a pile driver to a pile of mass m, which is driven into the ground vertically. If the resistive force is proportional to the velocity v, find the displacement x at any time $t > \epsilon$, where ϵ is very small.

(1°) The differential equation is

$$m\frac{dv}{dt} + rv = f(t) \quad (0 < t < \epsilon), \tag{1}$$

$$= 0 \quad (t > \epsilon),$$

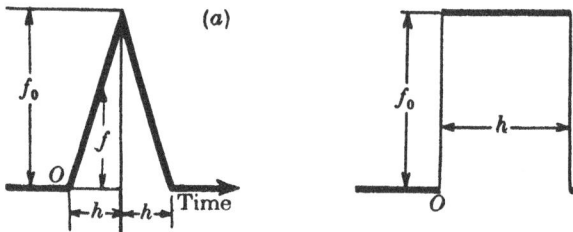

Fig. 50. Triangular and rectangular impulses.
The area represents the 'strength'.

where r is the resistive force per unit velocity, i.e. the mechanical resistance. Taking $v(t) \Rightarrow \phi(p)$ and applying (3) § 8·45 we get

$$m\phi + r\phi/p = A_0 \tag{2}$$

or $$\phi = (A_0/m)\, p/(p + r/m), \tag{3}$$

this being the transform solution. Inverting (3) by the Mellin theorem

$$v = \frac{(A_0/m)}{2\pi i} \int_{Br_1} \frac{e^{zt}\, dz}{(z + r/m)}. \tag{4}$$

The integrand of (4) has a simple pole at $z = -r/m$, so we obtain

$$v = \frac{A_0}{m}\, e^{-rt/m}. \tag{5}$$

The velocity attains the value A_0/m at $t = \epsilon \simeq 0$ and decays exponentially thereafter, since no energy is supplied when $t > \epsilon$.

The displacement at any time $t_1 > \epsilon$ is given by

$$x = \int_0^{t_1} v\,dt = \frac{A_0}{r}\left[e^{-rt/m}\right]_0^{t_1}$$

$$= \frac{A_0}{r}(1 - e^{-rt_1/m}). \qquad (6)$$

($2°$) By aid of § 10·31, the transform equation is

$$mp\phi + r\phi = A_0p, \qquad (7)$$

which is identical with (2). The transform p on the right-hand side of (7) complies with the Laplace integral, but not with the Mellin theorem (see [128]). The inversion of ϕ in (7) should come apparently from the Laplace integral, i.e. it is the solution of the integral equation $p\displaystyle\int_0^\infty e^{-pt}g(t)\,dt = \dfrac{A_0p}{(mp+r)}$ for $g(t)$. However, this transform can be inverted also by the Mellin theorem, since the corresponding contour integral converges.

In the preceding example if m is replaced by L, r by R, v by I and $f(t)$ by $E(t)$, we have the case of an inductance L of resistance R to which a voltage impulse $\displaystyle\int_0^\epsilon E(t)\,dt = A_0$ is applied. The current at $t > \epsilon$ is given by

$$I = \frac{A_0}{L}e^{-Rt/L}, \qquad (8)$$

while the quantity of electricity (analogous to the displacement) is

$$Q = \frac{A_0}{R}(1 - e^{-Rt/L}). \qquad (9)$$

10·23. Example. A generator of resistance R is connected in series with an uncharged capacitance C at a time epoch when the p.d. is a maximum. What is the current at time $t > 0$?

The wave form of the generator is portrayed in Fig. 51 (a), and is expressed by $E = E_0 \cos \omega t$. The differential equation of the system illustrated in Fig. 51 (b) is

$$RI + \frac{1}{C}\int_0^t I\,dt = E_0 \cos \omega t. \qquad (1)$$

By (2) §8·424 and (3) §8·45, with $I \Rightarrow \phi(p)$ the transform version of (1) is

$$(R + 1/pC)\,\phi = \frac{E_0 p^2}{(p^2 + \omega^2)}, \tag{2}$$

so

$$I \Rightarrow \phi = \frac{(E_0/R)\,p^3}{(p^2 + \omega^2)\,(p + 1/CR)}. \tag{3}$$

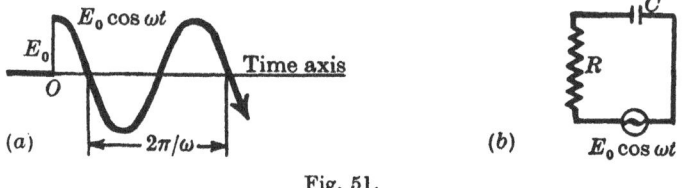

Fig. 51.

Inverting (3) by the Mellin theorem, we get

$$I = \frac{(E_0/R)}{2\pi i} \int_{Br_1} \frac{e^{zt} z^2\,dz}{(z - i\omega)\,(z + i\omega)\,(z + a)}, \tag{4}$$

where $a = 1/CR$. Evaluating (4) at the three simple poles $z = \pm i\omega,\, -a$, yields

$$I = \frac{E_0}{R}\left[\frac{a^2 e^{-at}}{a^2 + \omega^2} - \frac{\omega\, e^{+i\omega t}}{2i(a + i\omega)} + \frac{\omega\, e^{-i\omega t}}{2i(a - i\omega)}\right] \tag{5}$$

$$= \frac{E_0}{R}\left[\frac{a^2 e^{-at}}{a^2 + \omega^2} - \frac{\omega}{a^2 + \omega^2}(a \sin \omega t - \omega \cos \omega t)\right] \tag{6}$$

$$= \frac{E_0}{R}\left[\frac{e^{-t/CR}}{1 + \omega^2 C^2 R^2} - \frac{\omega C R}{\sqrt{(1 + \omega^2 C^2 R^2)}} \sin(\omega t - \tan^{-1} \omega C R)\right]. \tag{7}$$

At $t = 0$ the current rises immediately to the value E_0/R. When R is extremely small the first term in (7) represents initially a current impulse of amplitude E_0/R (see §9·25).

10·31. General type of impulse. Let (a) $y = f(t)$, a real function of t representing an impulse of short duration $t = 0$ to h, be single-valued and continuous between its discontinuities, where it is finite; (b) the discontinuities, maxima, and minima be limited in number; (c) $\int_0^h f(t)\,dt = S$, a constant, this being the strength of the impulse. Then as $h \to 0$, the transform of $f(t) \to Sp$ [133].

For

$$\phi(p) = p \int_0^h e^{-pt} f(t) \, dt, \tag{1}$$

and when h is small

$$\phi(p) \simeq p \int_0^h f(t) \, dt = Sp, \tag{2}$$

so that when $h \to 0$,

$$\phi(p) \to Sp \rightleftharpoons f(t). \tag{3}$$

Consequently if the duration of an impulse is finite, but very short, we may write $f(t) \rightleftharpoons Sp$, approximately. Additional details and a rigorous proof will be found in [236]. It is convenient to write

$$I(t) \rightleftharpoons p, \tag{4}$$

where $I(t)$ is the impulsive function of infinite amplitude but unit strength. For an impulse of strength S, we have

$$SI(t) \rightleftharpoons Sp. \tag{5}$$

Dimensions. The index in e^{-pt} is dimensionless, so if t represents time, p has dimension t^{-1}. Hence in (4) and (5) $I(t)$ has dimension t^{-1}. If $SI(t)$ represents a force (mechanical, electrical, etc.), S has dimensions force × time.

10·32. Example. Solve

$$d^4 y / dt^4 - a^4 y = SI(t), \tag{1}$$

if $y_0 = y_1 = y_2 = y_3 = 0$, i.e. quiescence initially.

Taking $y(t) \rightleftharpoons \phi(p)$, the transform equation for (1) is

$$(p^4 - a^4) \phi = Sp, \tag{2}$$

so

$$\phi = Sp / (p^2 + a^2)(p^2 - a^2). \tag{3}$$

$$= \frac{S}{2a^2} \left[\frac{p}{(p^2 - a^2)} - \frac{p}{(p^2 + a^2)} \right], \tag{4}$$

and on inversion using Appendix 10

$$y = (S/2a^3)(\sinh at - \sin at). \tag{5}$$

Alternatively, by the Mellin theorem

$$y = \frac{S}{2\pi i} \int_{Br_1} \frac{e^{zt}\, dz}{(z^2 - a^2)(z^2 + a^2)}. \tag{6}$$

The integrand has simple poles at $z = \pm a,\ \pm ia$. The contribution from the pole at $z = a$ is $S\, e^{at}/4a^3$. Writing $-a,\ \pm ia$ in turn for a, and adding, we get

$$y = \frac{S}{2a^3}\left[\frac{(e^{at} - e^{-at})}{2} - \frac{(e^{iat} - e^{-iat})}{2i}\right] \tag{7}$$

$$= (S/2a^3)\,(\sinh at - \sin at). \tag{8}$$

If this is inserted in the left-hand side of (1), it vanishes, so the solution must be considered to *apply after* the impulse has occurred. In the absence of $SI(t)$, the solution would be

$$y = A \sinh at + B \cosh at + C \sin at + D \cos at, \tag{9}$$

where $A,\ B,\ C,\ D$ are arbitrary constants dependent upon the initial conditions. For those given above, $A = B = C = D = 0$, so the solution is then $y \equiv 0$.

10·33. Solution for $SI(t)$ when that for $H(t)$ is known [236]. If the current in a circuit, initially quiescent, due to a p.d. $H(t)$ is

$$f(t) \Rightarrow \phi(p), \tag{1}$$

that due to $SI(t)$ is $f_1(t) \Rightarrow Sp\phi(p).$ $\qquad\qquad$ (2)

By (4) § 8·43 $Sf'(t) \Rightarrow Sp[\phi(p) - f(0)],$ $\qquad\qquad$ (3)

so $Sp\phi(p) \Leftarrow S[f'(t) + I(t)f(0)] = f_1(t).$ $\qquad$ (4)

If $f(0) = 0$, $f_1(t) = Sf'(t),$ $\qquad\qquad\qquad\qquad$ (5)

but if $f(0) \neq 0$, there is a current impulse $SI(t)f(0)$.

When the variable is $a\sqrt{(t^2 - x^2/v^2)} = au,$

$$S\left[\frac{d}{dt}f(au) + I(t - x/v)f(au)\right]_{t=x/v} \Rightarrow Sp\phi(p) \Leftarrow f_1(au), \tag{6}$$

where $f(au) \Rightarrow \phi(p)$.

Terms in the solution for $H(t)$ which correspond to non-zero initial conditions are independent of either $H(t)$ or $SI(t)$. They must be added to the above solutions to obtain the complete solutions.

10·41. Repeated impulses of infinite amplitude and strength S

[133]. From (5) § 10·31 the transform of the zeroth period (Fig. 52 (a)) is given by

$$\phi_0(p) = Sp, \tag{1}$$

so from (3) § 8·61, $\qquad \phi(p) = Sp/(1 - e^{-ph}). \tag{2}$

This cannot be inverted by the Mellin theorem, because the contour integral is divergent. Suppose, however, that the impulses are applied to the circuit shown in Fig. 52 (b), which is quiescent initially, S_1 being closed at $t = 0$. Then if the current $I \rightleftharpoons \varphi(p)$, the transform equation for the circuit is

$$(pL + R)\varphi = Sp/(1 - e^{-ph}), \tag{3}$$

so $\qquad\qquad\qquad \varphi = \dfrac{(S/L)p}{(p + a)(1 - e^{-ph})}, \tag{4}$

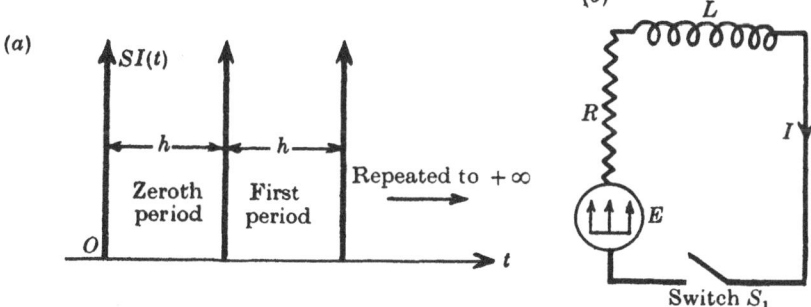

Fig. 52. (a) Impulse of infinite amplitude, but strength S, repeated at interval h. (b) Schematic diagram of LR circuit to which repeated impulse is applied.

with $a = R/L$. Applying the Mellin theorem, the current is given by

$$I = \frac{(S/L)}{2\pi i} \int_{Br_1} \frac{e^{zt}\,dz}{(z + a)(1 - e^{-zh})}, \tag{5}$$

the integral being convergent in virtue of the additional factor $(z + a)$ in the denominator. The integrand has simple poles at $z = -a$, $2n\pi i/h$, n integral, and a limit point of poles at infinity. It is shown in § 4·54, that for a similar integral, the contribution round a large semicircle on the left of Br_1, which meets the imaginary axis between the mth and $(m + 1)$th poles $\to 0$ as m and the radius $\to \infty$. Thus we proceed to evaluate the integral as in § 3·237.

For the pole at $z = -a$, the contribution is

$$e^{-at}/(1 - e^{ah}). \tag{6}$$

For the remaining singularities, we have

$$\sum_{n=-\infty}^{\infty} \left[\frac{e^{zt}}{(z+a)\dfrac{d}{dz}(1-e^{-zh})} \right]_{zh=2n\pi i} = \sum_{n=-\infty}^{\infty} \left[\frac{e^{zt}}{h(z+a)\,e^{-zh}} \right]_{zh=2n\pi i} \tag{7}$$

$$= \sum_{n=1}^{\infty} \left[\frac{e^{2n\pi ti/h}}{ah+2n\pi i} + \frac{e^{-2n\pi ti/h}}{ah-2n\pi i} \right] + \frac{1}{ah} \tag{8}$$

$$= \frac{1}{ah} + 2\sum_{n=1}^{\infty} \left[\frac{ah\cos(2n\pi t/h) + 2n\pi \sin(2n\pi t/h)}{a^2 h^2 + 4n^2 \pi^2} \right] \tag{9}$$

$$= \frac{1}{ah} + 2\sum_{n=1}^{\infty} \left[\frac{\cos\{(2n\pi t/h) - \tan^{-1}(2n\pi/ah)\}}{\sqrt{(a^2 h^2 + 4n^2 \pi^2)}} \right]. \tag{10}$$

Hence from (5), (6), (10), when $t > 0$ the current is given by

$$I = (S/L)\Bigg\{(L/Rh) + [e^{-Rt/L}/(1-e^{Rh/L})] + (2L/h)$$
$$\times \sum_{n=1}^{\infty} \left[\frac{\cos\{(2n\pi t/h) - \tan^{-1}(2n\pi L/Rh)\}}{\sqrt{(R^2 + 4n^2 \pi^2 L^2/h^2)}} \right]\Bigg\}. \tag{11}$$

The first term in (11), namely S/Rh, represents the unidirectional current, the second is the transient, while the 'sigma' term represents the Fourier expansion of the steady alternating current, which is continuous. The fundamental frequency is $1/h$.

If $R \to 0$ in (11), we obtain

$$I = \frac{S}{L}\left[\frac{t}{h} + \left\{ \frac{1}{2} + \sum_{n=1}^{\infty} \frac{\sin(2n\pi t/h)}{n\pi} \right\} \right]. \tag{12}$$

(12) represents the piecewise continuous 'staircase' function illustrated in Figs. 49 (a), 53 (a). The first term St/Lh gives the straight line through the lower corners of the stairs. The portion above this line is the piecewise continuous periodic 'saw-tooth' function shown in Fig. 53 (b), whose Fourier expansion is represented by the part in { } in (12). The staircase and saw-tooth functions are finitely discontinuous at $t = nh$, and the latter function is given by its Fourier expansion at all *other* points. In accordance with the last paragraph of § 10·31, the dimensions of S are voltage × time.

10·42. Alternate impulses reversed [133]. Referring to Fig. 54 (a), if alternate impulses are reversed, the transform of the zeroth period is

$$\phi_0(p) = Sp(1-e^{-ph}), \tag{1}$$

so by (3) § 8·61

$$\phi(p) = Sp(1-e^{-ph})/(1-e^{-2ph}) = Sp/(1+e^{-ph}). \tag{2}$$

Then for the circuit of Fig. 52 (b), we get

$$\varphi = \frac{(S/L)\,p}{(p+a)\,(1+e^{-ph})},\qquad (3)$$

so

$$I = \frac{(S/L)}{2\pi i}\int_{Br_1}\frac{e^{zt}dz}{(z+a)\,(1+e^{-zh})}.\qquad (4)$$

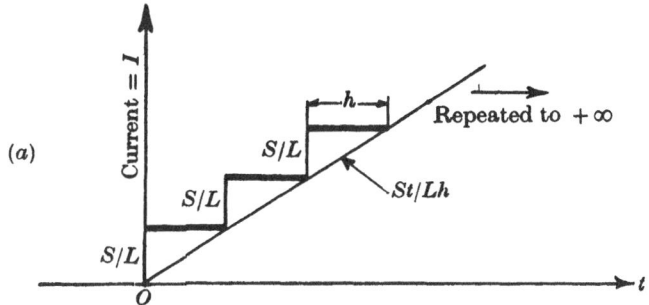

(a)

(b)

Mean value $= S/2L$

Fig. 53. (a) The piecewise continuous 'staircase' function, which has finite discontinuities at $t = nh$, n a positive integer. (b) The piecewise continuous 'saw-tooth' function.

The integrand has simple poles at $z = -a$, $(2n+1)\,\pi i/h$, n integral, and a limit point of poles at infinity. Proceeding as in § 10·41, we obtain

$$I = \frac{S}{L}\left\{\frac{e^{-Rt/L}}{(1+e^{Rh/L})}+\frac{2L}{h}\sum_{n=1}^{\infty}\frac{\cos\left[(2n+1)\,\pi t/h-\tan\,(2n+1)\,\pi L/Rh\right]}{\sqrt{[R^2+(2n+1)^2\,\pi^2 L^2/h^2]}}\right\}.\qquad (5)$$

There is no term representing a unidirectional current as in (11) § 10·41, this being due to absence of a pole at the origin in (4), which is to be expected on physical grounds, since alternate impulses are reversed. Hence during the steady state, the 'sigma' term represents the Fourier expansion of the current which is continuous. If $R \to 0$ in (5), we have

$$I = \frac{S}{L}\left\{\frac{1}{2}+\frac{2}{\pi}\sum_{n=1}^{\infty}\frac{\sin\,(2n+1)\,\pi t/h}{(2n+1)}\right\},\qquad (6)$$

which is the Fourier expansion of the periodic piecewise continuous 'Morse' dot function illustrated in Fig. 54 (b). It is finitely discontinuous at $t = nh$, being represented by (6) except at these points (see [236], p. 135).

10·511. Frequency spectra of impulses. A periodic continuous or piecewise continuous function may be represented symbolically by a Fourier series—as exemplified in previous sections—which comprises a set of *discrete* frequency components usually having different amplitudes. The graphical representation in Fig. 55 (*d*) is known as a 'line' spectrum.

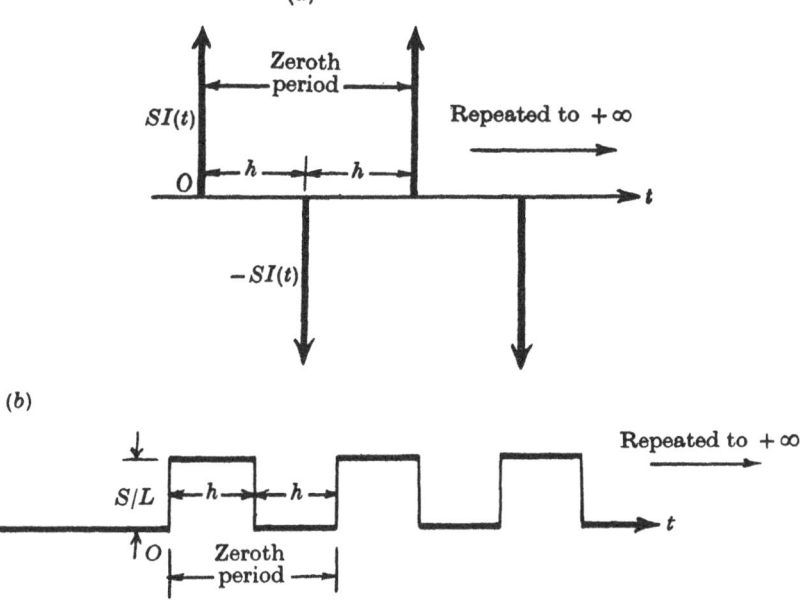

Fig. 54. (*a*) Alternate positive and negative impulses of infinite amplitude, but strength S_1 repeated at interval $2h$. (*b*) The piecewise continuous 'Morse dot' function. Finite discontinuities occur at $t = nh$.

In § 10·512 et seq. it is shown that all angular frequencies in the range $\omega \geqslant 0$ are present in the spectrum of an impulse, so it must be a 'band' type (see Figs. 55 (*e*), 57).

In acoustics, the audible part of the band spectrum of an impulsive sound depends upon the frequency range of the listener's auditory system. The maximum range for young people is approximately 20–25,000 cycles per second. Since the frequency band of an impulsive sound is infinite in extent, it includes the range of audibility of the human ear. Hence if two bodies having supersonic natural frequencies collide, e.g. two small steel spheres, the collision is audible, being perceived as a click.

The sound emitted by a single note on a pianoforte may be represented approximately by a combination of line and band spectra, as illustrated in Fig. 55 (a). The line spectrum pertains to the fundamental and overtones of the strings, while the hammer blow is responsible mainly for the band spectrum. Strictly the motion of the strings is damped and, therefore, non-periodic. Consequently the corresponding spectra are narrow resonance type curves instead of lines, so the complete spectrum is really a band type. By aid of Fourier's integral theorem, the frequency spectrum of an impulse of known mathematical form may be derived as shown below.

10·512. Fourier's theorem. Let $f(t)$ be a function of t having period $2h$ and the following properties: (i) it is finite and single-valued, (ii) it has a limited number of finite discontinuities, and of maxima and minima in the interval $t = (0, 2h)$. Then $f(t)$ may be expressed in the well-known form

$$f(t) = b_0 + \sum_{n=1}^{\infty} b_n \cos \frac{n\pi t}{h} + a_n \sin \frac{n\pi t}{h} \qquad (1)$$

$$= b_0 + \sum_{n=1}^{\infty} A_n \cos(\omega_n t - \theta_n), \qquad (2)$$

where $\omega_n = n\pi/h$, $\theta_n = \tan^{-1}(a_n/b_n)$, $A_n = \sqrt{(a_n^2 + b_n^2)}$, and

$$b_0 = \frac{1}{2h} \int_0^{2h} f(\lambda) \, d\lambda, \text{ the mean value over period } 2h; \qquad (3)$$

$$b_n = \frac{1}{h} \int_0^{2h} f(\lambda) \cos(n\pi\lambda/h) \, d\lambda; \qquad (4)$$

$$a_n = \frac{1}{h} \int_0^{2h} f(\lambda) \sin(n\pi\lambda/h) \, d\lambda, \qquad (5)$$

λ being a dummy variable replacing t.

Suppose $f(t)$ takes the form of a series of pulses repeated at interval $2h$, as illustrated in Fig. 55 (b). Then using the above formulae, we find that

$$f(t) = E_0 \frac{h_1}{h} + \frac{2E_0}{\pi} \sum_{n=1}^{\infty} \frac{\sin{(n\pi h_1/h)}}{n} \cos\left\{\frac{n\pi(t-h_1)}{h}\right\}. \qquad (6)$$

We shall investigate the effect of keeping h_1 constant and increasing h, so that ultimately when $h \to +\infty$, there is a solitary pulse as in Fig. 55 (c). In (6)

$$A_n = (2E_0/n\pi)\sin{(n\pi h_1/h)} = (2E_0/n\pi)\sin{\omega_n h_1}. \qquad (7)$$

Now $\omega_n = n\pi/h$, and if ω_n is a fixed interval of ω, n increases directly with increase in h. Thus if $h_1/h = \frac{1}{4}$, and $\omega_4 h = \pi$, apart from $n = 0$, there will be four coefficients $A_1, ..., A_4$ in $\omega_4 = 4\pi/h$, as illustrated in Fig. 55 (d). With $h_1/h = \frac{1}{64}$, and $\omega_{64} h = \pi$, there will be sixty-four coefficients $A_1, ..., A_{64}$ in the same interval. In virtue of (7), their amplitudes decrease steadily with increase in h. As $h \to +\infty$, $h_1/h \to 0$, and the number of coefficients in a finite interval $\to \infty$. The coefficients are now of infinitesimal order of magnitude, and the band or continuous frequency spectrum of Fig. 55 (e) results. This leads directly to Fourier's integral theorem, and by its aid the band spectra of various functions may be found as shown in subsequent sections.

10·513. Fourier's integral theorem. Substituting from (3)–(5) into (1) § 10·512 gives

$$f(t) = \frac{1}{2h}\int_0^{2h} f(\lambda)\,d\lambda + \frac{1}{h}$$

$$\times \sum_{n=1}^{\infty}\int_0^{2h} f(\lambda)\{\cos\omega_n t\cos\omega_n\lambda + \sin\omega_n t\sin\omega_n\lambda\}\,d\lambda \qquad (1)$$

$$= \frac{1}{2h}\int_0^{2h} f(\lambda)\,d\lambda + \frac{1}{h}\sum_{n=1}^{\infty}\int_0^{2h} f(\lambda)\cos{[\omega_n(t-\lambda)]}\,d\lambda. \qquad (2)$$

Put $h(\omega_n - \omega_{n-1}) = h\Delta\omega = \pi$, and the second integral in (2) becomes

$$\frac{1}{\pi}\sum_{n=1}^{\infty}\Delta\omega\int_0^{2h} f(\lambda)\cos{[\omega_n(t-\lambda)]}\,d\lambda. \qquad (3)$$

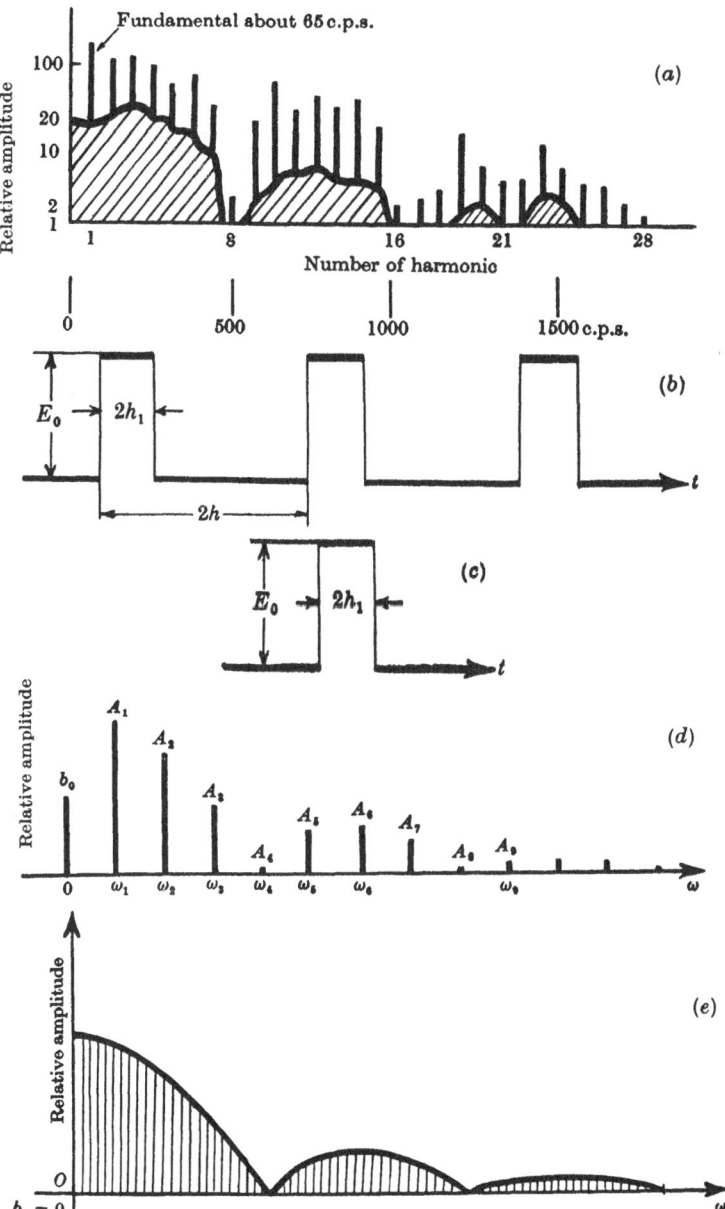

Fig. 55. (a) Band and line spectra for note on pianoforte two octaves below middle C. (b) Impulses of duration $2h_1$ repeated at interval $2h$, $h_1 < h$. (c) Fig. 55(b) when h_1 is constant and $h \to +\infty$. (d) Line spectrum for Fig. 55(b), with $h_1/h = \frac{1}{4}$. (e) Band spectrum for Fig. 55(b), when $h_1/h \to 0$, h_1 constant, $h \to +\infty$.

Now let $\Delta\omega \to d\omega$ such that $h\,d\omega = \pi$, i.e. $h \to +\infty$, and write ω for ω_n. The first integral in (2) vanishes if (5) converges, and (3) gives

$$f(t) = \frac{1}{\pi} \int_0^\infty d\omega \int_0^\infty f(x) \cos\left[\omega(t-x)\right] dx, \qquad (4)$$

on writing x for λ. The lower limit in the ω-integral is zero, since $n = h\omega/\pi$, and, therefore, $\omega \to 0$ as $h \to \infty$ for $n = 1$. The result at (4) is known as Fourier's integral theorem for the range $x = (0, +\infty)$, and it is valid if the two integrals converge. Since $|\cos\left[\omega(t-x)\right]| \leqslant 1$, it follows that the first integral converges if

$$\int_0^\infty |f(x)|\,dx \qquad (5)$$

does so, although in some cases this condition may be too stringent.

Similarly for the range $x = (-\infty, +\infty)$, we derive

$$f(t) = \frac{1}{\pi} \int_0^\infty d\omega \int_{-\infty}^\infty f(x) \cos\left[\omega\,(t-x)\right] dx. \qquad (6)$$

If Re denotes the real part, (6) may be expressed in the alternative form

$$f(t) = \frac{1}{\pi} Re \left\{ \int_0^\infty d\omega \int_{-\infty}^\infty e^{i\omega(t-x)} f(x)\,dx \right\} \qquad (7)$$

$$= \frac{1}{\pi} Re \left\{ \int_0^\infty e^{i\omega t}\,d\omega \int_{-\infty}^\infty e^{-i\omega x} f(x)\,dx \right\}. \qquad (8)$$

Here
$$F(i\omega) = \int_{-\infty}^\infty e^{-i\omega x} f(x)\,dx \qquad (9)$$

is defined to be the *Fourier transform* of $f(x)$ for the range $x = (-\infty, +\infty)$. For the range $x = (0, +\infty)$

$$f(t) = \frac{1}{\pi} Re \left\{ \int_0^\infty e^{i\omega t}\,d\omega \int_0^\infty e^{-i\omega x} f(x)\,dx \right\} \qquad (10)$$

$$= \frac{1}{\pi} Re \left\{ \int_0^\infty e^{i\omega t} F(i\omega)\,d\omega \right\}, \qquad (11)$$

where
$$F(i\omega) = \int_0^\infty e^{-i\omega x} f(x)\,dx \qquad (12)$$

is the *half-range Fourier transform*. The amplitude of any component of frequency ω is

$$|F(i\omega)|\,d\omega, \tag{13}$$

so that $|F(i\omega)|$ is an *index* of the relative amplitudes of the components of the frequency spectrum of $f(t)$.

When $f(t)$ has certain properties given in reference [241], the integral theorem for the *half range* $t = (0, +\infty)$ may be expressed as follows:

If

$$F(i\omega) = \int_0^\infty e^{-i\omega x} f(x)\,dx, \tag{14}$$

then

$$f(t) = \frac{1}{\pi}\,Re\left\{\int_0^\infty e^{i\omega t}\,F(i\omega)\,d\omega\right\}, \tag{15}$$

and vice versa, provided that (14) converges *uniformly* with regard to ω in the range $(0, +\infty)$, and (15) does likewise with regard to t. This implies restrictions on $f(t)$, and some functions do not have Fourier transforms in the above sense, e.g. $H(t)$, $\sin \alpha t$, $\cos \alpha t$. For these functions (14) diverges.

10·514. Frequency spectrum derived from p-multiplied Laplace transform. If in (14) § 10·513 we write p for $i\omega$, we get

$$F(p) = \int_0^\infty e^{-px} f(x)\,dx = \frac{\phi(p)}{p}, \tag{1}$$

provided (14) § 10·513 converges as stipulated. Then

$$F(i\omega) = [\phi(p)/p]_{p=i\omega}. \tag{2}$$

The uniform convergence entails a *continuous* frequency spectrum, which may have zeros as shown later. In *some* instances when (14) § 10·513 diverges, the spectrum may be found as *a limiting case*. For example

$$\int_0^l e^{-i\omega x} H(x)\,dx, \tag{3}$$

which corresponds to Heaviside's unit function, diverges since it does not tend to a definite value as $l \to +\infty$. Now for all

practical purposes, if $\epsilon > 0$, but *extremely small*, $e^{-\epsilon t} H(t) \simeq H(t)$.
Then by (14) § 10·513

$$F(i\omega) = \int_0^\infty e^{-(i\omega+\epsilon)x}\, dx = 1/(\epsilon + i\omega). \tag{4}$$

When $\epsilon \to 0$, $|F(i\omega)| \to 1/\omega$, which gives the spectrum of $H(t)$ *regarded as a limiting case*. The index in (4) may be written $-i(\omega - i\epsilon)x$, so we have introduced a complex or generalized angular frequency $(\omega - i\epsilon)$. Hence to derive the spectrum in *cases of this type*, divide the p-multiplied Laplace transform by p, substitute $(\epsilon + i\omega)$ for p, calculate the modulus and let $\epsilon \to 0$.

10·52. Example. Derive the frequency spectrum of the rectangular type of impulse in Fig. 50 (b).
$f(t)$ is defined thus:

$$f(t) = f_0 \left.\begin{matrix} \\ \\ \end{matrix}\right\} \begin{matrix} 0 < t < h \\ = 0 \end{matrix} \quad \begin{matrix} 0 < t < h \\ t < 0,\ h < t. \end{matrix} \tag{1}$$

We take the *strength* of the impulse as $f_0 h = S$, a constant, so that $f_0 = S/h$. Then in (14) § 10·513 the range of the x-integral is $x = (0, h)$, so the Fourier transform is

$$F(i\omega) = \frac{S}{h} \int_0^h e^{-i\omega x}\, dx = \frac{iS}{\omega h}(e^{-i\omega h} - 1) \tag{2}$$

$$= (S/\omega h)\,[i(\cos \omega h - 1) + \sin \omega h]. \tag{3}$$

Hence

$$|F(i\omega)| = (S/\omega h)\,\sqrt{[(1 - \cos \omega h)^2 + \sin^2 \omega h]}$$

$$= (S/\omega h)\,|\sqrt{2(1 - \cos \omega h)}| \tag{4}$$

$$= S\left\{ \frac{|\sin(\tfrac{1}{2}\omega h)|}{\tfrac{1}{2}\omega h} \right\}. \tag{5}$$

The spectrum corresponding to (5) is shown in Fig. 56 (a) for three values of h. As h decreases, the interval between consecutive zeros increases, and the first part of the graph takes the form of the broken line. As $h \to 0$ in (5), with ω, finite $\{\ \} \to 1$, so in the limit all frequencies in the range $0 \leqslant \omega < \infty$ have the same relative amplitude S, i.e. the spectrum is uniform. An impulse of the type $I(t)$ cannot, of course, occur in practice. But its use in analysis often gives an adequate approximation to one of finite amplitude and relatively short duration h. If h in (5) is finite when $\omega \to \infty$, then $|F(i\omega)| \to 0$. In practice the range of ω is finite and conventionally we may consider the spectrum of $I(t)$ to be uniform.

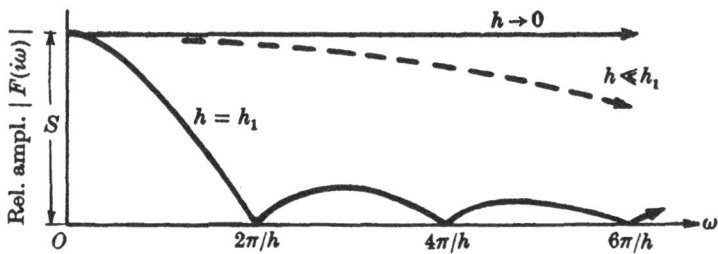

Fig. 56(a). Frequency spectrum of rectangular impulse or Morse dot of Fig. 50(b), with various h.

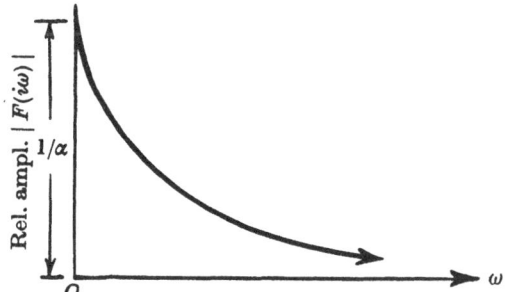

Fig. 56(b). Frequency spectrum of $e^{-\alpha t}$, $t \geqslant 0$.

Inversion of Fourier transform. Inserting the right-hand side of (2) into (15) § 10·513, we get

$$f(t) = Re\left\{\frac{iS}{\pi h}\int_0^\infty [e^{-i\omega(h-t)} - e^{i\omega t}]\frac{d\omega}{\omega}\right\} \tag{6}$$

$$= \frac{S}{\pi h}\int_0^\infty [\sin\omega(h-t) + \sin\omega t]\frac{d\omega}{\omega} \tag{7}$$

$$\begin{aligned} &= (S/\pi h)\,(\tfrac{1}{2}\pi + \tfrac{1}{2}\pi) = S/h = f_0\\ &= (S/\pi h)\,(-\tfrac{1}{2}\pi + \tfrac{1}{2}\pi) \qquad\;\; = 0 \end{aligned}\left.\begin{aligned}&\\&\end{aligned}\right\}\quad\begin{aligned}&0 < t < h,\\&t < 0, h < t,\end{aligned} \tag{8}$$

as at (1). The integrals in (7) were obtained from 3°, § 3·3.

10·53. Spectrum of $e^{-\alpha t}$, $t \geqslant 0$, $\alpha > 0$. By (2) § 10·514 and (10) Appendix 10,

$$F(i\omega) = [1/(p+\alpha)]_{p=i\omega}, \tag{1}$$

so

$$|F(i\omega)| = 1/\sqrt{(\alpha^2+\omega^2)}, \tag{2}$$

which is portrayed graphically in Fig. 56 (b). When α is small enough, the spectrum of $H(t)$—given in § 10·514—is approximated.

Inversion of (1). Using a rationalized denominator, by (15) § 10·513 we have

$$f(t) = \frac{1}{\pi} Re\left\{ \int_0^\infty e^{i\omega t}\left(\frac{\alpha - i\omega}{\alpha^2+\omega^2}\right) d\omega \right\} = \frac{1}{\pi} \int_0^\infty \left[\frac{\alpha\cos\omega t + \omega\sin\omega t}{(\alpha^2+\omega^2)}\right] d\omega. \tag{3}$$

Now

$$\int_0^\infty \left(\frac{\omega\sin\omega t}{\alpha^2+\omega^2}\right) d\omega \Rightarrow \int_0^\infty \left(\frac{\omega^2}{\alpha^2+\omega^2}\right)\left(\frac{p}{p^2+\omega^2}\right) d\omega = I, \tag{4}$$

where the transform of $\sin\omega t$ has been used. Then

$$I = \frac{p}{p^2-\alpha^2} \int_0^\infty \left[\frac{p^2}{p^2+\omega^2} - \frac{\alpha^2}{\alpha^2+\omega^2}\right] d\omega = \frac{p}{p^2-\alpha^2} \frac{\pi(p-\alpha)}{2} \tag{5}$$

$$= \pi p/2(p+\alpha) \Subset \tfrac{1}{2}\pi e^{-\alpha t}. \tag{6}$$

Hence

$$\frac{1}{\pi} \int_0^\infty \left(\frac{\omega\sin\omega t}{\alpha^2+\omega^2}\right) d\omega = \tfrac{1}{2}e^{-\alpha t}. \tag{7}$$

We have also

$$\int_0^\infty \frac{\cos\omega t}{\alpha^2+\omega^2} d\omega \Rightarrow p^2 \int_0^\infty \frac{d\omega}{(\alpha^2+\omega^2)(p^2+\omega^2)} = I_1, \tag{8}$$

so

$$I_1 = \frac{p^2}{p^2-\alpha^2} \int_0^\infty \left[\frac{1}{\alpha^2+\omega^2} - \frac{1}{p^2+\omega^2}\right] d\omega = \frac{p^2}{p^2-\alpha^2}\left(\frac{1}{\alpha} - \frac{1}{p}\right)\frac{\pi}{2} \tag{9}$$

$$= \pi p/2\alpha(p+a) \Subset (\pi/2\alpha) e^{-\alpha t}, \tag{10}$$

so

$$\frac{\alpha}{\pi} \int_0^\infty \frac{\cos\omega t}{(\alpha^2+\omega^2)} d\omega = \tfrac{1}{2}e^{-\alpha t}. \tag{11}$$

Hence by (3), (7), (11) $$f(t) = e^{-\alpha t}, \tag{12}$$

as required. In the foregoing analysis the artifice in (4) was used to permit ready evaluation of the infinite integrals concerned. For justification of this procedure see [236].

10·54. Spectrum of $e^{-\alpha t} \cos \omega_0 t$, $t \geqslant 0$. Using (17) Appendix 10, and (2) § 10·514

$$F(i\omega) = (p+\alpha)/[(p+\alpha)^2 + \omega_0^2]_{p=i\omega}, \tag{1}$$

so $\quad |F(i\omega)| = \sqrt{(\omega^2 + \alpha^2)}/\sqrt{[(\omega_0^2 - \omega^2 + \alpha^2)^2 + 4\omega^2\alpha^2]}.$ (2)

This is a band spectrum of the type in Fig. 62 (b). If $\alpha \ll \omega_0$, when $\omega = \omega_0$

$$|F(i\omega)| \simeq 1/[2\alpha(1 + \alpha^2/8\omega_0^2)], \tag{3}$$

so that as $\alpha \to 0$, $|F(i\omega)| \to \infty$, and a line spectrum which corresponds to $\cos \omega_0 t$ is obtained. In this instance, by (12) § 10·513, the Fourier transform for the range $t > 0$ is

$$F(i\omega) = \int_0^\infty e^{-(i\omega+\alpha)x} \cos \omega_0 x \, dx. \tag{4}$$

This integral diverges unless $\alpha > 0$, so when $\alpha \to 0$ the line spectrum is a limiting case.

10·55. Relationship between spectrum of impulse and current amplitude in LC circuit.

Laplace transform of first half-period of $E_0 \sin \alpha t$. The impulse is defined as follows:

$$E(t) = E_0 \sin \alpha t \left.\begin{array}{l} \\ \end{array}\right\} \quad \begin{array}{l} 0 \leqslant t \leqslant \pi/\alpha, \\ t < 0, \ t > \pi/\alpha. \end{array} \tag{1}$$
$$= 0$$

Now

$$E_0 \sin \alpha t \Rightarrow E_0 \alpha p/(p^2 + \alpha^2) \quad \text{and} \quad E_0 \sin \alpha(t - \pi/\alpha) \Rightarrow E_0 \alpha p \, e^{-p\pi/\alpha}/(p^2 + \alpha^2).$$

Adding these yields for the first half-period

$$E(t) \Rightarrow E_0 \alpha p(1 + e^{-p\pi/\alpha})/(p^2 + \alpha^2). \tag{2}$$

Circuital equation. This is (see Fig. 61 (a) with $E(t)$ for $I(t)$)

$$L\frac{dI}{dt} + \frac{1}{C}\int_0^t I \, dt = E(t), \tag{3}$$

so with $I \Rightarrow \phi$, the transform equation, for initial quiescence, is

$$Lp\phi + \phi/pC = E_0 \alpha p(1 + e^{-p\pi/\alpha})/(p^2 + \alpha^2). \tag{4}$$

Taking $\omega_1^2 = 1/LC$ and resolving into partial fractions, we obtain

$$\phi = \frac{E_0 \alpha/L}{\alpha^2 - \omega_1^2} \left[\frac{p^2}{p^2 + \omega_1^2} - \frac{p^2}{p^2 + \alpha^2} \right] (1 + e^{-p\pi/\alpha}). \tag{5}$$

Inverting (4) by aid of Appendix 10 and § 8·51, for $t > \pi/\alpha$ the current is

$$I = \frac{E_0 \alpha/L}{\alpha^2 - \omega_1^2} [(\cos \omega_1 t - \cos \alpha t) + \{\cos \omega_1 (t - \pi/\alpha) - \cos \alpha (t - \pi/\alpha)\}] \quad (6)$$

$$= \{E_0 \alpha/L(\alpha^2 - \omega_1^2)\} [\cos [\omega_1 (t - \pi/\alpha)] + \cos \omega_1 t] \quad (7)$$

$$= \{2E_0 \alpha/L(\alpha^2 - \omega_1^2)\} \cos (\omega_1 \pi/2\alpha) \cos [\omega_1 (t - \pi/2\alpha)], \quad (8)$$

so the amplitude is *proportional* to $2E_0 \alpha \mid \cos (\omega_1 \pi/2\alpha)/(\alpha^2 - \omega_1^2) \mid$.   (9)

We shall now show that (9) is identical with the relative amplitude $\mid F(i\omega) \mid$ when $\omega = \omega_1$ and $t > \pi/a$.

Spectrum of $E(t)$. From (2) § 10·514 and (2) above,

$$F(i\omega) = E_0 \alpha [(1 + e^{-p\pi/\alpha})/(p^2 + \alpha^2)]_{p=i\omega} = \{(E_0 \alpha)/(\alpha^2 - \omega^2)\} (1 + e^{-i\omega\pi/\alpha}),$$

so $\mid F(i\omega) \mid = \mid \{E_0 \alpha/(\alpha^2 - \omega^2)\} \sqrt{[\{1 + \cos (\omega\pi/\alpha)\}^2 + \sin^2 (\omega\pi/\alpha)]} \mid$ (10)

$$= \left| \frac{2E_0 \alpha \cos (\omega\pi/2\alpha)}{(\alpha^2 - \omega^2)} \right|. \quad (11)$$

This is identical with (9) for $\omega = \omega_1$. Hence in the case of the LC *loss-free* circuit, the current amplitude is equal to $\mid F(i\omega_1) \mid L^{-1}$, ω_1 being the natural frequency of the circuit. This result shows the relevance of the frequency spectrum of the impulse (Fig. 57), in relation to the current response of the circuit. The reader may ascertain the effect of introducing a small resistance, on the response. When $R \to 0$, the expression should reduce to (8).

Particular case. When $\omega = \alpha$, $\mid F(i\omega) \mid$ has the indeterminate form $0/0$. To ascertain its value, we differentiate the numerator and denominator independently and put $\omega = \alpha$. Thus

$$\mid F(i\omega) \mid_{\omega=\alpha} = 2E_0 \alpha \left| \frac{d}{d\omega} \cos (\omega\pi/2\alpha) \middle/ \frac{d}{d\omega} (\alpha^2 - \omega^2) \right|_{\omega=\alpha}$$

$$= \frac{2E_0 \alpha\pi}{2\alpha} \left| \frac{\sin (\omega\pi/2\alpha)}{2\omega} \right|_{\omega=\alpha} = \frac{\pi E_0}{2\alpha},$$

so $\mid F(i\alpha) \mid/E_0 = \pi/2\alpha$.   (12)

Since $\cos (\omega\pi/2\alpha) = 0$ for $\omega\pi/2\alpha = (2n+1)\pi/2$, or $\omega = (2n+1)\alpha$, the spectrum has zeros corresponding to $n = 1, 2, 3, \ldots$, i.e. $\omega = 3\alpha, 5\alpha, 7\alpha, \ldots$. When $t > \pi/\alpha$, the current is zero if the natural frequency of the circuit has one of these values of ω.

10·56. Frequency spectrum of current in circuit subjected to unit impulse $I(t)$. For initial quiescence, if the Laplace transform solution for the current is $\phi(p)$, the spectrum is given by

$$\mid F(i\omega) \mid = \mid \phi(i\omega) \mid/\omega. \quad (1)$$

Referring to Fig. 62 the transform equation is

$$(Lp + R + 1/Cp)\phi = p, \tag{2}$$

so $\qquad \phi/p = 1/(Lp + R + 1/Cp) = 1/Z(p), \tag{3}$

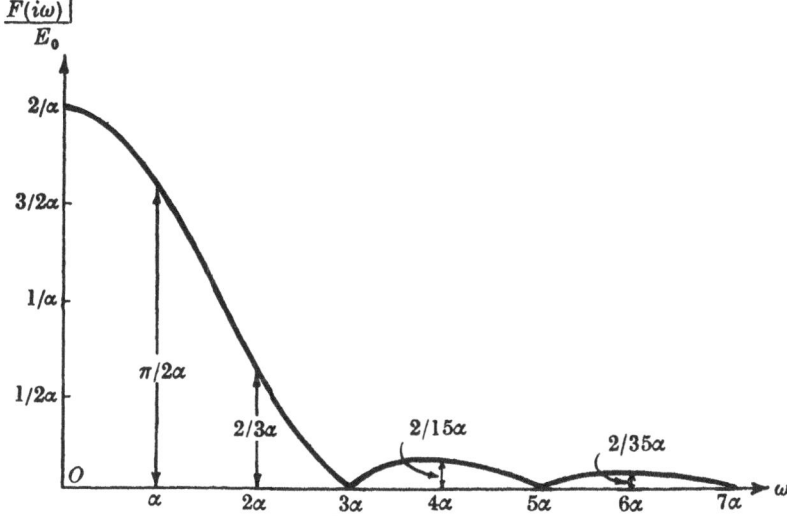

Fig. 57. Band spectrum for first half-period of $E_0 \sin \alpha t$.

where $Z(p)$ is the transform impedance, and $1/Z(p)$ the transform admittance ([236], p. 59). If the current $I = f(t)$, then

$$\int_0^\infty e^{-pt} f(t)\, dt = \frac{\phi}{p}, \tag{4}$$

so $\qquad F(i\omega) = \int_0^\infty e^{-i\omega x} f(x)\, dx = \frac{\phi(i\omega)}{i\omega}, \tag{5}$

provided the integral converges. Thus from (3), (5) it follows that

$$|F(i\omega)| = |\phi(i\omega)|/\omega = 1/|Z(i\omega)|. \tag{6}$$

Hence the spectrum of the current due to $I(t)$—which latter has a uniform spectrum—is given by the reciprocal of the modulus of the steady state circuital impedance.

10·57. Examples. Since the spectrum of $I(t)$ is uniform, the response obtained by applying an impulse of very short duration to an electrical or other system, initially quiescent, is an index of the frequency response. A number of circuits is considered below.

(i) For a resistance R, $1/|Z(i\omega)| = 1/R$, a constant, so the spectrum is uniform like that of $I(t)$. The current is an impulse $I(t)/R$ as shown in Fig. 58. In practice, no resistance is devoid of capacitance and inductance, so the current-time curve would resemble that for a highly-damped LCR circuit.

(ii) For an inductance L, $1/|Z(i\omega)| = 1/\omega L$, so the spectrum curve is a rectangular hyperbola which $\rightarrow +\infty$ as $\omega \rightarrow 0$, and $\rightarrow 0$ as $\omega \rightarrow +\infty$ (Fig. 59). Moreover, the spectrum is essentially a line type, in virtue of the infinity at $\omega = 0$. The current is unidirectional and constant, i.e. zero frequency. This case is analogous to that of a mass m on a smooth horizontal plane, subjected to an impulsive blow. Thereafter the mass moves with constant velocity.

(iii) For a capacitance and resistance in series,

$$1/|Z(i\omega)| = \omega C/\sqrt{(\omega^2 C^2 R^2 + 1)},$$

which gives the band spectrum of Fig. 60. If $R = 0$, the spectrum is a straight line through the origin having slope C. The current for $R > 0$ is

$$I = (1/R)[I(t) - e^{-t/CR}/CR],$$

where $I(t)/R$ represents a positive current impulse which charges C instantaneously. Thereafter C commences to discharge, so the current reverses and is represented by the second term on the right-hand side.

(iv) For the LC circuit of Fig. 61 (a),

$$1/|Z(i\omega)| = \omega C/|\omega^2 LC - 1|.$$

In virtue of the infinity when $\omega^2 = 1/LC = \omega_0^2$, the spectrum is essentially a line type at the resonance point $\omega = \omega_0$. The undamped current has this frequency.

(v) For the LCR circuit of Fig. 62 (a),

$$1/|Z(i\omega)| = 1/\sqrt{[R^2 + (\omega L - 1/\omega C)^2]},$$

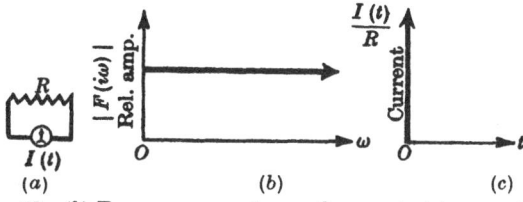

Fig. 58. (b) Frequency spectrum of current; (c) current-time curve,
when $I(t)$ is applied to circuit (a).

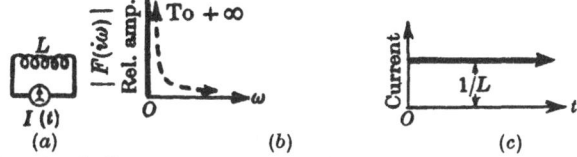

Fig. 59. (b) Frequency spectrum of current; (c) current-time curve,
when $I(t)$ is applied to circuit (a).

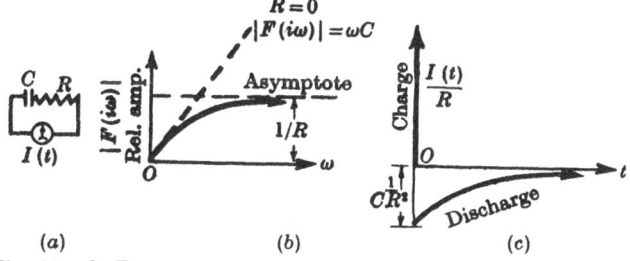

Fig. 60. (b) Frequency spectrum of current; (c) current-time curve,
when $I(t)$ is applied to circuit (a).

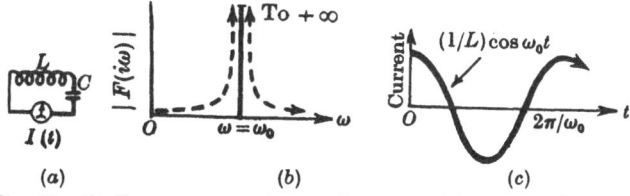

Fig. 61. (b) Frequency spectrum of current; (c) current-time curve,
when $I(t)$ is applied to circuit (a).

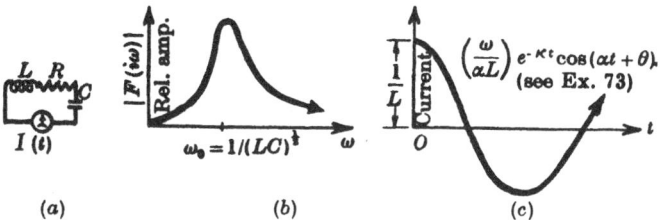

Fig. 62. (b) Frequency spectrum of current; (c) current-time curve,
when $I(t)$ is applied to circuit (a).

so the spectrum is given by the familiar selectivity or tuning curve of the circuit, shown in Fig. 62 (b). If the circuit is 'oscillatory', the current is a damped cosine curve, as in Fig. 62 (c) (see Prob. 73, p. 329).

(vi) For the low-pass filter of Fig. 86 (b) with $Z_0 = 0$ and terminated to avoid reflexion at the far end, replacing E by $I(t) \Rightarrow p$ in (1) § 13·521, we get

$$\frac{\phi_m(p)}{p} = \frac{\sqrt{(C/L)}}{\sqrt{(1+p^2a^2)}\{pa + \sqrt{(1+p^2a^2)}\}^{2m}}. \qquad (1)$$

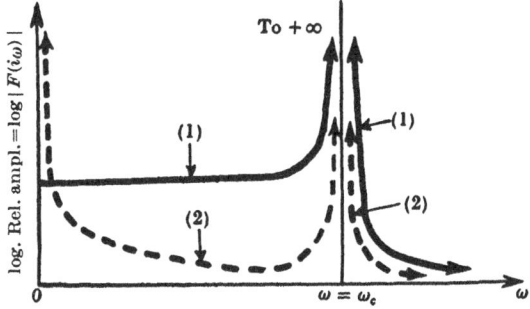

Fig. 63. Frequency spectra of current in low-pass filter of Fig. 86(b), with $z_0 = 0$, when (1) $I(t)$, and (2) $H(t)$ are applied to the input.

Now $1/a = \omega_c$, the cut-off frequency, so we obtain

$$F(i\omega) = \sqrt{(C/L)}/\sqrt{[1-(\omega/\omega_c)^2]}\{(i\omega/\omega_c)+\sqrt{[1-(\omega/\omega_c)^2]}\}^{2m}, \quad (2)$$

and, therefore,

$$|F(i\omega)| = \sqrt{(C/L)}/\sqrt{[1-(\omega/\omega_c)^2]} \quad (\omega < \omega_c),$$
$$= \sqrt{(C/L)}/\sqrt{[(\omega/\omega_c)^2-1]}\{\omega/\omega_c + \sqrt{[(\omega/\omega_c)^2-1]}\}^{2m}$$
$$(\omega > \omega_c). \quad (3)$$

Logarithmically* the relative amplitude of the spectrum is almost constant from $\omega = 0$ to a point near cut-off, as illustrated in Fig. 63, curve (1). As $\omega \to \omega_c$ from either side, the relative amplitude $\to \infty$, since a 'resonant' condition is approached. When $\omega > \omega_c$, the relative amplitude decreases rapidly with increase in ω.

* Decibel basis.

The transform of the current in the mth inductance of the filter is given by $\phi_m(p)$ in (1) above. Thus by aid of (50) Appendix 10,

$$I_m = (2/L) J_{2m}(\omega_c t), \tag{4}$$

which may be plotted from tables of Bessel functions [234]. The current-time curves for $m = 0, 1$ are depicted in Fig. 64.

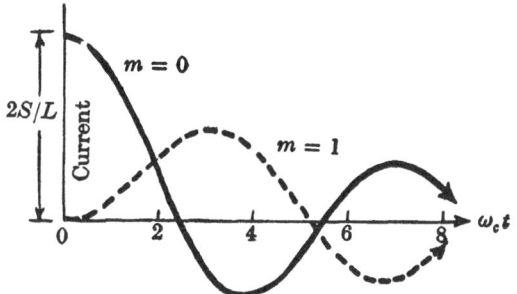

Fig. 64. Current-time curve for zeroth and first meshes
of filter (see Fig. 86(b)), when $SI(t)$ is applied.

The reader should deal with the high pass-filter in Fig. 87 (a), taking $Z_0 = 0$. At $t = 0$ the uncharged capacitances act as short circuits, so the current starts as an impulse $I(t)\sqrt{(C/L)}$. Thereafter the capacitances commence to discharge through the inductances.

Spectrum when $H(t)$ is applied to circuit. The transform of the applied p.d. is now unity instead of p, so the spectrum may be obtained if the formulae corresponding to $I(t)$ are divided by ω. That for the low-pass filter is shown in Fig. 63, curve (2).

PART III

TECHNICAL APPLICATIONS OF PARTS I AND II

XI

ELECTRICAL CIRCUITS: VIBRATIONAL SYSTEMS: AEROPLANE DYNAMICS: DEFLECTION OF BEAMS

11·11. Simple inductive circuit. This comprises a resistance R in series with an inductance L as shown diagrammatically in Fig. 65. Find the current at any time after the application of (1°) a constant p.d. $E = E_0$, (2°) a p.d. $E = E_0 \cos \omega t$, the circuit being quiescent initially.

(1°) The differential equation is

$$L\frac{dI}{dt} + RI = E. \tag{1}$$

With $I \Rightarrow \phi(p)$, the transform version of (1) for $E = E_0$ is

$$\phi = (E_0/L)/(p+a), \tag{2}$$

where $a = R/L$. By (11) Appendix 10, the inverse of (2) is

$$I = (E_0/R)(1 - e^{-at}). \tag{3}$$

Alternatively, by the Mellin theorem

$$I = \frac{E_0/L}{2\pi i} \int_{Br_1} \frac{e^{zt}\,dz}{z(z+a)}. \tag{4}$$

The integrand has simple poles at $z = 0$ and $z = -a$, so by §§ 3·21, 4·53,

$$I = (E_0/L)[1/a - e^{-at}/a] \tag{5}$$

$$= (E_0/R)(1 - e^{-Rt/L}), \tag{6}$$

which is a very well-known formula.

(2°) We substitute the transform of $E_0 \cos \omega t$ for E_0 in (2), thereby obtaining

$$I = \frac{(E_0/L)p^2}{(p+a)(p^2+\omega^2)}. \tag{7}$$

Inverting (7) by the Mellin theorem, we have

$$I = \frac{(E_0/L)}{2\pi i} \int_{Br_1} \frac{e^{zt} z \, dz}{(z+a)(z^2+\omega^2)}$$

$$= \frac{(E_0/L)}{2\pi i} \int_{Br_1} \frac{e^{zt} z \, dz}{(z+a)(z-i\omega)(z+i\omega)}. \tag{8}$$

The integrand of (8) has three simple poles at $z = -a, \; +i\omega, \; -i\omega$, so, by § 4·53,

$$I = (E_0/L)\left[\frac{-ae^{-at}}{a^2+\omega^2} + \frac{i\omega e^{i\omega t}}{2i\omega(a+i\omega)} + \frac{i\omega e^{-i\omega t}}{2i\omega(a-i\omega)} \right] \tag{9}$$

$$= \frac{E_0}{\sqrt{(R^2+\omega^2 L^2)}}\left[\cos(\omega t - \alpha) - \frac{R e^{-Rt/L}}{\sqrt{(R^2+\omega^2 L^2)}} \right], \tag{10}$$

where $\alpha = \tan^{-1}(\omega L/R)$.

The first term in (10) represents the alternating current of constant amplitude $E_0/\sqrt{(R^2+\omega^2 L^2)}$, $\sqrt{(R^2+\omega^2 L^2)}$ being the impedance of the circuit at frequency ω. The second term in (10) represents a transient which dies away exponentially with increase in time. $\alpha = \tan^{-1}(\omega L/R)$ is the angle of lag of the current behind the applied p.d., it being due to the presence of inductance L. When $t = 0$ in (10), the two terms are equal but of opposite sign, so the current is zero.

11·12. Application of the product theorem. The method of solution in 2° § 11·11 is probably the briefest and best. There is, however, another method which may be considered. It is outlined in Appendix 8, and depends upon the product theorem, which can be applied when the unit function solution (2) § 11·11 is known. By (6) § 11·11 the current corresponding to unit applied p.d. is

$$A(t) = \frac{1}{R}(1 - e^{-Rt/L}). \tag{1}$$

The function $A(t)$ which corresponds to $f_1(t)$ in (1) § 3 Appendix 8 is called the *indicial admittance* of the circuit in Fig. 65. Knowing this, the solution of the problem for an arbitrary applied p.d., $f_2(t)$, can be found by aid of any of the formulae (4)–(9) § 3 Appendix 8. Choosing (4), the solution for an applied p.d. $E_0 \cos \omega t$ is

$$I = \frac{E_0}{R} \frac{d}{dt} \int_0^t (1 - b \, e^{a\lambda}) \cos \omega \lambda \, d\lambda, \tag{2}$$

where $a = R/L$ and $b = e^{-Rt/L} = e^{-at}$. Using a well-known standard integral, (2) gives

$$I = \frac{E_0}{R}\frac{d}{dt}\int_0^t \cos \omega\lambda\, d\lambda$$

$$-\frac{E_0}{R}\frac{d}{dt}\left\{\frac{1}{a^2+\omega^2}(\omega \sin \omega t + a \cos \omega t - a\, e^{-at})\right\} \qquad (3)$$

$$= \frac{E_0}{R}\frac{d}{dt}\left\{\left[\frac{1}{\omega}-\frac{\omega}{a^2+\omega^2}\right]\sin \omega t - \frac{a\cos \omega t}{a^2+\omega^2}+\frac{a\,e^{-at}}{a^2+\omega^2}\right\} \qquad (4)$$

$$= \frac{E_0 a}{R(a^2+\omega^2)}\frac{d}{dt}\left\{\frac{\sqrt{(a^2+\omega^2)}}{\omega}\sin\,[\omega t - \tan^{-1}(\omega/a)]+e^{-at}\right\} \qquad (5)$$

$$= \frac{E_0 a}{R(a^2+\omega^2)}\{\sqrt{(a^2+\omega^2)}\cos\,[\omega t - \tan^{-1}(\omega/a)]-a\,e^{-at}\}. \qquad (6)$$

When $a = R/L$ is substituted in (6), formula (10) § 11·11 is reproduced.

11·21. Modified inductive circuit. A p.d. of the form in Fig. 66 (b) is applied to the circuit of Fig. 66 (a), initially quiescent.

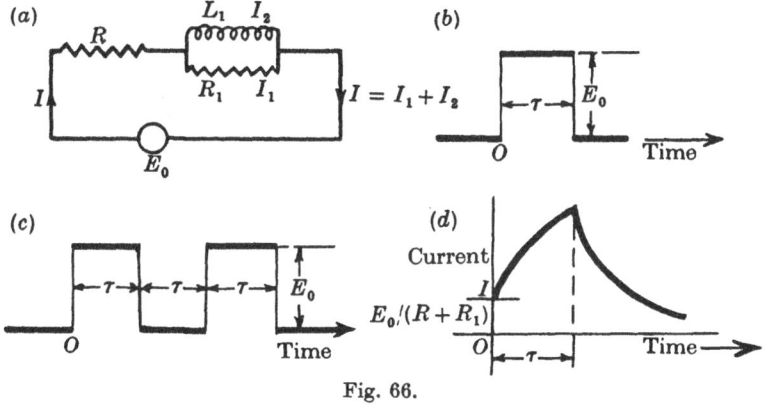

Fig. 66.

What is the current I at any time $t > 0$? In transform notation, with $I \Rightarrow \phi(p)$, $I_1 \Rightarrow \phi_1(p)$, $I_2 \Rightarrow \phi_2(p)$, the p.d. across $L_1 R_1$ is

$$E_0 - R\phi = R_1\phi_1 = pL_1\phi_2, \qquad (1)$$

also

$$\phi = \phi_1 + \phi_2. \qquad (2)$$

From (1)

$$\phi_1 = (E_0 - R\phi)/R_1 \qquad (3)$$

and
$$\phi_2 = (E_0 - R\phi)/pL_1. \tag{4}$$

Substituting from (3) and (4) into (2), we get

$$\phi = E_0(pL_1 + R_1)/[pL_1(R_1 + R) + RR_1] \tag{5}$$

$$= \frac{E_0}{\alpha}\left(\frac{pL_1 + R_1}{p + \beta}\right), \tag{6}$$

where $\alpha = L_1(R_1 + R)$ and $\beta = RR_1/L_1(R_1 + R)$. (6) may be inverted by (10), (11) Appendix 10.

Alternatively by the Mellin theorem

$$I = \frac{E_0}{\alpha}\frac{1}{2\pi i}\int_{Br_1}\frac{e^{zt}(zL_1 + R_1)\,dz}{z(z + \beta)}. \tag{7}$$

The integrand has two simple poles at $z = 0$ and at $z = -\beta$, so

$$I = \frac{E_0}{\alpha}\left[\frac{R_1}{\beta} - \frac{e^{-\beta t}(R_1 - \beta L_1)}{\beta}\right] \tag{8}$$

$$= E_0\left[\frac{1}{R} - \frac{R_1}{R(R + R_1)}e^{-RR_1 t/(R_1 + R)L_1}\right]. \tag{9}$$

Formula (9) gives the current from $t \geqslant 0$ to $t \leqslant \tau$. If at $t = \tau$ we imagine a p.d. $-E_0$ to be applied in series with E_0, the effect is equivalent to replacing the source of supply by a short circuit. We treat the system as though it were quiescent at $t = \tau$, so the current due to $-E_0$ is obtained from (9) by writing $-E_0$ for E and $(t - \tau)$ for t. Thus

$$I_\tau = -E_0\left[\frac{1}{R} - \frac{R_1}{R(R + R_1)}e^{-RR_1(t - \tau)/(R_1 + R)L_1}\right]. \tag{10}$$

Since the system is a linear one, the two solutions (9) and (10) can be superimposed (added), so that when $t > \tau$, the current is

$$I = \frac{E_0 R_1}{R(R + R_1)}e^{-RR_1 t/(R_1 + R)L_1}[e^{RR_1\tau/(R_1 + R)L_1} - 1]. \tag{11}$$

The current from $t = 0$ onwards is shown in Fig. 66 (d).

At any time after $t = \tau$, the current could be found also by using the shift theorem in § 8·51. The transform of the applied p.d. representing the dot of the Morse telegraph code in Fig. 66 (b) is $E_0(1-e^{-p\tau})$. Substituting this in (6) and using the Mellin theorem

$$I = \frac{E_0}{\alpha}\frac{1}{2\pi i}\int_{Br_1}\frac{[e^{zt}-e^{z(t-\tau)}](zL_1+R_1)\,dz}{z(z+\beta)},\qquad(12)$$

which yields (9) plus (10), i.e. formula (11).

If the applied p.d. were two dots of the Morse code as in Fig. 66 (c), the transform version would be

$$E_0(1-e^{-p\tau}+e^{-2p\tau}-e^{-3p\tau}),$$

so that when $t>3\tau$,

$$I = \frac{E_0}{\alpha}\frac{1}{2\pi i}\int_{Br_1}\frac{[e^{zt}-e^{z(t-\tau)}+e^{z(t-2\tau)}-e^{z(t-3\tau)}](zL_1+R_1)\,dz}{z(z+\beta)}\qquad(13)$$

$$= -\frac{E_0}{\alpha\beta}[e^{-\beta t}-e^{-\beta(t-\tau)}+e^{-\beta(t-2\tau)}-e^{-\beta(t-3\tau)}]\,(R_1-\beta L_1),\qquad(14)$$

the integrand having no pole at the origin. Thus for $t>3\tau$,

$$I = \frac{E_0 R_1}{R(R_1+R)}\,e^{-\beta t}[e^{3\beta\tau}-e^{2\beta\tau}+e^{\beta\tau}-1].\qquad(15)$$

11·22. Oscillatory electrical circuit. An LCR circuit of the type in Fig. 62 (a) has a switch replacing $I(t)$. When $t<0$ the switch is open and C has a charge Q_0. What is the current I at any time after closing the switch?

The circuital differential equation may be written

$$\frac{L\,dI}{dt}+RI+\frac{1}{C}\int_0^t I\,dt+\frac{Q_0}{C}=0,\qquad(1)$$

or

$$\frac{dI}{dt}+2\kappa I+\omega_0^2\int_0^t I\,dt = -Q_0\omega_0^2,\qquad(2)$$

where $\kappa=R/2L$, $\omega_0^2=1/LC$. Since $I=0$ when $t=0$, there is no initial conditions term, so with $I\Rightarrow\phi$, the transform version of (2) is

$$(p+2\kappa+\omega_0^2/p)\,\phi = -Q_0\omega_0^2,\qquad(3)$$

or

$$\phi = -Q_0\omega_0^2 p/(p^2+2\kappa p+\omega_0^2)\qquad(4)$$

$$= -Q_0\omega_0^2 p/[(p+\kappa)^2+\alpha^2],\qquad(5)$$

where $\alpha^2 = (\omega_0^2 - \kappa^2) > 0$ for an oscillatory discharge. Using (3) § 8·411, and writing $(p - \kappa)$ for p in the denominator of (5) gives

$$e^{\kappa t} I \Rightarrow - Q_0 \omega_0^2 p / (p^2 + \alpha^2). \tag{6}$$

The right-hand side may be inverted either by the Mellin theorem, or by aid of Appendix 10. Using the latter, we have

$$e^{\kappa t} I = - Q_0(\omega_0^2/\alpha) \sin \alpha t, \tag{7}$$

so

$$I = - Q_0(\omega_0^2/\alpha) e^{-\kappa t} \sin \alpha t, \tag{8}$$

the minus sign indicating discharge of C, which commences at $t = 0$.

For the critical case, when oscillation just ceases, $\alpha = 0$, then (6) gives

$$e^{\kappa t} I \Rightarrow - Q_0 \omega_0^2 / p \,\Subset\, - Q_0 \omega_0^2 t, \tag{9}$$

so

$$I = - Q_0 \omega_0^2 t e^{-\kappa t}, \tag{10}$$

which has a minimum value at $t = 1/\kappa$.

For the aperiodic case $\omega_0^2 < \kappa^2$, so $\alpha < 0$. Writing $- \beta^2 = (\omega_0^2 - \kappa^2)$ for α^2 in (6) yields

$$e^{\kappa t} I = - Q_0(\omega_0^2/\beta) \sinh \beta t, \tag{11}$$

so

$$I = - Q_0(\omega_0^2/\beta) e^{-\kappa t} \sinh \beta t. \tag{12}$$

11·3. Torsional oscillations of motor-generator combination. Fig. 67 is a simplified diagram illustrating an electrical motor-generator combination, the armatures (or rotors) of the two machines being coupled mechanically by a shaft. At time $t = 0$ current is supplied to the motor, which starts to rotate and take the generator armature with it. Owing to compliance of the coupling shaft, the two armatures are displaced angularly relative to each other. Assuming that the torque exerted by the motor is constant for a short time, find the behaviour of the system during this interval.

Let f be the tangential force at radius b which causes a torque $fb = T$, by the motor; I_1, I_2 be the respective moments of inertia of the two armatures about their axes; θ_1, θ_2 the angular displacements of the motor and generator respectively from θ_1, $\theta_2 = 0$ at $t = 0$; τ the torque to twist the coupling shaft (of negligible moment of inertia) through one radian (if elastic throughout this angular range).

Since the applied torque is constant, it has the form of Heaviside's unit function (for the short interval under consideration). The differential equations of the system for $t > 0$ are

$$I_1 \frac{d^2\theta_1}{dt^2} + \tau(\theta_1 - \theta_2) = T \text{ (torque due to motor)}, \tag{1}$$

and

$$I_2 \frac{d^2\theta_2}{dt^2} = \tau(\theta_1 - \theta_2) \text{ [torque at generator end of shaft]}. \tag{2}$$

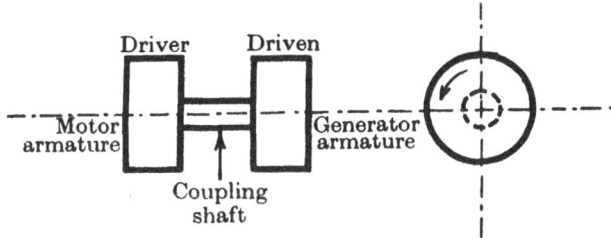

Fig. 67.

Substituting the left-hand side of (2) for $\tau(\theta_1 - \theta_2)$ in (1), taking $\theta_1(t) \Rightarrow \phi_1(p)$, $\theta_2(t) \Rightarrow \phi_2(p)$, and applying (6) § 8·43 with

$$f(0) = f'(0) = 0$$

for quiescent initial conditions, we get

$$I_1 p^2 \phi_1 = T - I_2 p^2 \phi_2, \tag{3}$$

or

$$\phi_1 = (T - I_2 p^2 \phi_2)/I_1 p^2. \tag{4}$$

Substituting for ϕ_1 from (4) into the transform version of (2), i.e. $I_2 p^2 \phi_2 = \tau(\phi_1 - \phi_2)$, we get

$$\phi_2(I_2 p^2 + \tau) = \tau(T - I_2 p^2 \phi_2)/I_1 p^2,$$

so

$$\phi_2 = \frac{K}{p^2(p^2 + a^2)}, \tag{5}$$

where $K = T\tau/I_1 I_2$, and $a^2 = \tau(I_1 + I_2)/I_1 I_2$.

Since $\theta_2(0) = 0$, by (5) § 8·43 the angular velocity of the generator is

$$\frac{d\theta_2}{dt} \equiv p\phi_2 = \frac{K}{p(p^2+a^2)} \tag{6}$$

$$= \frac{K}{a^2}\left[\frac{1}{p} - \frac{p}{p^2+a^2}\right], \tag{7}$$

(7) may be inverted by using (3), (4) Appendix 10. Alternatively by the Mellin theorem

$$\dot{\theta}_2 = \frac{K}{2\pi i}\int_{Br_1} \frac{e^{zt}\,dt}{z^2(z^2+a^2)}. \tag{8}$$

The integrand of (8) has a double pole at the origin, and the residue is found by the method of (3) § 3·241. Thus the contribution from this double pole is

$$K\frac{d}{dz}\left(\frac{e^{zt}}{z^2+a^2}\right)_{z=0} = K\,e^{zt}\left[\frac{t}{z^2+a^2} - \frac{2z}{(z^2+a^2)^2}\right]_{z=0} \tag{9}$$

$$= Kt/a^2 = Tt/(I_1+I_2). \tag{10}$$

For the poles at $z = \pm ia$, we have the contribution

$$K\left[\frac{-e^{iat}}{2a^3i} + \frac{e^{-iat}}{2a^3i}\right] = -\frac{T\sqrt{(I_1 I_2)}}{\sqrt{\tau}(I_1+I_2)^{\frac{3}{2}}}\sin at. \tag{11}$$

Adding (10) and (11),

$$\dot{\theta}_2 = \frac{T}{(I_1+I_2)}\left[t - \sqrt{\left(\frac{I_1 I_2}{\tau(I_1+I_2)}\right)}\sin t \sqrt{\left(\frac{\tau(I_1+I_2)}{I_1 I_2}\right)}\right]. \tag{12}$$

Formula (12) shows that the angular velocity of the generator armature can be considered to have two parts, one which increases linearly with time, the other which represents an oscillation of angular frequency

$$\omega = \sqrt{\left(\frac{\tau(I_1+I_2)}{I_1 I_2}\right)}. \tag{13}$$

This oscillation is due to the interaction of the shaft torque compliance and the moments of inertia of the two armatures. Frictional and magnetic losses would damp the oscillation out rapidly. It will be understood that in practice the motor torque decreases with increase in time, ultimately assuming a value required to overcome losses and windage on no-load. If in the present case the torque were suddenly removed after a time-interval t_1, the form of T would be that shown in Fig. 66 (b). The angular velocity for $t > t_1$ would be given by (12) minus its value with $(t - t_1)$ written for t as in § 11·21.

11·41. Influence of gun recoil on the motion of an aeroplane [103]. Referring to Fig. 68 (a) a large calibre gun is mounted on the fuselage at a distance of 7 ft. behind the centre of gravity. It fires 40 shots per minute vertically upwards, the recoil per shot being 10^3 lb. wt. lasting for 0·3 sec. and causing an average force of 200 lb. wt. acting vertically downwards. Given the data below, the problem is to find θ the angular pitching velocity about the Y-axis due to gunfire, on the assumption that no corrective influence is exerted by the pilot.

Gross weight ...	...	4700 lb.
Mass	...	146·2 slugs
Span	...	38 ft.
Length	...	27 ft. 7 in.
Max. speed	...	172 m.p.h.

X, Y, Z are forces in the directions of the x-, y-, z-axes, respectively.

U, V, W are velocities in the directions of these axes; u, v, w are incremental velocities due to the disturbance.

M_u, M_w, M_q are moments about Y-axis due to velocities u, w, and to pitching, respectively.

k_B is radius of gyration about Y-axis; q is angular velocity about Y-axis due to the disturbance.

$X_u = \partial X/\partial u = -0·0739$

$X_w = \partial X/\partial w = 0·0935$

$X_q = \partial X/\partial q = 0$

$Z_u = \partial Z/\partial u = -0·323$

$Z_w = \partial Z/\partial w = -2·28$

$Z_q = \partial Z/\partial q = 0$

Z_g (vertical force due to gun)
$\quad = 200/146·2 = 1·37$

$M_u = \partial M/\partial u = 0$

$M_w = \partial M/\partial w = -1·4$

$M_q = \partial M/\partial q = -160$

M_g (moment due to gun fire)
$\quad = 1400/146·2 = 9·56$

$k_B^2 = 38·6$

U (forward velocity)
$\quad = 252$ ft. sec.$^{-1}$

In accordance with aeronautical practice the forces and moments are given per unit mass of the aeroplane, the unit of mass being 1 slug $= 32·2$ lb. ft.$^{-1}$ sec.^{2} in Britain, but 32·17 lb. ft.$^{-1}$ sec.^{2} in U.S.A.

11·42. Equations of motion. Since X_q, Z_q and M_q are zero, the equations of motion along the x-axis, the z-axis, and about the y-axis (pitching) are, respectively,

$$
\left.
\begin{aligned}
&\frac{\partial u}{\partial t} - u\frac{\partial X}{\partial u} - w\frac{\partial X}{\partial w} + g\theta = 0, \\[2mm]
&\frac{\partial w}{\partial t} - u\frac{\partial Z}{\partial u} - w\frac{\partial Z}{\partial w} - U\frac{\partial \theta}{\partial t} = Z_g \\[1mm]
&\text{(force due to gun recoil),} \\[2mm]
&-w\frac{\partial M}{\partial w} + k_B^2\frac{\partial^2\theta}{\partial t^2} - \frac{\partial M}{\partial q}\frac{\partial \theta}{\partial t} = M_g \\[1mm]
&\text{(moment due to gun recoil).}
\end{aligned}
\right\}
\tag{1}
$$

With $\theta(t) \Rrightarrow \phi(p)$, the transform version of (1) is

$$
\left.
\begin{aligned}
(p - X_u)\,u - X_w w + g\phi &= 0, \\
-Z_u u + (p - Z_w)\,w - Up\phi &= Z_g, \\
-M_w w + (k_B^2 p^2 - M_q p)\,\phi &= M_g.
\end{aligned}
\right\}
\tag{2}
$$

Expressing the solution for ϕ from (2) in determinantal form, we obtain

$$
\phi = \begin{vmatrix} 0 & p-X_u & -X_w \\ Z_g & -Z_u & p-Z_w \\ M_g & 0 & -M_w \end{vmatrix} \Bigg/ \begin{vmatrix} +g & -X_w & p-X_u \\ -Up & p-Z_w & -Z_u \\ k_B^2 p^2 - M_q p & -M_w & 0 \end{vmatrix}
\tag{3}
$$

$$
= \phi_1(p)/\phi_2(p).
\tag{4}
$$

By expanding the determinants in (3), we find that

$$
\begin{aligned}
\phi_1(p) = Z_g[M_w p - M_w X_u] \\
+ M_g[p^2 - p(X_u + Z_w) + (X_u Z_w - Z_u X_w)],
\end{aligned}
\tag{5}
$$

$$
\begin{aligned}
\phi_2(p) = p(p - X_u)\,[(p - Z_w)\,(k_B^2 p - M_q) - M_w U] \\
+ Z_u[M_w g - X_w(k_B^2 p^2 - M_q p)].
\end{aligned}
\tag{6}
$$

Substituting the given numerical values in (5) and (6), we get

$$
\phi_1(p) = 9{\cdot}56p^2 + 20{\cdot}6p + 1{\cdot}76
\tag{7}
$$

and $\quad \phi_2(p) = 38{\cdot}6p^4 + 250{\cdot}8p^3 + 736{\cdot}3p^2 + 57{\cdot}8p + 14{\cdot}54.$ $\qquad$ (8)

The roots of $\phi_2(p)$, which must have negative real parts for stability of the aeroplane, are, approximately,

$$\alpha = -3{\cdot}21 + 2{\cdot}89i,$$

$$\beta = -3{\cdot}21 - 2{\cdot}89i,$$

$$\gamma = -0{\cdot}037 + 0{\cdot}14i,$$

and $\quad\quad\quad\quad \delta = -0{\cdot}037 - 0{\cdot}14i.$

In solving (8) to obtain these roots, the accuracy of γ and δ depends upon small differences between relatively large quantities. To get more accurate values it is necessary to use seven-figure logarithms. The present numerical values will serve, however, to illustrate the analysis.

By the Mellin theorem, the pitching angle is

$$\theta = \frac{1/38{\cdot}6}{2\pi i} \int_{Br_1} \frac{e^{zt}(9{\cdot}56z^2 + 20{\cdot}6z + 1{\cdot}76)\,dz}{z(z-\alpha)(z-\beta)(z-\gamma)(z-\delta)}. \tag{9}$$

The evaluation at each of the five poles $z = 0, \alpha, \beta, \gamma, \delta$ is effected in the usual manner. From (8) the product $38{\cdot}6\alpha\beta\gamma\delta$ is obviously $14{\cdot}54$, so the residue at $z = 0$ is $1{\cdot}76/14{\cdot}54 \simeq 0{\cdot}121$. The approximate final result is

$$\theta = 0{\cdot}121 - e^{-3{\cdot}21t}(0{\cdot}004\cos 2{\cdot}89t + 0{\cdot}014\sin 2{\cdot}89t)$$

$$- e^{-0{\cdot}037t}(0{\cdot}12\cos 0{\cdot}14t - 0{\cdot}17\sin 0{\cdot}14t). \tag{10}$$

When t is small we take z large in accordance with §§ 4·61, 4·62 and (9) gives

$$\theta \simeq \frac{1/38{\cdot}6}{2\pi i} \int_{Br_1} \frac{e^{zt}\,9{\cdot}56dz}{z^3}$$

$$= 0{\cdot}124t^2, \tag{11}$$

so the angular displacement-time curve is parabolic at the start. From (10) the angular pitching velocity is

$$\dot\theta = \frac{d\theta}{dt} = e^{-3{\cdot}21t}[-0{\cdot}028\cos 2{\cdot}89t + 0{\cdot}056\sin 2{\cdot}89t]$$

$$+ e^{-0{\cdot}037t}[0{\cdot}028\cos 0{\cdot}14t + 0{\cdot}01\sin 0{\cdot}14t]. \tag{12}$$

The angular pitching velocity $\dot{\theta}$ attains a maximum value as shown in Fig. 68 (b). There are two components, that having the smaller damping being the more important. It is preferable that the motion should be aperiodic. Although this is not so, the

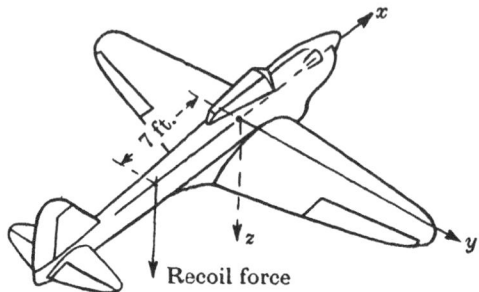

Fig. 68(a). Diagrammatic view of aeroplane and principal axes.

Fig. 68(b).

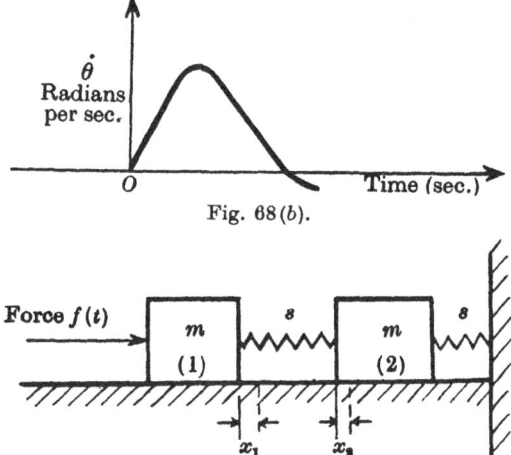

Fig. 68(c). Schematic diagram of a pair of masses m, on smooth horizontal plane, coupled by springs of stiffness s, and driven by a force $f(t)$ applied to the first mass. The motion is along a common linear axis.

damping is such that the recoil oscillation has subsided about one second after the shot is fired. The final path of the aeroplane differs according as the gun is fired or not. But the accuracy of aim depends upon the rate of change in the direction of motion, i.e. upon $\dot{\theta}$, and this is probably too small to have any appreciable effect.

11·51. Deflexion of beams. The standard method given in engineering and other texts for over half a century is often simpler and more direct than that using Laplace transform. However, we shall give two examples illustrating the latter method. Additional applications will be found in [269, 270].

The differential equation for the downward deflexion ξ at any point x on a horizontal uniform beam carrying a load $w(x)$—a function of x having a transform—is

$$EI\, d^4\xi/dx^4 = w(x), \qquad (1)$$

the modulus of elasticity E, and the appropriate moment of inertia of the cross-section I, being constant. Here we use x for t, and take $\xi(x) \Rightarrow \phi(p)$, $w(x) \Rightarrow \psi(p)$. The four conditions at $x = 0$, namely,

$$\xi(0) = \xi_0, \quad \xi'(0) = \xi_1, \quad \ldots, \quad \xi'''(0) = \xi_3,$$

replace those corresponding to $t = 0$. Only two of these will be *known*. The other two are derived from the boundary conditions, usually those at the end of the beam remote from $x = 0$, after the formal solution for $\xi(x)$ has been obtained.

The transform equation for (1) is by §9·21

$$p^4\phi = (\psi/EI) + p^4\xi_0 + p^3\xi_1 + p^2\xi_2 + p\xi_3, \qquad (2)$$

so

$$\phi = (\psi/p^4EI) + \xi_0 + \xi_1/p + \xi_2/p^2 + \xi_3/p^3. \qquad (3)$$

11·52. Example. Determine $\xi(x)$ for a horizontal uniform beam length l hinged freely* at each end, and carrying a uniformly distributed load w per unit length.

The known conditions at $x = 0$ are $\xi_0 = 0$ (zero deflexion), $\xi_2 = 0$ (zero bending moment), while $\psi = $ w, a constant. Then (3) §11·51 gives

$$\phi = (\text{w}/p^4EI) + \xi_1/p + \xi_3/p^3, \qquad (1)$$

so by inversion

$$\xi(x) = (\text{w}/24EI)\, x^4 + \xi_1 x + \xi_3 x^3/6. \qquad (2)$$

We now use the boundary conditions at $x = l$ to determine the unknowns ξ_1, ξ_3. Then $\xi(l) = 0$, and $\xi''(l) = 0$, so from (2) we get

$$(\text{w}/24EI)\, l^3 + \xi_3 l^3/6 + \xi_1 = 0, \qquad (3)$$

and

$$\xi_3 = -(\text{w}/2EI)\, l. \qquad (4)$$

(3) and (4) yield

$$\xi_1 = (\text{w}/24EI)\, l^3. \qquad (5)$$

Substituting from (4), (5) into (2) leads to

$$\xi(x) = (\text{w}/24EI)\{x(x^3 - 2x^2l + l^3)\}. \qquad (6)$$

* Horizontally movable hinges are implied so that the reactions at the supports are always vertical and statically determinate.

For the central deflexion (6) gives the well-known result

$$\xi(\tfrac{1}{2}l) = 5Wl^3/384EI, \tag{7}$$

$W = wl$ being the total load.

11·53. Example. Determine $\xi(x)$ for a uniform horizontal beam length l built in at both ends and carrying concentrated loads W_1, W_2 at $x = h_1, h_2$, respectively.

In a short range $x = (h_1, h_1 + \Delta x)$, suppose that the load is uniformly distributed, such that $w_1 \Delta x = W_1$, w_1 being the load per unit length. Then

$$w_1(x) = w_1[H(x - h_1) - H(x - h_1 - \Delta x)] \Rightarrow W_1 e^{-ph_1}(1 - e^{-p\Delta x})/\Delta x, \tag{1}$$

and with $w_1(x) \Rightarrow \psi_1(p)$, we get

$$\psi_1 = \frac{W_1 e^{-ph_1}}{\Delta x}\left[p\Delta x - \frac{(p\Delta x)^2}{2!} + \dots\right]. \tag{2}$$

When $\Delta x \to 0$, $\psi_1 = W_1 p\, e^{-ph_1} \Leftarrow W_1 I(x - h_1),$ (3)

so the concentrated load is equivalent analytically to W_1 times the impulsive function* at $x = h_1$. A similar argument applies to W_2 at $x = h_2$.

The known conditions at $x = 0$ are $\xi_0 = 0$ (zero deflexion), $\xi_1 = 0$ (zero slope), while $\psi = W_1 p e^{-ph_1} + W_2 p e^{-ph_2}$. Substituting into (3) §11·51, gives

$$\phi = \{(W_1 e^{-ph_1} + W_2 e^{-ph_2})/p^3 EI\} + \xi_2/p^2 + \xi_3/p^3. \tag{4}$$

By inversion

$$\xi(x) = (1/6EI)\{W_1(x - h_1)^3 H(x - h_1) + W_2(x - h_2)^3 H(x - h_2)\}$$
$$+ \xi_2 x^2/2 + \xi_3 x^3/6. \tag{5}$$

To determine ξ_2, ξ_3 we use the boundary conditions $\xi(l) = 0$, $\xi'(l) = 0$. Thus from (5), we have

$$(1/3EI)\{W_1(l - h_1)^3 + W_2(l - h_2)^3\} + \xi_2 l^2 + \tfrac{1}{3}\xi_3 l^3 = 0, \tag{6}$$

and $(1/2EI)\{W_1(l - h_1)^2 + W_2(l - h_2)^2\} + \xi_2 l + \tfrac{1}{2}\xi_3 l^2 = 0.$ (7)

Solving for ξ_2, ξ_3 yields

$$\xi_2 = (1/l^2 EI)\{W_1 h_1(l - h_1)^2 + W_2 h_2(l - h_2)^2\}, \tag{8}$$

and $\xi_3 = -(1/l^3 EI)\{W_1(l - h_1)^2(l + 2h_1) + W_2(l - h_2)^2(l + 2h_2)\}.$ (9)

Substituting from (8), (9) into (5), and rearranging, we obtain finally

$$\xi(x) = (W_1/6EI)\{(x - h_1)^3 H(x - h_1) + x^2(l - h_1)^2[3h_1 l - x(l + 2h_1)]/l^3\}$$
$$+ (W_2/6EI)\{(x - h_2)^3 H(x - h_2) + x^2(l - h_2)^2[3h_2 l - x(l + 2h_2)]/l^3\}. \tag{10}$$

* The variable is x, not t, so an impulsive *blow* is not in question.

It is seen that the two expressions for the deflexions due to W_1, W_2 are identical in form. This follows from the fact that for small ξ the loaded beam is a linear system, so the deflexions due to individual loads may be superimposed. For loads $W_3, \ldots,$ at $h_3, \ldots,$ it is necessary merely to add expressions as in (10), using the appropriate subscripts.

If $W_1 = W$, $W_2 = 0$, $h_1 = \frac{1}{2}l$, (10) degenerates to the well-known formula for the central deflexion, namely,

$$\xi(\tfrac{1}{2}l) = Wl^3/192EI. \tag{11}$$

11·61. Mechanical system having two degrees of freedom.

Fig. 68 (c) represents a loss-free mechanical system comprising identical masses m and coupling springs of negligible mass, each having stiffness s. Considered separately, each (m, s) pair has a natural angular frequency $\omega = \sqrt{(s/m)}$, there being one degree of freedom, since the motion of m may be specified by one co-ordinate only, i.e. x. But when the two pairs are coupled together as shown, there are two degrees of freedom, two coordinates x_1, x_2, being needed, one for each mass. One mass cannot move without affecting the other, owing to the coupling springs. There are now *two* natural frequencies ω_1, ω_2, both of which differ from ω, and as will be shown later, $\omega_1 > \omega > \omega_2$. When the system oscillates freely after being disturbed, the motion of either mass is a function of ω, ω_1 and ω_2.

The differential equations for the respective masses are:

$$m\ddot{x}_1 + s(x_1 - x_2) = f(t), \tag{1}$$

and
$$m\ddot{x}_2 + s(x_2 + x_2 - x_1) = 0. \tag{2}$$

Taking

$$x_1(t) \Rightarrow \phi_1(p), \quad x_2(t) \Rightarrow \phi_2(p), \quad f(t) \Rightarrow \phi(p), \quad \omega^2 = s/m,$$

and assuming quiescence initially, the transform equations corresponding to (1), (2) are

$$p^2\phi_1 + \omega^2(\phi_1 - \phi_2) = \phi/m, \tag{3}$$

and
$$p^2\phi_2 + \omega^2(2\phi_2 - \phi_1) = 0. \tag{4}$$

From (4)
$$\phi_2 = \omega^2\phi_1/(p^2 + 2\omega^2), \tag{5}$$

and on inserting this into (3), we get

$$[(p^2+\omega^2)-\omega^4/(p^2+2\omega^2)]\,\phi_1 = \phi/m, \tag{6}$$

so

$$\phi_1 = \phi(p^2+2\omega^2)/m(p^4+3p^2\omega^2+\omega^4). \tag{7}$$

Now $(p^4+3p^2\omega^2+\omega^4)$ may be written $(p^2+\omega_1^2)\,(p^2+\omega_2^2)$, where ω_1, ω_2 are the natural frequencies, so we have to find the roots of

$$p^4+3p^2\omega^2+\omega^4 = 0. \tag{8}$$

Thus

$$p^2 = -\tfrac{1}{2}(3\pm\sqrt{5})\,\omega^2, \tag{9}$$

so

$$\omega_1^2 = \tfrac{1}{2}(3+\sqrt{5})\,\omega^2, \quad \omega_2^2 = \tfrac{1}{2}(3-\sqrt{5})\,\omega^2, \tag{10}$$

which gives

$$\omega_1 = 1{\cdot}618\omega, \quad \omega_2 = 0{\cdot}618\omega, \tag{11}$$

and, therefore,

$$\omega_1 > \omega > \omega_2. \tag{12}$$

11·62. Displacement of first mass. From (7) §11·61

$$\phi_1 = \frac{\phi(p^2+2\omega^2)}{m(p^2+\omega_1^2)\,(p^2+\omega_2^2)}, \tag{1}$$

and if we take $f(t) = SI(t)$, the impulse function of strength S—which has dimensions (force × time)—we get

$$\phi_1 = \frac{(S/m)\,p(p^2+2\omega^2)}{(p^2+\omega_1^2)\,(p^2+\omega_2^2)}. \tag{2}$$

Inversion of (2). By the Mellin theorem,

$$x_1 = \frac{(S/m)}{2\pi i}\int_{Br_1}\frac{e^{zt}\,(2\omega^2+z^2)\,dz}{(z-i\omega_2)(z+i\omega_2)(z-i\omega_1)(z+i\omega_1)}. \tag{3}$$

The contribution from the pole at $z = i\omega_2$ is

$$\frac{(S/m)\,(2\omega^2-\omega_2^2)\,e^{i\omega_2 t}}{2i\omega_2(\omega_1^2-\omega_2^2)}. \tag{4}$$

For the pole at $z = -i\omega_2$, change i to $-i$ in (4), and we get

$$-\frac{(S/m)\,(2\omega^2-\omega_2^2)\,e^{-i\omega_2 t}}{2i\omega_2(\omega_1^2-\omega_2^2)}. \tag{5}$$

Adding (4), (5) yields for the poles at $z = \pm i\omega_2$

$$\frac{(S/m)\,(2\omega^2-\omega_2^2)}{\omega_2(\omega_1^2-\omega_2^2)}\sin\omega_2 t. \tag{6}$$

The contribution from the poles at $z = \pm i\omega_1$ may be obtained from (6) by interchanging ω_1, ω_2, which gives

$$-\frac{(S/m)\,(2\omega^2 - \omega_1^2)}{\omega_1(\omega_1^2 - \omega_2^2)}\sin\omega_1 t. \tag{7}$$

The sum of (6) and (7) is the value of (3), so the displacement of the first mass is

$$x_1 = \frac{(S/m)}{(\omega_1^2 - \omega_2^2)}\left[\left(\frac{2\omega^2 - \omega_2^2}{\omega_2}\right)\sin\omega_2 t - \left(\frac{2\omega^2 - \omega_1^2}{\omega_1}\right)\sin\omega_1 t\right]. \tag{8}$$

(8) represents a combination of two harmonic motions having frequencies ω_1, ω_2. Since $\omega_1 > \omega_2$, the component of lower frequency has the greater amplitude.

As an exercise the reader may verify that the displacement of the second mass is

$$x_2 = \frac{(S/m)\,\omega^2}{(\omega_1^2 - \omega_2^2)}\left[\frac{\sin\omega_2 t}{\omega_2} - \frac{\sin\omega_1 t}{\omega_1}\right]. \tag{9}$$

By aid of (11) § 11·61 we find that

$$(\omega_1^2 - \omega_2^2) = 2\cdot236\omega^2, \quad (2\omega^2 - \omega_1^2) = -0\cdot618\omega^2,$$
$$(2\omega^2 - \omega_2^2) = 1\cdot618\omega^2. \tag{10}$$

Substituting from (11) § 11·61 and (10) above into (8) leads to

$$x_1 = (S/\omega m)\,[(1\cdot171\sin(0\cdot618\omega t) + 0\cdot171\sin(1\cdot618\omega t)], \tag{11}$$

so the ratio of the two amplitudes is about 6·85/1.

Finally we remark that the uniform band frequency spectrum of the applied impulse $SI(t)$—see § 10·52—has been converted into a spectrum having two lines at ω_1, ω_2, the natural frequencies of the system. For a loss-free system having n degrees of freedom, there would be n lines, one corresponding to each natural frequency.

XII

RADIO AND TELEVISION RECEIVERS

12·11. Radio receiver. Fig. 69 (a) is a diagram of the audio-frequency stages of a radio-broadcasting receiver. The detector valve (not shown) passes on a p.d. $-E_g$ to the grid and cathode of the amplifying valve V_1. The latter is resistance-capacitance coupled to the power or output valve V_2, from which the loud speaker is operated via a step-down transformer T. Find the p.d. across the grid and cathode of V_2 when $-E_g$ has the form of Heaviside's unit function, the circuit being quiescent initially.

12·12. Simplified circuit. First we draw the equivalent circuit Fig. 69 (b), in which R_a is the anode resistance of V_1, $mE_g = E$ is the p.d. in the equivalent anode circuit, m being the magnification factor of the valve. The positive sign is due to a phase change of π introduced by the valve. Capacitances associated with the input circuits to V_1 and V_2 are neglected in order to avoid complicating the analysis at this stage. The p.d. across AB is applied to the branches R_1 and C_1R_2 in parallel, so with $I \Rightarrow \phi$, $I_1 \Rightarrow \phi_1$, $I_2 \Rightarrow \phi_2$

$$E - R_a\phi = R_1\phi_1 = (R_2 + 1/pC_1)\,\phi_2. \tag{1}$$

The current through the valve is equal to the sum of the two branch currents, so

$$I = I_1 + I_2, \tag{2}$$

and, therefore, $\phi = \phi_1 + \phi_2$.
From (1)

$$\phi = (E/R_a) - (1 + pC_1R_2)\,\phi_2/pC_1R_a, \tag{3}$$

and $\quad \phi_1 = (1 + pC_1R_2)\,\phi_2/pC_1R_1. \tag{4}$

Substituting from (3), (4) into (2), we get

$$\phi_2\left\{1 + \frac{1 + pC_1R_2}{pC_1}\left(\frac{R_1 + R_a}{R_1R_a}\right)\right\} = \frac{E}{R_a}, \tag{5}$$

which after a little reduction gives

$$\phi_2 = Ep/\alpha(p+\beta), \tag{6}$$

where $\quad \alpha = R_a + R_2(1 + R_a/R_1),$

and $\quad \beta = (1 + R_a/R_1)/C_1[R_a + R_2(1 + R_a/R_1)].$

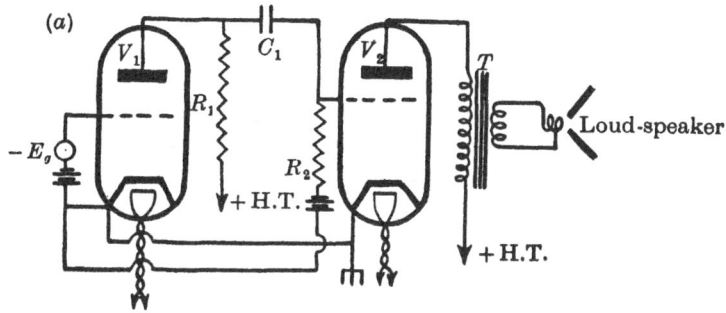

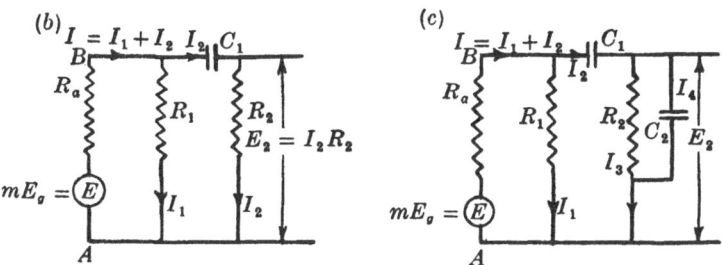

Fig. 69. (a) Audio-frequency part of radio receiver. (b) Simplified circuit equivalent to (a) when ω is not too large. (c) Circuit equivalent to (a) for audible frequencies.

(6) may be inverted using (10) Appendix 10. Alternatively, by the Mellin theorem, the current

$$I_2 = \frac{E}{\alpha} \frac{1}{2\pi i} \int_{Br_1} \frac{e^{zt}\,dz}{(z+\beta)} = \frac{E\,e^{-\beta t}}{\alpha}. \tag{7}$$

Hence the p.d. across the input to V_2 in Fig. 69 (a) is

$$E_2 = I_2 R_2 = \frac{E\,e^{-(1+R_a/R_1)\,t/C_1[R_a+R_2(1+R_a/R_1)]}}{(1+R_a/R_1 + R_a/R_2)}. \tag{8}$$

From (8), when the p.d. $-E_g$ is applied to V_1, that to V_2 immediately assumes the value

$$E_2 = E/(1 + R_a/R_1 + R_a/R_2)$$

(see Fig. 70 (a)). E_2 then starts to decay exponentially due to the charging of capacitance C_1 through the resistances R_a, R_2, in which process R_1 also plays an incidental part. The larger C_1 the more slowly does the p.d. E_2 decay with increase in time, so that if C_1 were large enough, the extremely low audio-frequencies (bass register) would be amplified adequately. The reverse would be true if C_1 were small enough to cause a rapid decay in E_2. The precipitate rise at $t = 0$ signifies that high audio frequencies would be amplified adequately.

12·21. Inclusion of input impedance of V_2. To effect simplification, the input impedance of V_2 was ignored in the preceding analysis, but in practice it is an all-important item, and one which determines the upper frequency response of the audio amplifier. The impedance in question would be high enough to be left out of account were it not for capacitance of the grid to cathode of V_2 plus wiring, and also the feeding back of energy from the anode to the grid circuit. With a screened-grid valve, where the grid to anode capacitance is a tiny fraction of a micro-microfarad, the feed-back effect in audio-frequency amplifiers is negligible. At the moment we are concerned with triodes, and the input of V_2 alone can be represented approximately by a capacitance C_2 as shown in Fig. 69 (c).

To determine I_2 for an impressed p.d. $-E_g$ as before, we use Fig. 69 (c). The circuital equations, however, may be derived from (5) § 12·12 if R_2 is replaced by the transform of R_2 and C_2 in parallel, namely, $R_2/(1 + pC_2R_2)$ (see [236], p. 59). This gives

$$\phi_2\left[1 + \frac{1 + pR_2(C_1 + C_2)}{pC_1(1 + pC_2R_2)}\left(\frac{R_1 + R_a}{R_1R_a}\right)\right] = \frac{E}{R_a}. \tag{1}$$

The input p.d. to V_2 is now I_3R_2, so we have to find I_3. With $I_3 \Rightarrow \phi_3$, $I_4 \Rightarrow \phi_4$,

$$\phi_3 R_2 = \phi_4/pC_2, \quad \text{or} \quad \phi_4 = pC_2R_2\phi_3, \tag{2}$$

and

$$\phi_2 = \phi_3 + \phi_4. \tag{3}$$

From (2) and (3) $\phi_3 = \phi_2/(1 + pC_2R_2).$ (4)

Using (4) in (1), we find after some reduction that

$$E_2 = I_3R_2 \Rightarrow (E/\gamma)\,p/(p^2 + \delta p + \theta),\qquad (5)$$

where $\gamma = C_2R_a;$

$$\delta = [(R_1 + R_a)(C_1 + C_2)/C_1C_2R_1R_a] + 1/C_2R_2,$$

and $\theta = (R_1 + R_a)/C_1C_2R_1R_2R_a.$

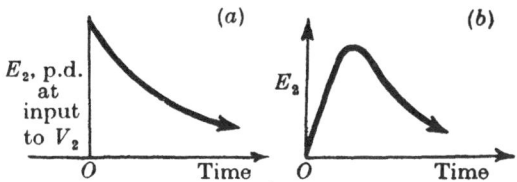

Fig. 70. The initial rate of rise in (b) is E/γ.

If $-\xi$ and $-\psi$ are the roots of $p^2 + \delta p + \theta = 0$, then $\delta = (\xi + \psi)$, $\theta = \xi\psi$, and (5) may be written

$$E_2 \Rightarrow \frac{E}{\gamma(\psi - \xi)}\left[\frac{p}{p+\xi} - \frac{p}{p+\psi}\right].\qquad (6)$$

Inverting (6) by (10) Appendix 10 gives

$$E_2 = \frac{E}{\gamma(\psi - \xi)}(e^{-\xi t} - e^{-\psi t}).\qquad (7)$$

Alternatively, by applying the Mellin theorem to (5), we get

$$E_2 = \frac{E/\gamma}{2\pi i}\int_{Br_1}\frac{e^{zt}\,dz}{(z + \xi)(z + \psi)}\qquad (8)$$

which yields the right-hand side of (7). For stability of the amplifier ξ, ψ must be real and > 0.

12·22. Initial shape of input p.d. to V_2. To determine the form of E_2 near the origin of time, we make z large in (8) § 12·21, in accordance with §§ 4·61, 4·62, and neglect ξ, ψ, so

$$E_2 \simeq \frac{E/\gamma}{2\pi i}\int_{Br_1}\frac{e^{zt}\,dz}{z^2} = \frac{Et}{\gamma}.\qquad (1)$$

Hence the input to V_2 starts to rise linearly with time, and not precipitately in accordance with the p.d. $-E_g$ at $t = 0$ (see Fig. 70 (b)). Consequently the influence of the input capacitance C_2 is to retard the initiation of the response, thereby introducing distortion. The impedance of C_2 decreases with rise in frequency $(Z = 1/\omega C_2)$, and at high audio frequency, the shunting effect of C_2 on R_2 causes a reduction in amplification. As shown in § 10·514, the applied p.d. $-E_g$ comprises an infinite frequency spectrum ranging from zero to infinity. Owing to the fall in amplification with rise in frequency, the initial rate of rise of p.d. is reduced.

12·3. Differentiating (7) § 12·21 with respect to t, the maximum value of E_2 occurs when

$$t = \frac{1}{(\psi - \xi)} \log \frac{\psi}{\xi}. \tag{1}$$

The form of the E_2/time curve is illustrated in Fig. 70 (b) from which it is evident (on comparing the initial slope with that in Fig. 70 (a)) that C_2 must be as small as possible if transients with higher frequency components are to be amplified with negligible distortion.

12·41. Sinusoidal input p.d. to V_1. We come now to the case where mE_g in Fig. 69 (c) takes the form $E_0 \sin \omega t$. The transform solution is obtained if E in (5) § 12·21 is replaced by the transform of $E_0 \sin \omega t$, namely $E_0 \omega p/(p^2 + \omega^2)$. Thus by the Mellin theorem

$$E_2 = \frac{(E_0 \omega/\gamma)}{2\pi i} \int_{Br_1} \frac{e^{zt} z \, dz}{(z^2 + \omega^2)(z + \xi)(z + \psi)}. \tag{1}$$

The integrand of (1) has four simple poles at $z = \pm i\omega, -\xi, -\psi$. Evaluating at these poles in accordance with § 3·21, we obtain

$$E_2 = \frac{(E_0 \omega/\gamma)}{(\xi - \psi)} \left[\frac{\xi e^{-\xi t}}{(\omega^2 + \xi^2)} - \frac{\psi e^{-\psi t}}{(\omega^2 + \psi^2)} \right]$$

$$+ (E_0 \omega/\gamma) \left[\frac{i\omega e^{i\omega t}}{2i\omega(\xi + i\omega)(\psi + i\omega)} + \frac{(-i\omega) e^{-i\omega t}}{(-2i\omega)(\xi - i\omega)(\psi - i\omega)} \right]$$

$$= \text{transient p.d.} + \text{steady p.d.} \tag{2}$$

The second part of (2) gives for the sine-wave p.d.,

$$E_{2s} = \frac{(E_0\omega/\gamma)\left[(\xi\psi - \omega^2)\cos\omega t + \omega(\psi + \xi)\sin\omega t\right]}{(\xi^2 + \omega^2)(\psi^2 + \omega^2)} \quad (3)$$

$$= \frac{(E_0\omega/\gamma)\sin(\omega t + \alpha)}{\sqrt{\{(\omega^2 + \xi^2)(\omega^2 + \psi^2)\}}}, \quad (4)$$

where $\tan\alpha = (\xi\psi - \omega^2)/\omega(\psi + \xi)$. When $t = 0$, the sum of (4) and the first term in (2) is zero, as it should be. The transient p.d. is given by the sum of two exponential terms, and it rises precipitately to the value

$$\frac{E_0\omega(\omega^2 - \xi\psi)}{(\omega^2 + \xi^2)(\omega^2 + \psi^2)\gamma}$$

at $t = 0$. To offset this, the sine-wave component is displaced by a phase angle α. From § 12·21 we have $\delta = (\xi + \psi)$ and $\theta = \xi\psi$, so from above

$$\tan\alpha = \frac{-\omega^2 + \theta}{\omega\delta} = \frac{-\omega^2 C_1 C_2 R_1 R_2 R_a + R_1 + R_a}{\omega[(R_1 + R_a)(C_1 + C_2)R_2 + C_1 R_1 R_a]}. \quad (5)$$

At very low frequencies the phase angle is positive; it vanishes with the numerator, while at moderate and high audio frequencies it is negative. When $C_1 = 0$, α is positive. In practice $C_1 \gg C_2$, $R_2 \gg R_1$ and R_a, and when ω is large enough at the higher audio frequencies, (5) may be written with adequate accuracy

$$\tan\alpha = -\omega C_2 \bigg/ \left(\frac{1}{R_1} + \frac{1}{R_a}\right) = -\omega C_2\left(\frac{R_1 R_a}{R_1 + R_a}\right), \quad (6)$$

the bracketed factor representing the resistance of R_1 and R_a in parallel. From (6) the phase angle increases numerically with rise in frequency. When C_2 is zero or small enough to be neglected, e.g. if V_2 is a screened-grid valve, the phase angle can be ignored.

12·42. Influence of C_2 upon amplification.

We note that the denominator of (4) § 12·41 may be expressed in the form

$$K = \gamma\sqrt{\{\omega^2(\psi + \xi)^2 + (\omega^2 - \xi\psi)^2\}} = \gamma\sqrt{\{\omega^2\delta^2 + (\omega^2 - \theta)^2\}} \quad (1)$$

$$= \omega R_a\sqrt{\left\{\left(\frac{1}{R_1} + \frac{1}{R_a}\right)^2 + \omega^2 C_2^2\right\}}, \quad (2)$$

when the foregoing approximations and conditions at the higher audio frequencies are used. Then (4) § 12·41 may be written

$$E_{2s} = E_0 \sin(\omega t + \alpha) \Big/ R_a \sqrt{\left\{ \left(\frac{1}{R_1} + \frac{1}{R_a} \right)^2 + \omega^2 C_2^2 \right\}}, \qquad (3)$$

which indicates that the input p.d. to V_2 and, therefore, the amplification due to V_1, falls away as the frequency rises. If $\omega^2 C_2^2$ is negligible, $\alpha = 0$ and (3) degenerates to

$$E_{2s} = \frac{E_0 \sin \omega t}{1 + (R_a/R_1)}. \qquad (4)$$

The amplification due to V_1 is now given by $m/[1 + (R_a/R_1)]$, whereas with C_2 present, it is $m \Big/ R_a \sqrt{\left\{ \left(\frac{1}{R_1} + \frac{1}{R_a} \right)^2 + \omega^2 C_2^2 \right\}}$, and decreases with rise in frequency.

12·51. Output transformer of radio receiver [123].

The circuit to be analysed is illustrated in Fig. 71 (a), while the equivalent circuit is shown in Fig. 71 (b). To effect simplification we shall confine our attention to moderate and high audio frequencies, where to a first approximation the loud-speaker impedance can be represented by a resistance in series with an inductance.

In Fig. 71 (b), R_a = anode resistance of power valve plus A.C. resistance of transformer primary; L_1, L_2 = inductance of primary and secondary windings, respectively; R = resistance of speaker in operation plus A.C. resistance of secondary; L = inductance of speaker plus effect of magnetic leakage which is assumed small; M = mutual inductance between L_1 and L_2; I_1, I_2 = currents in windings. The self and mutual capacitances of the windings are taken to be negligible owing to sectionization.

12·52. Input p.d. of form $H(t)$.

First we consider the case where the input E_g has the form of Heaviside's unit function. The problem is to find an expression for the secondary current at any time $t>0$, the circuit being quiescent initially. With $I_1 \Rightarrow \phi_1$, $I_2 \Rightarrow \phi_2$, for the primary circuit we have

$$pL_1\phi_1 + pM\phi_2 + R_a\phi_1 = -E; \tag{1}$$

and for the secondary circuit

$$p(L_2+L)\phi_2 + pM\phi_1 + R\phi_2 = 0. \tag{2}$$

From (2) $\qquad \phi_1 = -[R+p(L+L_2)]\phi_2/pM. \tag{3}$

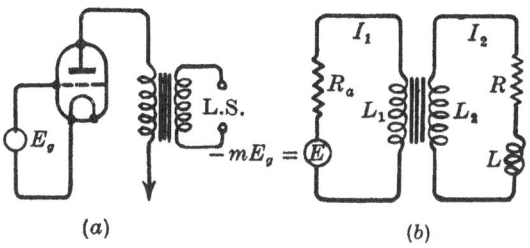

(a) (b)

Fig. 71. (a) Output circuit of radio receiver.
(b) Equivalent circuit of (a).

Substituting from (3) into (1), we get

$$pM\phi_2 - (pL_1+R_a)[R+p(L+L_2)]\phi_2/pM = -E. \tag{4}$$

By hypothesis the magnetic leakage is negligible, so $M^2 = L_1L_2$, and after reduction (4) yields

$$\phi_2 = \frac{E(n_2/n_1)p}{L(p+a)(p+b)}. \tag{5}$$

where $n_2/n_1 = \sqrt{(L_2/L_1)}$ = ratio of secondary to primary turns, and

$$\left.\begin{array}{c}a\\b\end{array}\right\} = \frac{\alpha}{2} \pm \sqrt{\{(\tfrac{1}{4}\alpha^2) - \beta\}}, \quad \alpha = \frac{R}{L} + \frac{R_a}{L_1} + \frac{R_aL_2}{LL_1}, \quad \beta = RR_a/LL_1.$$

Resolving (5) into partial fractions, we get

$$\phi_2 = \frac{E(n_2/n_1)}{(a-b)L}\left[\frac{p}{p+b} - \frac{p}{p+a}\right], \tag{6}$$

and so, by (10) Appendix 10

$$I_2 = \frac{E(n_2/n_1)}{(a-b)L}(e^{-bt} - e^{-at}). \tag{7}$$

Alternatively, by the Mellin theorem, (5) gives

$$I_2 = \frac{E(n_2/n_1)}{L} \frac{1}{2\pi i} \int_{Br_1} \frac{e^{zt}\,dz}{(z+a)\,(z+b)} \qquad (8)$$

$$= \frac{E(n_2/n_1)}{L} \left[\frac{e^{-at}}{b-a} + \frac{e^{-bt}}{a-b} \right] \qquad (9)$$

$$= \frac{E(n_2/n_1)}{(a-b)\,L} [e^{-bt} - e^{-at}], \qquad (10)$$

as at (7).

12·53. To illustrate the type of curve represented by (10) § 12·52 we take the following practical data and calculate the secondary current I_2: $R_a = 10^3$ ohms; $R = 10$ ohms; $L_1 = 20$ henrys; $L_2 = 0·08$ henry; $L = 10^{-3}$ henry; $n_2/n_1 = 16$. Inserting these numerical values in (10) § 12·52, we obtain

$$I_2 = 4·52 \times 10^{-3} E(e^{-35·7t} - e^{-1·4 \times 10^4 t}), \qquad (1)$$

a relationship shown graphically in Fig. 72 (a). The secondary current is not identical in form with that of the p.d. applied to the grid of the power valve. It rises to a maximum after a time lapse of $4·27 \times 10^{-4}$ sec., and then decays asymptotically to zero according to the law $e^{-35·7t}$.

12·61. Input p.d. of form $E_0 e^{-\alpha t} \sin \omega t$. To simplify the analysis it is preferable to discuss first the case when L is zero. For the unit function type of p.d., (4) § 12·52 gives

$$\phi_2 = pME \Big/ (R_a L_2 + RL_1) \left(p + \frac{RR_a}{R_a L_2 + RL_1} \right). \qquad (1)$$

The transform solution of the present problem is obtained if E is replaced by the transform of $E_0 e^{-\alpha t} \sin \omega t$. Thus we get

$$\phi_2 = Kp^2/(p+c)\,[(p+\alpha)^2 + \omega^2], \qquad (2)$$

where $K = \dfrac{\omega M E_0}{R_a L_2 + RL_1} = \dfrac{\omega E_0 (n_2/n_1)}{R + R_a (n_2/n_1)^2}$ and $c = \dfrac{RR_a}{R_a L_2 + RL_1}$.

Inverting (2), we obtain

$$I_2 = \frac{K}{2\pi i}\int_{Br_1} \frac{e^{zt} z\,dz}{(z+c)(z+\alpha-i\omega)(z+\alpha+i\omega)} \tag{3}$$

$$= K\left[\frac{-c\,e^{-ct}}{(\alpha-c)^2+\omega^2}+\frac{e^{(i\omega-\alpha)t}(i\omega-\alpha)}{2i\omega(c-\alpha+i\omega)}+\frac{e^{-(i\omega+\alpha)t}(i\omega+\alpha)}{2i\omega(c-\alpha-i\omega)}\right] \tag{4}$$

$$= \frac{K/\omega}{\omega^2+(c-\alpha)^2}\{\sqrt{(A^2+B^2)}\,e^{-\alpha t}\sin[\omega t+\tan^{-1}(B/A)]-\omega c\,e^{-ct}\},$$

$$\tag{5}$$

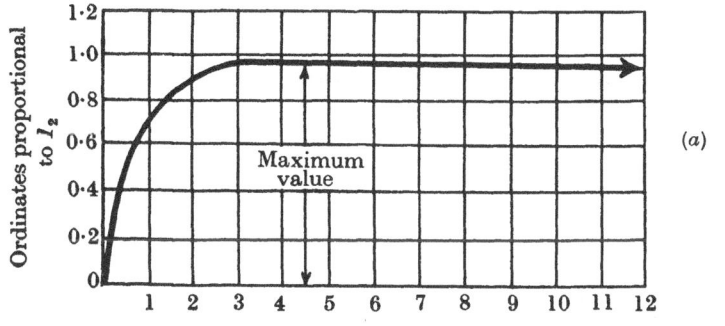

Fig. 72 (a).

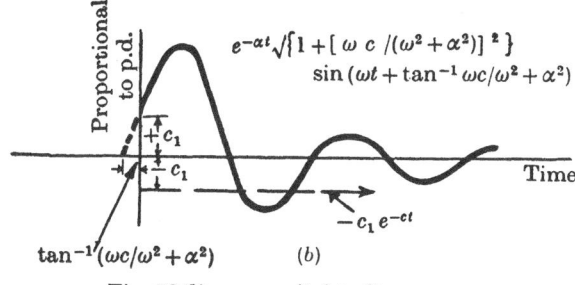

Fig. 72 (b). $c_1 = \omega c/(\omega^2+\alpha^2)$.

where $A = (\alpha^2+\omega^2-\alpha c)$, and $B = \omega c$. If $\alpha \gg c$, this being a likely practical condition, (5) becomes

$$(K/\omega)\left[\sqrt{\left\{1+\left(\frac{\omega c}{\omega^2+\alpha^2}\right)^2\right\}}\,e^{-\alpha t}\sin\left\{\omega t+\tan^{-1}\frac{\omega c}{\omega^2+\alpha^2}\right\}-\frac{\omega c\,e^{-ct}}{\omega^2+\alpha^2}\right].$$

$$\tag{6}$$

The current-time curve corresponding to (6) is shown in Fig. 72 (b). Distortion of the wave form caused by the output circuit is represented in (6) by the phase angle

$$\tan^{-1}\left[\omega c/(\omega^2+\alpha^2)\right],$$

and the exponential term. So long as c is sufficiently small, these quantities will also be small and, therefore, unimportant. Since R is fixed usually, both L_2 and L_1 must be large enough, while R_a, the anode resistance, should be comparatively small, as in the case of a power triode.

12·62. Secondary current when $L \neq 0$. When $L \neq 0$, the transform of the secondary current is

$$\phi_2 = \frac{E_0(n_2/n_1)\,\omega}{L}\,p^2 \Big/ (p+a)\,(p+b)\,[(p+\alpha)^2+\omega^2], \qquad (1)$$

the inverse of which may be written

$$I_2 = \frac{E_0(n_2/n_1)\,\omega}{L}\,[A_1\,e^{-at}+B_1\,e^{-bt}+C_1\,e^{-\alpha t}\sin{(\omega t+\theta)}]. \qquad (2)$$

12·63. Initial part of current curve. When t is small, we make p large and neglect a, b, α, ω, so (1) § 12·62 yields

$$I_2 = \frac{E_0(n_2/n_1)\,\omega}{L}\,\frac{1}{2\pi i}\int_{Br_1}\frac{e^{zt}\,dz}{z^3} = \frac{E_0(n_2/n_1)\,\omega t^2}{2L}. \qquad (1)$$

The initial slope of (1) is zero, whereas that of the impressed transient is ωE_0, the difference being due to the presence of L in the secondary circuit. The two exponential terms in (2) § 12·62 do not contribute anything of importance acoustically. The oscillatory part of the reproduced transient is given by the third term in (2) § 12·62. The influence of $L>0$ is to reduce the external multiplier and to introduce a larger phase angle θ. On comparison with the case where $L=0$, it is seen that the difference in wave form between the impressed and reproduced transients (using a properly designed transformer) depends mainly upon the value of L, which should, therefore, be as small as possible to avoid appreciable distortion.

12·71. Reproduction by a television amplifier [122].

A circuit used in a television receiver is illustrated in Fig. 73 (a), while the equivalent circuit is shown in Fig. 73 (b). The problem

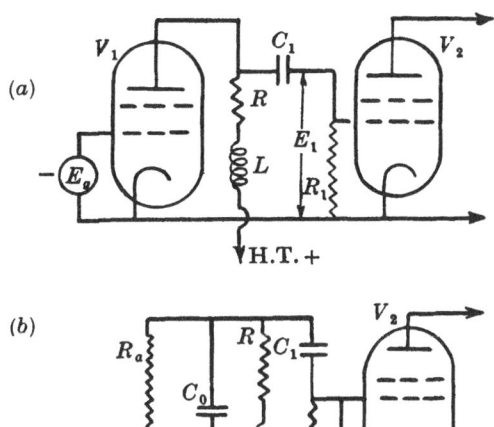

Fig. 73. (a) Illustrating coupling circuit between two screened-grid valves. (b) Equivalent circuit of (a). C_0 = anode to cathode capacitance + leads; C_2 = grid to cathode capacitance + leads.

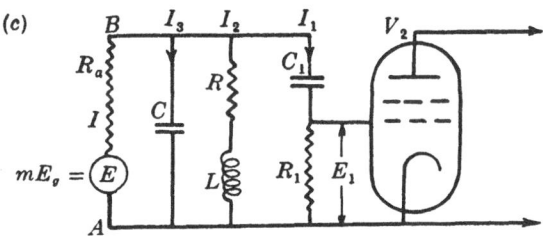

Fig. 73. (c) Circuit equivalent to Fig. 73 (a), (b) when $C_1 \gg C_2$ and $R_1 \gg 1/\omega C_1$; $C = C_0 + C_2$; E_1 = output p.d. from V_1 = input p.d. to V_2.

is to find the response of this circuit, quiescent initially, when the applied p.d. $-E_g$ takes the form of Heaviside's unit function. In practice $C_1 \gg C_2$ and $R_1 \gg 1/\omega C_1$ over the frequency range required, so C_2 may be taken in parallel with C_0. Fig. 73 (b) can then be replaced by Fig. 73 (c); where $C = C_0 + C_2$, which is the sum of the following capacitances: (a) anode to cathode of V_1, (b) grid to cathode of V_2, (c) leads and wiring associated with (a), (b). The inductance L, whose self-capacitance and capacitance to earth are negligible, offsets the shunting effect of C at the upper or radio-frequency end of the range.

12·72. Analysis of circuit of Fig. 73 (c). R_a represents the internal A.C. resistance of the valve V_1, while $mE_g = E$ is the product of the amplification factor and the p.d. applied to the grid, with reversed sign owing to phase reversal by the valve. The p.d. across AB, namely, $E - IR_a$, is applied to the three branches C, RL and $C_1 R_1$ in parallel. Thus with $I \Rightarrow \phi$, $I_1 \Rightarrow \phi_1$, $I_2 \Rightarrow \phi_2$, $I_3 \Rightarrow \phi_3$ we have

$$E - \phi R_a = \phi_3/pC = \phi_2(R+pL) = \phi_1\{R_1 + (1/pC_1)\}. \quad (1)$$

The valve current is equal to the sum of the three branch currents, so

$$I = I_1 + I_2 + I_3 \quad \text{and} \quad \phi = \phi_1 + \phi_2 + \phi_3. \quad (2)$$

From (1) we obtain

$$\phi = \frac{E}{R_a} - \left(\frac{1+pC_1 R_1}{pC_1 R_a}\right)\phi_1, \quad (3)$$

$$\phi_2 = \frac{(1+pC_1 R_1)\phi_1}{pC_1(R+pL)}, \quad (4)$$

and

$$\phi_3 = (1+pC_1 R_1)\frac{C}{C_1}\phi_1. \quad (5)$$

Substituting from (3), (4) and (5) in (2), we get

$$\phi_1\left\{1 + (1+pC_1 R_1)\left[\frac{1}{pC_1 R_a} + \frac{1}{pC_1(R+pL)} + \frac{C}{C_1}\right]\right\} = \frac{E}{R_a}, \quad (6)$$

which, after simplification, gives

$$I_1 \Rightarrow \phi_1 = \frac{E}{LCR_1 R_a}\left[\frac{p(R+pL)}{p^3 + \alpha p^2 + \beta p + \gamma}\right], \quad (7)$$

where
$$\alpha = \frac{R}{L} + \frac{1}{R_1}\left(\frac{1}{C} + \frac{1}{C_1}\right) + \frac{1}{CR_a},$$

$$\beta = \frac{1}{LC}\left(1 + \frac{R}{R_a} + \frac{R}{R_1}\right) + \frac{1}{C_1 R_1}\left(\frac{R}{L} + \frac{1}{CR_a}\right),$$

and
$$\gamma = \frac{R + R_a}{LCC_1 R_1 R_a}.$$

The following numerical data will be used:
$R_a = 5 \times 10^5$ ohms; $R_1 = 10^6$ ohms; $R = 5 \times 10^3$ ohms; $L = 5 \times 10^{-4}$ henry; $C_1 = 10^{-1}$ microfarad; $C = 30$ micro-microfarads; m of valve 2000. We find, with adequate approximation, that

$$\alpha = R/L = 10^7, \quad \beta = 1/LC = 6\cdot667 \times 10^{13}$$

and
$$\gamma = 1/LCC_1 R_1 = 6\cdot667 \times 10^{14}.$$

To invert (7), we must first solve the cubic equation

$$\sigma^3 + \alpha\sigma^2 + \beta\sigma + \gamma = (\sigma + a)(\sigma + b)(\sigma + c) = 0. \tag{8}$$

From above $\alpha \lll \beta \ll \gamma$, and the first root of (8) is obtained with adequate accuracy by putting $\beta\sigma + \gamma = 0$, which gives

$$a = \gamma/\beta = 1/C_1 R_1 = 10. \tag{9}$$

The remaining roots of (8) are found on dividing by $(\sigma + a)$, which yields

$$\left.\begin{matrix} b \\ c \end{matrix}\right\} = (R/2L) \pm i\sqrt{\{(1/LC) - (R^2/4L^2)\}}. \tag{10}$$

12·73. Inversion of (7) §12·72. Applying the Mellin theorem, we have

$$E_1 = I_1 R_1 = \frac{E}{LCR_a}\frac{1}{2\pi i}\int_{Br_1}\frac{e^{zt}(R + zL)\,dz}{(z + a)(z + b)(z + c)}. \tag{1}$$

The integrand of (1) has three simple poles at $z = -a$, $-b$ and $-c$. Evaluating the residues at these poles (see §§ 3·21, 4·53), we obtain

$$E_1 = \frac{E}{LCR_a}\left[\frac{(R - aL)e^{-at}}{(b - a)(c - a)} + \frac{(R - bL)e^{-bt}}{(a - b)(c - b)} + \frac{(R - cL)e^{-ct}}{(a - c)(b - c)}\right]. \tag{2}$$

Using the numerical data in §12·72, we find that $R \gg aL$, $Re(b) \gg a$, $Re(c) \gg a$, so to an adequate degree of approximation (2) may be written

$$
E_1 \simeq \frac{E}{LCR_a} \left[\frac{R e^{-t/C_1 R_1}}{bc} + \frac{(R - bL) e^{-(R/2L + i\omega_0)t}}{b(b - c)} \right.
$$
$$
\left. - \frac{(R - cL) e^{-(R/2L - i\omega_0)t}}{c(b - c)} \right], \qquad (3)
$$

where $\omega_0 = \sqrt{\{(1/LC) - (R^2/4L^2)\}}$.

When the numerical values of b, c and ω_0 are used, (3) gives

$$
E_1 \simeq \frac{ER}{R_a} \left[e^{-t/C_1 R_1} - \frac{e^{-Rt/2L}}{CR\omega_0} \sin(\omega_0 + \theta) \right], \qquad (4)
$$

where $\theta = \tan^{-1} \left\{ CR\omega_0 \Big/ \left(\frac{CR^2}{2L} - 1 \right) \right\}$, and $\sin \theta = CR\omega_0$. The output p.d. consists of two parts: (i) an exponential decay component, (ii) a highly damped sine wave of frequency about one megacycle per second. The beginning of the reproduced p.d. is illustrated in curve 1 of Fig. 74. Owing mainly to the charge to capacitance C in Fig. 73 (c), the p.d. does not attain the value ER/R_a until after a lapse of time of $2·04 \times 10^{-7}$ sec. Having reached this value, it overshoots by 15 per cent, owing to the oscillation arising from the LCR circuit. This oscillation is super-imposed upon the main p.d., which would be of rectangular form in the absence of distortion due to the inter-valve coupling circuit.

The time to reach the p.d. $E_1 = ER/R_a$ is found easily from (4). When $\sin(\omega_0 t + \theta)$ has its first zero, t is extremely small and $e^{-t/C_1 R_1} \simeq 1$. We have, therefore, $\omega_0 t + \theta = \pi$, so

$$
6·46 \times 10^6 t + 1·824 = \pi,
$$

which gives $t \simeq 2·04 \times 10^{-7}$, or about $\frac{1}{5}$th microsecond.

12·74. Approximation to (1) §12·73 when t is very small.

When t is of the order of a few hundredths of a micro-second, the main contribution to the value of (1) § 12·73 may be found by taking z very large (see § 4·61 et seq.). Consequently a, b, c and R are negligible, and we may write

$$E_1 \simeq \frac{E}{CR_a} \frac{1}{2\pi i} \int_{Br_1} \frac{e^{zt} dz}{z^2} = \frac{Et}{CR_a}. \tag{1}$$

From (1) it follows that the initial portion of the reproduced p.d. curve 1 of Fig. 74 is linear.

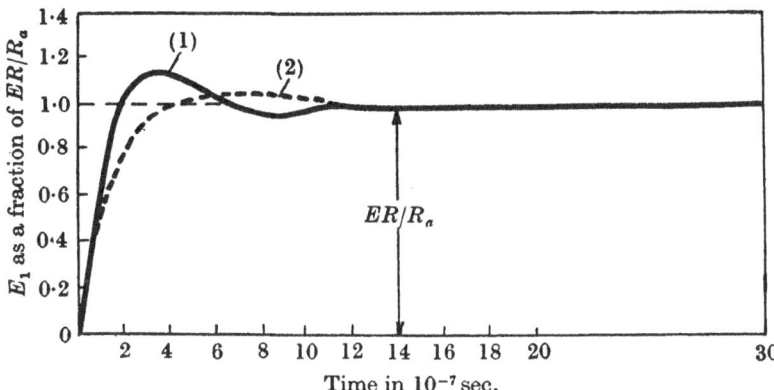

Fig. 74. Curve 1: Reproduced p.d. using the circuit of Fig. 73(a). Curve 2: Reproduced p.d. using the circuit of Fig. 73(a) and a damping resistance R_3 across L to make the LCR circuit aperiodic. In each case, the p.d. across R_1 is plotted.

By differentiating (4) § 12·73 and equating to zero we can find the time which elapses between the application of the p.d. $-E_g$ to the grid of the valve and the occurrence of the first maximum of the p.d. across the output resistance R_1 (Fig. 73 (c)).
Thus

$$\frac{dE_1}{dt} = \frac{ER}{R_a} \left\{ -\frac{e^{-t/C_1 R_1}}{C_1 R_1} + \frac{e^{-Rt/2L}}{CR\sqrt{(1-CR^2/4L)}} [\sin(\omega_0 t + \theta - \phi)] \right\}, \tag{2}$$

where
$$\phi = \tan^{-1} 2L\omega_0/R.$$

Inserting in (2) the numerical data given above, and equating to zero, we obtain with $\tau = 10^7 t$

$$\sin(0.646\tau + 0.913) \simeq 1.185 \times 10^{-6} e^{\frac{1}{2}\tau}. \tag{3}$$

As an approximation assume the right-hand side of (3) to be zero, then

$$\sin(0{\cdot}646\tau + 0{\cdot}913) = 0, \quad \text{or} \quad 0{\cdot}646\tau + 0{\cdot}913 = \pi,$$

which gives $\tau = 3{\cdot}45$ and, therefore,

$$t = 3{\cdot}45 \times 10^{-7} \simeq \tfrac{1}{3} \text{ microsecond.} \tag{4}$$

If we put $\tau = 3{\cdot}45$ in the right-hand side of (3) its value is of order 7×10^{-6}, so the preceding approximation is justifiable.

If after time t_1 when $E_1 \simeq ER/R_a$, the source $-E_g$ in Fig. 73 (a) is replaced by a short circuit, there is a transient oscillation as discussed above, the condenser C_1 discharges through the remainder of the circuit. The p.d. across R_1 is therefore reduced to zero and ultimately becomes slightly negative [according to the value of $(e^{-t_1/C_1 R_1} - 1)$].

Using § 10·33, the reader should investigate the behaviour of the amplifier when the applied p.d. is $-SI(t)$.

The solution of more complicated circuit problems will be found in [225].

XIII

PARTIAL LINEAR DIFFERENTIAL EQUATIONS: ELECTRICAL TRANSMISSION LINES: ELECTRICAL WAVE FILTERS

13·11. Method of solution. This is a slight modification of the procedure used in Chapter IX and depends upon application of the Mellin inversion theorem. Consider the linear partial differential equation

$$\frac{\partial^2\theta}{\partial x^2} - \frac{\partial\theta}{\partial t} = 0, \tag{1}$$

which occurs in problems in heat and moisture diffusion through rectangular slabs, also in the transmission of electrical energy in a uniform unloaded cable (negligible inductance and leakance). Suppose that θ, which is a function of both x and t, is zero as $t \to -0$, for any value of $x \geqslant 0$. This corresponds to quiescence initially. For boundary conditions take (i) $\theta \to 0$, $x \to +\infty$, $t > 0$, (ii) $\theta = \theta_0$, a constant, when $x = 0$, $t > 0$. Multiply (1) by $p\,e^{pt}$, where $R(p) > 0$, and integrate with respect to t from $t = 0$ to $+\infty$. Then we get

$$p\int_0^\infty e^{-pt}\left(\frac{\partial^2\theta}{\partial x^2}\right)dt - p\int_0^\infty e^{-pt}\left(\frac{\partial\theta}{\partial t}\right)dt = 0. \tag{2}$$

Assuming that the order of differentiation and integration in the first member of (2) may be changed, and taking the second by parts, we have

$$\frac{\partial^2}{\partial x^2}\left[p\int_0^\infty e^{-pt}\theta(x,t)\,dt\right] - p[e^{-pt}\theta(x,t)]_{t=0}^\infty$$
$$- p\left[p\int_0^\infty e^{-pt}\theta(x,t)\,dt\right] = 0. \tag{3}$$

Let $p\int_0^\infty e^{-pt}\theta(x,t)\,dt = \phi(x,p)$, i.e. $\theta \rightleftharpoons \phi$, then substituting the initial condition $\theta = 0$, $t = 0$ in the second member of (3), if $e^{-pt}\theta(x,t) \to 0$ as $t \to +\infty$, we get

$$d^2\phi/dx^2 - p\phi = 0. \tag{4}$$

Since ϕ is the Laplace transform of θ^*, (4) may be regarded as the transform equation corresponding to (1). Its solution with two arbitrary constants A, B—which are independent of x, but are functions of p in a general sense—is

$$\phi = A\,e^{-x\sqrt{p}} + B\,e^{x\sqrt{p}}. \tag{5}$$

We have now to express the boundary conditions in terms of ϕ. Using the definition above (4), when $\theta = 0$, $\phi = 0$, so the first condition is $\phi \to 0$, $x \to +\infty$, $t > 0$. When $\theta = \theta_0$, $p\int_0^\infty e^{-pt}\theta_0 dt = \theta_0$, so the second condition is $\phi = \theta_0$, $x = 0$, $t > 0$. Using the first condition in (5) entails $B = 0$, while the second gives $A = \theta_0$. Hence

$$\phi = \theta_0\,e^{-x\sqrt{p}}, \tag{6}$$

which leads to the integral equation for $\theta(x, t)$, namely,

$$p\int_0^\infty e^{-pt}\,\theta(x, t)\,dt = \theta_0\,e^{-x\sqrt{p}}. \tag{7}$$

ϕ in (6) fulfils the conditions for validity of the Mellin inversion theorem, so the solution of (7) in terms of x, t, is

$$\theta = \frac{\theta_0}{2\pi i}\int_{Br_1} e^{zt - x\sqrt{z}}\frac{dz}{z} \tag{8}$$

$$= \theta_0\,\mathrm{erfc}\,(x/2\sqrt{t}), \tag{9}$$

by (6) §5·221. This result could have been obtained also from (6) by using (33) Appendix 10. The graphs of $\mathrm{erf}\,u$ and $\mathrm{erfc}\,u$ are given in Fig. 33. For a fixed x and increasing time, the graph must be read backwards from u infinite towards $u = 0$; but for fixed t and increasing distance, it is read in the usual way. In the first case θ increases, but decreases in the second, as is to be expected on physical grounds.

13·12. Proof that (9) §13·11 is the appropriate solution of (1) §13·11.
We have to demonstrate that (9) § 13·11 satisfies:
  (i) the differential equation (1) § 13·11;
  (ii) the initial condition;
  (iii) the boundary conditions.

* With respect to t.

(i) $$\theta = \theta_0 \operatorname{erfc} u = \frac{2\theta_0}{\sqrt{\pi}} \int_u^{\infty} e^{-v^2}\, dy,$$

with $u = x/2\sqrt{t}$. Then

$$\frac{\partial \theta}{\partial x} = \frac{\partial \theta}{\partial u}\frac{\partial u}{\partial x} = -\frac{2\theta_0}{\sqrt{\pi}}\frac{e^{-x^2/4t}}{2\sqrt{t}} = -\frac{\theta_0\, e^{-x^2/4t}}{\sqrt{(\pi t)}}, \tag{1}$$

and

$$\frac{\partial^2 \theta}{\partial x^2} = \frac{\theta_0\, x\, e^{-x^2/4t}}{2\sqrt{\pi t^{\frac{3}{2}}}}. \tag{2}$$

Also $$\frac{\partial \theta}{\partial t} = \frac{\partial \theta}{\partial u}\frac{\partial u}{\partial t} = -\frac{2\theta_0\, e^{-x^2/4t}}{\sqrt{\pi}}\left(\frac{-x}{4t^{\frac{3}{2}}}\right) = \frac{\theta_0\, x\, e^{-x^2/4t}}{2\sqrt{\pi t^{\frac{3}{2}}}}. \tag{3}$$

The identity of (2), (3) proves that (9) § 13·11 satisfies (1) §13·11.

(ii) When $x > 0$, $\operatorname{erfc}(x/2\sqrt{t}) \to 0$, as $t \to +0$—see Fig. 33—so $\theta = 0$, which satisfies the initial condition.

(iii) Since $\operatorname{erfc} u \to 0$ as $u \to +\infty$, $\theta \to 0$ when $x \to +\infty$, $0 < t < \infty$, which satisfies the first boundary condition. For $x = 0$, $t > 0$, $\operatorname{erfc}(x/2\sqrt{t}) = 1$—see Fig. 33—and (9) § 13·11 gives $\theta = \theta_0$, so the second boundary condition is satisfied.

Hence (9) § 13·11 is the appropriate solution of (1) § 13·11. The reader may check also that it is permissible to invert the order of differentiation and integration in (2) § 13·11 (see [236] for additional checks). In [236] the transform method is proved to be valid for solving ordinary linear differential equations having constant coefficients. It is not possible, however, to establish a *general* proof for partial differential equations, so each solution must be verified independently.

13·13. Non-quiescent initial condition. To illustrate this we shall solve the equation for radial heat diffusion in a right circular cylinder whose length $\gg$ radius, so that end effect is negligible. Taking $\theta(r, t)$ for temperature, $r =$ radius, $k =$ diffusivity,* the appropriate equation is

$$\frac{\partial^2 \theta}{\partial r^2} + \frac{1}{r}\frac{\partial \theta}{\partial r} - \frac{1}{k}\frac{\partial \theta}{\partial t} = 0, \tag{1}$$

θ being a function of r and t.

* Defined in § 15·11.

Let the initial condition be $\theta(r, t) = \theta_0$, a constant for $r < a$, $t \to 0$, and the boundary condition be $\theta(r, t) = \theta_1$, a constant for $r = a$, $t > 0$, i.e. $0 < t < \infty$. Actually a second boundary condition is needed, namely, that the axial temperature is finite.

Multiply (1) by $p\,e^{-pt}$, where $R(p) > 0$, and integrate from $t = 0$ to $+\infty$. Then we get

$$p \int_0^\infty e^{-pt} \left(\frac{\partial^2 \theta}{\partial r^2} \right) dt + p \int_0^\infty e^{-pt} \left(\frac{1}{r} \frac{\partial \theta}{\partial r} \right) dt - \frac{p}{k} \int_0^\infty e^{-pt} \left(\frac{\partial \theta}{\partial t} \right) dt = 0.$$

(2)

If it is permissible to invert the order of differentiation and integration in the first two members of (2), after integrating the third member by parts, we obtain

$$\frac{\partial^2}{\partial r^2} \left[p \int_0^\infty e^{-pt} \theta(r, t)\, dt \right] + \frac{1}{r} \frac{\partial}{\partial r} \left[p \int_0^\infty e^{-pt} \theta(r, t)\, dt \right]$$

$$- \frac{p}{k} [e^{-pt} \theta(r, t)]_{t=0}^\infty - \frac{p}{k} \left[p \int_0^\infty e^{-pt} \theta(r, t)\, dt \right] = 0. \quad (3)$$

Writing

$$\phi(r, p) = p \int_0^\infty e^{-pt} \theta(r, t)\, dt, \qquad (4)$$

and substituting the initial condition $\theta = \theta_0$ as $t \to 0$ in the third member of (3), if $e^{-pt} \theta(r, t) \to 0$ as $t \to +\infty$, we get the transform equation

$$\frac{d^2\phi}{dr^2} + \frac{1}{r} \frac{d\phi}{dr} - \frac{p}{k} \phi = -\frac{p\theta_0}{k}. \qquad (5)^*$$

The formal solution of (5) is [234]

$$\phi = A J_0[ir \sqrt{(p/k)}] + B Y_0[ir \sqrt{(p/k)}] + \theta_0. \qquad (6)$$

Near the axis of the cylinder as $r \to 0$, $Y_0 \to -\infty$, and since θ and, therefore, ϕ must be finite, it follows that B is zero. At the surface of the cylinder $r = a$, and $\theta = \theta_1$, so by (4), $\phi = \theta_1$. Substituting $B = 0$, $r = a$, $\phi = \theta_1$ in (6) leads to

$$A = (\theta_1 - \theta_0)/J_0[ia \sqrt{(p/k)}]. \qquad (7)$$

Inserting this value of A, and $B = 0$ in (6), the transform solution of our problem is

$$\phi = (\theta_1 - \theta_0) \frac{J_0[ir \sqrt{(p/k)}]}{J_0[ia \sqrt{(p/k)}]} + \theta_0. \qquad (8)$$

---

* The term $-p\theta_0/k$ is due to the initial condition $\theta = \theta_0$ as $t \to 0$ (see [236]).

Hence by (4) and (8) we have the integral equation

$$p \int_0^\infty e^{-pt} \theta(r,t) \, dt = \theta_0 + (\theta_1 - \theta_0) \frac{J_0[ir\sqrt{(p/k)}]}{J_0[ia\sqrt{(p/k)}]}. \tag{9}$$

Since the right-hand side of (9) satisfies the conditions for validity of the Mellin theorem, by § 8·11 the solution of (9) is

$$\theta(r,t) = \frac{(\theta_1 - \theta_0)}{2\pi i} \int_{Br_1} \frac{e^{zt} J_0[ir\sqrt{(z/k)}] \, dz}{z J_0[ia\sqrt{(z/k)}]} + \theta_0 \tag{10}$$

$$= (\theta_1 - \theta_0)\left[1 - 2 \sum_{n=1}^\infty e^{-\alpha_n^2 kt/a^2} \frac{J_0(\alpha_n r/a)}{\alpha_n J_1(\alpha_n)}\right] + \theta_0, \tag{11}$$

by (5) § 4·73. Convergence of the *series* in (11) is examined in § 5, Appendix 9.

13·14. Proof that (11) § 13·13 is the appropriate solution of (1) § 13·13. We have now to give a demonstration similar to that in § 13·12.

(i) To show that (11) § 13·13 satisfies (1) § 13·13, we substitute it into the latter. Then

$$-\frac{1}{k}\frac{\partial\theta}{\partial t} = -2(\theta_1 - \theta_0) \sum_{n=1}^\infty e^{-\alpha_n^2 kt/a^2} \left[\frac{\alpha_n J_0(\alpha_n r/a)}{a^2 J_1(\alpha_n)}\right]; \tag{1}$$

$$\frac{1}{r}\frac{\partial\theta}{\partial r} = 2(\theta_1 - \theta_0) \sum_{n=1}^\infty e^{-\alpha_n^2 kt/a^2} \left[\frac{J_1(\alpha_n r/a)}{ra J_1(\alpha_n)}\right]; \tag{2}$$

$$\frac{\partial^2\theta}{\partial r^2} = 2(\theta_1 - \theta_0) \sum_{n=1}^\infty e^{-\alpha_n^2 kt/a^2} \left[\frac{\alpha_n J_0'(\alpha_n r/a)}{a^2 J_1(\alpha_n)}\right] \tag{3}$$

$$= 2(\theta_1 - \theta_0) \sum_{n=1}^\infty e^{-\alpha_n^2 kt/a^2} \left[\frac{\alpha_n J_0(\alpha_n r/a)}{a^2 J_1(\alpha_n)} - \frac{J_1(\alpha_n r/a)}{ra J_1(\alpha_n)}\right], \tag{4}$$

by aid of the recurrence formula

$$J_1'(u) = J_0(u) - J_1(u)/u \tag{5}$$

(see [234], p. 24, (21)). The differentiations are permissible in virtue of uniform convergence of the series involved (see Appendix 9). Since the sum of the right-hand members in (1), (2), (4) is zero, (11) § 13·13 satisfies (1) § 13·13.

(ii) By § 4·73 (below (5)) the initial condition $\theta = \theta_0$ as $t \to 0$ is satisfied;

(iii) When $r = a$, $J_0(\alpha_n r/a) = 0$, since $J_0(\alpha_n) = 0$. Thus the 'sigma' term in (11) § 13·13 vanishes for $t \geqslant 0$, and we get $\theta = \theta_1$, so the boundary condition is satisfied. Hence (11) § 13·13 is the appropriate solution of (1) § 13·13.

In the remainder of the text, checks of the above type are left as exercises for the reader. The t-solution may not always be suitable for this purpose. Then the integral form as at (8) § 13·11, (10) § 13·13 may be used. Differentiation of the contour integrals under the sign is usually valid if the resulting integral is uniformly convergent in the appropriate intervals of x, t, or r, t, as the case may be.

13·15. System subject to a variable force. The equations in §§ 13·11, 13·13 relate to systems having a constant impressed force, or its equivalent. Consider the equation

$$\frac{\partial^2 \xi}{\partial x^2} + \frac{1}{x}\frac{\partial \xi}{\partial x} - \left(a\frac{\partial^2 \xi}{\partial t^2} + b\frac{\partial \xi}{\partial t}\right) = \mathbf{f}\sin\omega t, \tag{1}$$

which refers to the diaphragm of a condenser microphone subjected to a sinusoidal sound pressure $\mathbf{f}$ of angular frequency ω.

We proceed as in § 13·13. Let the initial conditions be $\xi = \xi_0(x)$, $\xi' = \xi_1(x)$, $t \to 0$, where ξ_0, ξ_1 are arbitrary functions of x which represent the diaphragm displacement and velocity, respectively, as $t \to 0$. Then by (2) § 9·22, the transform of the 'initial conditions function' associated with the third and fourth terms on the left-hand side of (1), is given by the array

$$\phi_3(p) = \begin{array}{c} \\ \xi_0 \\ \xi_1 \end{array}\!\!\begin{array}{cc} p^2 & p \\ \left(\begin{array}{cc} -a & -b \\ 0 & -a \end{array}\right. & \end{array}\!\!\Big\} = -[\xi_0 a p^2 + (\xi_0 b + \xi_1 a)p]. \tag{2}$$

Taking $\phi(x,p) = p\displaystyle\int_0^\infty e^{-pt}\xi(x,t)\,dt$, replacing the right-hand side of (1) by its transform, writing p for $\partial/\partial t$, and adding (2) to the right-hand side, as in (5) § 13·13, we obtain the transform equation

$$\frac{d^2\phi}{dx^2} + \frac{1}{x}\frac{d\phi}{dx} - (ap^2 + bp)\,\phi = \frac{\mathbf{f}\omega p}{(p^2 + \omega^2)} - [\xi_0 a p^2 + (\xi_0 b + \xi_1 a)p]. \tag{3}$$

The formal solution of (3) is

$$\phi = AI_0(\lambda x) + BK_0(\lambda x) - \{f\omega/(p^2 + \omega^2)(ap + b)\} + \chi(x, p), \quad (4)$$

where $\lambda = \sqrt{(ap^2 + bp)}$, and χ is a particular integral corresponding to the initial conditions function (2). The arbitrary constants A, B are found as functions of p by inserting the boundary conditions in (4). Then $\phi(x, p)$ is inverted by the Mellin theorem, provided the requisite conditions are satisfied. Thus

$$\xi(x, t) = \frac{1}{2\pi i} \int_{Br_1} \frac{e^{zt}\phi(x, z)\,dz}{z}. \quad (5)$$

13·16. Recapitulation of §§13·11–13·15. The method of solving linear partial differential equations may be epitomized thus: (1°) Write ϕ for the dependent variable, e.g. θ, p for $\partial/\partial t$ on the left-hand side, and the transform of the applied force on the right-hand side. (2°) Add the transform of the initial conditions function, i.e. ϕ_3 from (2) § 9·22 to the right-hand side. (3°) Solve the resulting transform equation for ϕ. (4°) Determine the arbitrary constants in terms of p, using the boundary conditions. If ϕ complies with the conditions for validity of the Mellin theorem, it is then inverted using the complex integral. The function of (θ, t) so obtained, or the complex integral representing it, is now demonstrated to be the appropriate solution. Alternatively ϕ may be inverted by aid of a list of Laplace transforms as given in Appendix 10 or [235 a, b].

13·21. Electrical transmission lines. Let E and I represent, respectively, the potential difference and current at any point distant x from some convenient origin in a uniform transmission line having inductance **L**, capacitance **C**, resistance **R**, and leakance **G**, all assumed constant and reckoned per *unit length*. Referring to the schematic representation of Fig. 75 (a), the inductance and resistance of a length ∂x of the line are, respectively, **L** ∂x and **R** ∂x. Thus the potential difference across ∂x due to a current I is

$$-\partial E = (\mathbf{L}\,\partial I/\partial t + \mathbf{R}I)\,\partial x, \quad (1)$$

so
$$-\partial E/\partial x = (\mathbf{L}\partial/\partial t + \mathbf{R})\,I, \qquad (2)$$

the minus sign indicating that E decreases with increase in x.

Unless stated otherwise, we shall take quiescent initial conditions, i.e. as $t \to -0$, $E = 0$, $I = 0$, in all cable and filter problems in this chapter.

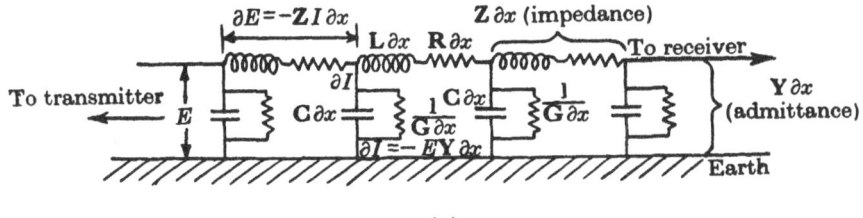

(a)

Fig. 75(a). Schematic diagram for cable having inductance, capacitance, resistance and leakance.

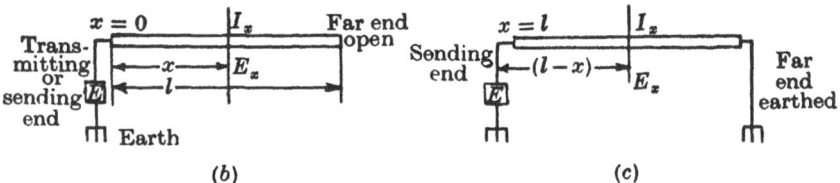

(b)   (c)

Fig. 75(b), (c). Schematic diagrams for cable circuit.

Writing $E \Rightarrow \varphi$, $I \Rightarrow \phi$, and p for $\partial/\partial t$, (2) gives
$$-\partial\varphi/\partial x = (\mathbf{L}p + \mathbf{R})\,\phi, \qquad (3)$$

where $(\mathbf{L}p + \mathbf{R})$ is the transform of the series impedance of the line per unit length.

Current flows from the central copper conductor of the line to earth, in virtue of capacitance and leakance. The values of these two for a length ∂x are $\mathbf{C}\partial x$ and $\mathbf{G}\partial x$, respectively. The current flow is
$$-\partial I = (\mathbf{C}\partial E/\partial t + \mathbf{G}E)\,\partial x, \qquad (4)$$

so
$$-\partial I/\partial x = (\mathbf{C}\partial/\partial t + \mathbf{G})\,E, \qquad (5)$$

the minus sign indicating that I decreases as x increases. The transform version of (5) is
$$-\partial\phi/\partial x = (\mathbf{C}p + \mathbf{G})\,\varphi, \qquad (6)$$

$(\mathbf{C}p + \mathbf{G})$ being the transform of the shunt admittance per unit length. From (3)

$$\phi = -\{1/(\mathbf{L}p + \mathbf{R})\}\,\partial\varphi/\partial x, \tag{7}$$

so (6) and (7) yield

$$d^2\varphi/dx^2 - \lambda^2\varphi = 0, \tag{8}$$

with $\lambda^2 = (\mathbf{L}p + \mathbf{R})(\mathbf{C}p + \mathbf{G})$. Replacing φ by E and p by $\partial/\partial t$ in (8), we get Heaviside's telegraph equation, namely,

$$\mathbf{L}\mathbf{C}\frac{\partial^2 E}{\partial t^2} + (\mathbf{C}\mathbf{R} + \mathbf{L}\mathbf{G})\frac{\partial E}{\partial t} + \mathbf{G}\mathbf{R}E = \frac{\partial^2 E}{\partial x^2}, \tag{9}$$

of which (8) is the transform version. The formal solution of (8) may be written in two ways, namely,

$$\varphi = A\,e^{-\lambda x} + B\,e^{\lambda x}, \tag{10}$$

and

$$\varphi = A_1\sinh\lambda x + B_1\cosh\lambda x. \tag{11}$$

The second form at (11) is the more convenient for certain boundary conditions as in § 13·24. When l, the length of the line, is very great, the current at the far end is so small that reflexion therefrom at a sufficiently remote point x can be neglected. Consequently the second solution in (10) is not required, so we have

$$\varphi = A\,e^{-\lambda x}, \tag{12}$$

which represents a wave propagated from the sending end.

By (7) and (12)

$$\phi = \frac{A\lambda\,e^{-\lambda x}}{(\mathbf{L}p + \mathbf{R})}, \tag{13}$$

while (12), (13) give

$$\frac{\varphi}{\phi} = \left\{\frac{\mathbf{L}p + \mathbf{R}}{\mathbf{C}p + \mathbf{G}}\right\}^{\frac{1}{2}} = Z_c(p), \tag{14}$$

this being known as the transform of the surge or 'characteristic' impedance of the line.

13·22. Unloaded submarine telegraph cable.
The so-called unloaded submarine telegraph cable, devised about a century ago, is the simplest type of line. Many thousands of miles of such cables are still in use, owing largely to the ingenuity of cable engineers in devising the 'regenerator' system of

transmission. At the speeds of signalling in vogue, the inductance and leakance of an unloaded cable can be neglected, so we have to consider the influence of resistance $\mathbf{R}$ and capacitance $\mathbf{C}$ only.

13·23. P.D. and current in very long unloaded cable.

For an unloaded cable $\lambda = \sqrt{(\mathbf{C}\mathbf{R}p)}$, and if, owing to its great length, the effect of reflection from the far end at an intermediate point distant x from the transmitter is negligible, (12) § 13·21 gives

$$\varphi = A\, e^{-b\sqrt{p}}, \qquad (1)$$

where $b = x\sqrt{(\mathbf{C}\mathbf{R})}$. If at $t = 0$, a battery of voltage E_0 (usually about 40) is applied at the sending end $x = 0$, $A = E_0$ in (1), so

$$\varphi = E_0\, e^{-b\sqrt{p}}, \qquad (2)$$

this being the transform for the p.d. at any point distant x from the battery ($x < l$). Using (33) Appendix 10, the inverse of (2) is

$$E = E_0\, \mathrm{erfc}\,(b/2\sqrt{t}). \qquad (3)$$

Alternatively, by the Mellin theorem

$$E = \frac{E_0}{2\pi i} \int_{Br_1} e^{zt - b\sqrt{z}}\frac{dz}{z} \qquad (4)$$

$$= E_0\, \mathrm{erfc}\,(b/2\sqrt{t}) = E_0\, \mathrm{erfc}\,\{x\sqrt{(\mathbf{C}\mathbf{R})}/2\sqrt{t}\}, \qquad (5)$$

by (6) § 5·221. The form of the E-t curve may be gleaned from the erfc u curve in Fig. 33, read backwards for t increasing.

From (2) § 13·21, since $\mathbf{L}$ is negligible,

$$I = -\frac{1}{\mathbf{R}}\frac{\partial E}{\partial x},$$

so from (4) by differentiating with respect to x, by aid of (4) § 5·222, we get

$$I = E_0\sqrt{(\mathbf{C}/\pi \mathbf{R}t)}\, e^{-x^2\mathbf{C}\mathbf{R}/4t}. \qquad (6)$$

At the sending end $x = 0$, and from (5) the current into this end of the cable is

$$I_0 = E_0\sqrt{(\mathbf{C}/\pi \mathbf{R}t)}. \qquad (7)$$

When $t = 0$ the current at $x = 0$ is infinite momentarily, owing to the uncharged cable acting as a short-circuit by virtue of its capacitance. When $t = 0$ and $x > 0$, by (6) $I = 0$, but when $t = \frac{1}{2}x^2\mathbf{C}\mathbf{R}$ it attains a maximum value.

The thermal analogue of this case is a uniform semi-infinite rod of homogeneous material, whose longitudinal surface is prevented from losing heat. At $t = 0$, one end is raised to a temperature θ_0 above its equilibrium value prior to this time. θ_0 corresponds to the applied voltage E_0, and q the quantity of heat passing any transverse section per unit time is analogous to I the cable current. The solution for a semi-infinite slab under the same conditions is identical with that for the rod. q per unit area is analogous to I per unit area, i.e. the heat flux and current densities, respectively.

13·24. Short unloaded cable. In a relatively short cable or line whose far end is not terminated by the characteristic or surge impedance, reflection occurs from that end, and its effect must be taken into account. This aspect of the subject is important not only in connection with cables, but also in the analogous problem of heat transmission in a large slab of thin material. We use the solution at (11) § 13·21, and consider the case of a short cable earthed at its distant end.

To simplify the analysis we take the origin $x = 0$ there. Then the p.d. is zero at $x = 0$, owing to the cable being earthed, but it is E_0 at the sending end where $x = l$. Substituting these conditions in (11) § 13·21, we have $B_1 = 0$, so

$$A_1 = E_0/\sinh \lambda l, \tag{1}$$

and $$\varphi = E_0 \sinh \lambda x/\sinh \lambda l, \tag{2}$$

there being no additional term in virtue of quiescence initially. From (2) § 13·21, remembering that $L \simeq 0$, and that E now increases with increase in x, since the latter is measured from the far end, we have $I = \dfrac{1}{R}\dfrac{\partial E}{\partial x}$, and from (3) § 13·21 $\phi = \dfrac{1}{R}\dfrac{\partial \varphi}{\partial x}$. Applying the latter to (2) we obtain

$$\phi = \frac{E_0 \lambda}{R}\frac{\cosh \lambda x}{\sinh \lambda l} = E_0 \sqrt{\left(\frac{C}{R}\right)} \sqrt{p}\,\frac{\cosh b \sqrt{p}}{\sinh a \sqrt{p}}, \tag{3}$$

where $a = l\sqrt{(CR)}$ and $b = x\sqrt{(CR)}$. Applying the Mellin theorem to (2) gives

$$E = \frac{E_0}{2\pi i}\int_{Br_1} \frac{e^{zt}\sinh b\sqrt{z}\,dz}{z\sinh a\sqrt{z}}. \tag{4}$$

The integrand has a simple pole at $z = 0$, a fact which will be realized more readily by expanding the hyperbolic functions. Thus

$$\frac{\sinh b \sqrt{z}}{z \sinh a \sqrt{z}} = \frac{bz^{\frac{1}{2}} + b^3 z^{\frac{3}{2}}/3! + b^5 z^{\frac{5}{2}}/5! + \dots}{z[az^{\frac{1}{2}} + a^3 z^{\frac{3}{2}}/3! + a^5 z^{\frac{5}{2}}/5! + \dots]}. \tag{5}$$

Dividing above and below by $\sqrt{z}$, and then letting $z \to 0$, yields b/az in the limit, since the other terms tend to zero with z. Thus (4) becomes

$$E = \frac{E_0 b/a}{2\pi i} \int_{Br_1} \frac{e^{zt} dz}{z} = \frac{E_0 b}{a}, \tag{6}$$

this being the contribution from the pole at the origin.

There are also poles due to the zeros of $\sinh a \sqrt{z}$, which occur when $a \sqrt{z} = \pm i n \pi$, $n = 1, 2, 3, \dots$, and there is a limit point at infinity. The value $n = 0$ is inadmissible, since $z = 0$ is a zero of $\sinh b \sqrt{z}$. At the poles $z = -n^2 \pi^2/a^2$, $n = 1, 2, 3, \dots$, and the negative integers give the same values of z as the positive integers. The contribution from these singularities is, by §§ 3·237, 4·72

$$E_0 \sum_{n=1}^{\infty} \left[\frac{e^{zt} \sinh b \sqrt{z}}{z \frac{d}{dz}(\sinh a \sqrt{z})} \right]_{z = -n^2 \pi^2/a^2} \tag{7}$$

$$= 2 \frac{E_0}{a} \sum_{n=1}^{\infty} \left[\frac{e^{zt} \sinh b \sqrt{z}}{\sqrt{z} \cosh a \sqrt{z}} \right]_{\substack{z = -n^2 \pi^2/a^2 \\ a\sqrt{z} = in\pi}}. \tag{8}$$

Inserting the values of z and $b \sqrt{z}$ into (8), the numerator in [ ] gives $i\, e^{-n^2 \pi^2 t/a^2} \sin (n\pi b/a)$, while the denominator yields

$$\frac{in\pi}{a} \cos n\pi = \frac{(-1)^n\, in\pi}{a}.$$

Substituting these values in (8) and adding (6), the p.d. at any point x, at time $t > 0$, is

$$E = E_0 \left\{ \frac{b}{a} + 2 \sum_{n=1}^{\infty} (-1)^n e^{-n^2 \pi^2 t/a^2} \frac{\sin (n\pi b/a)}{n\pi} \right\} \tag{9}$$

$$= E_0 \left[\frac{x}{l} + 2 \sum_{n=1}^{\infty} (-1)^n e^{-n^2 \pi^2 t/l^2 \text{ CR}} \frac{\sin (n\pi x/l)}{n\pi} \right]. \tag{10}$$

$l^2CR = lC . lR$ is the product of the total capacitance and conductor resistance of the line. At the sending end all the summation terms vanish, owing to the factor $\sin n\pi$, so $E = E_0$, while at the far end where $x = 0$, E is zero, since all terms in (10) vanish. Moreover, the boundary conditions are satisfied.

The thermal analogue in this case is the same as in § 13·23 except that the rod is short or the slab thin, so that reflection from the far end surface of constant temperature (lower than θ_0) cannot be neglected. The earthed end of the cable is at constant potential, which is analogous to constant temperature at the far end of the rod or slab.

13·25. Arrival current in unloaded cable. The current at any point of the cable may be obtained by inverting (3) § 13·24, which gives

$$I_{t>0} = \frac{E_0}{Rl}\left[1 + 2\sum_{n=1}^{\infty}(-1)^n e^{-n^2\pi^2 t/l^2 CR}\cos\left(n\pi x/l\right)\right]. \qquad (1)$$

When $t \to \infty$, $I \to E_0/R$, R being the *total* conductor resistance; so Ohm's law holds, as it should during the steady state. At the sending end of the cable $x = l$, and when $t = 0$ the current is infinite momentarily, as in the very long cable ((7) § 13·23). The arrival current at the far end of the cable is given by

$$I_{\substack{x=0\\t>0}} = (E_0/Rl)\{1 - 2[e^{-\pi^2 t/l^2 CR} - e^{-4\pi^2 t/l^2 CR} + e^{-9\pi^2 t/l^2 CR} - ...]\}, \qquad (2)$$

and its time relationship takes the form illustrated in Fig. 76. (2) is inapplicable when $t = 0$, since the series is oscillatory.

13·26. Cable earthed at transmitter after an interval t_1. Earthing the cable at the sending end when $t = t_1 > 0$ is equivalent to connecting a battery of voltage $-E_0$ in series with the existing battery. The effect of the battery voltage $-E_0$, considered to act alone, is to cause a reverse current

$$I_{\substack{x=0\\t>t_1}} = -(E_0/Rl)\{1 - 2[e^{-\pi^2(t-t_1)/l^2 CR} - e^{-4\pi^2(t-t_1)/l^2 CR} + ...]\}. \qquad (1)$$

Since the system is a linear one, the solutions (2) § 13·25 and (1) above may be superimposed, and at any time $t > t_1$ the arrival current is given by their sum. This is illustrated by the drooping portion of the curve in Fig. 76. Subsequent changes in battery voltage can be treated in like manner.

13·27. Practical data for uniform unloaded cable. In submarine telegraphy, the cable is used for the simultaneous transmission and reception of messages, i.e. duplex working. For this purpose a balanced bridge network (Fig. 77) is employed.

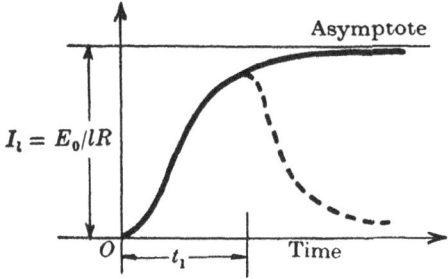

Fig. 76. Arrival curve (when $E = E_0 H(t)$) at earthed end of unloaded submarine cable. The broken curve shows the modification when E is reduced to zero at $t = t_1$.

The cable is connected to the point B, and an artificial line, having (as nearly as possible) the same electrical characteristics as the cable, is joined to D. The artificial line, comprising discrete resistances and capacitances, is kept in a special room at the cable station, its temperature being maintained at about 65° F. On connecting the battery to the duplex bridge, if the balance were perfect, there would be no p.d. across BD, so current would not pass through the receiving circuit connected to these points. A good, but imperfect balance can be obtained. Since the received signal strength decreases with increase in the speed of sending, a point is reached when the out-of-balance currents, due to transmission, render the received signals unreadable. The main disturbance is caused by the higher frequency components of the sending impulses, so it may be necessary to reduce their steepness by connecting a shunted inductance in series at the apex A, as shown in Fig. 77. The function of the arms AB, AD, is to 'shape' the transmitted signals. The capacitance curbs the low frequency components and obstructs unidirectional current, so the resistance is used as a by-pass. These arms offer considerable impedance to the incoming current from the far end of the cable, so this current flows to earth through the relatively low

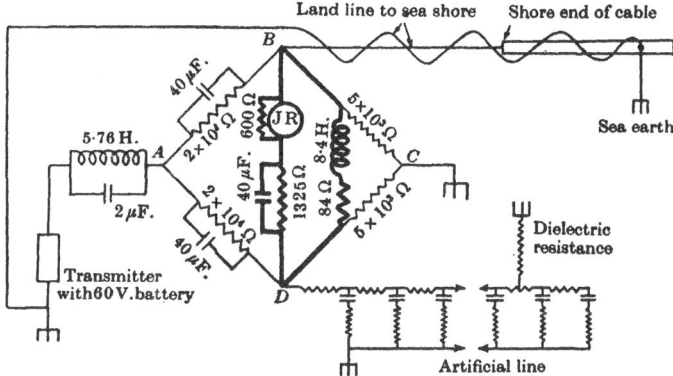

Fig. 77. Circuit diagram for duplex working with unloaded submarine cable. Length 1349 nauts, total conductor resistance 5833 ohms, total capacitance 489μF., total $CR = 2\cdot85$ sec., resistance of sea earth $10\cdot8$ ohms. J.R. is a jockey relay which passes the incoming signals to the land line connected to the central telegraph office.

impedance relay circuit. The relay is associated with a 'shaping' circuit comprising a shunted capacitance in series, and an inductance (magnetic shunt) with series resistance, connected in parallel therewith. The latter pair by-pass the low frequency components of the signal, but impede those of high frequency which, therefore, flow through the relay. The action of the shunted condenser in series with the relay is substantially the reverse of that of the magnetic shunt. Owing to the greater attenuation of the high than the low frequencies by the cable, as the signal passes along, it gradually loses the sharp profile it had at the transmitter, and near the receiver the form is rounded, as illustrated in Fig. 76. The low frequencies are then preponderant, so the shaping circuit is designed to increase the ratio of the high to the low frequency components of the current. The resistance arms BC, DC damp out surging, which may occur during transmission, due to interaction of capacitance and inductance.

Earthing the cable is an important feature. A twin-flex cable is run from the shore to the sea earth several miles out. One core is connected to the signal cable and the other to its outer steel armour, which makes contact with the sea water, as illustrated in Fig. 77.

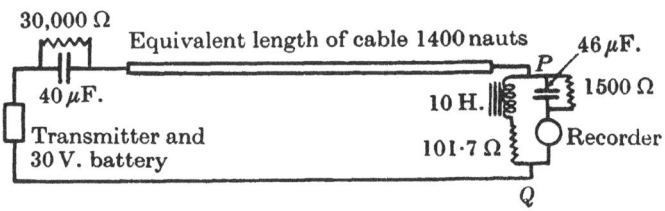

Fig. 78. Circuit diagram for tests with artificial line (see Fig. 87 (b)) simulating cable. Total resistance $R = 6725$ ohms, total $C = 442 \cdot 5 \mu$F., total CR of cable $= 2 \cdot 9$ sec.

13·28. Example of signal shaping. Two records are reproduced in which shaped and unshaped signals may be compared. The tests were made on an artificial line used as a simplex or one-way circuit, as shown schematically in Fig. 78. The unshaped signals were obtained on replacing the receiving circuit PQ by a moving-coil siphon recorder. The latter introduces some correction, but it is negligible compared with that due to the shaping circuit.

'Shaped' signals, representing the figures 5, 0, and the letters k, e, t, are shown in Fig. 79 (a). It is seen that they always return to a datum line (not indicated), being of uniform height in each *group*. The 'unshaped' signals of Fig. 79 (b), however, 'climb' well above the datum line. Excepting the letter t, the signals are almost shapeless. The resemblance of the profile between groups to that of Fig. 76 is evident. Climbing is due to the tendency of the current to attain an ultimate steady state, as indicated by the solid line in Fig. 76. The shaping circuit is designed to counteract this, and the major part of the low frequency current is by-passed by the inductive arm at PQ in

Fig. 78. Needless to remark, the mathematical analysis of the circuit depicted in Fig. 77 is too complicated for inclusion here.

In practice the recorder is used only for balancing the cable and the artificial line. During routine work the recorder is replaced by a 'jockey' relay (Fig. 77) which is associated with transfer of the incoming signals to the land-line link between the sea coast and the central telegraph office inland.

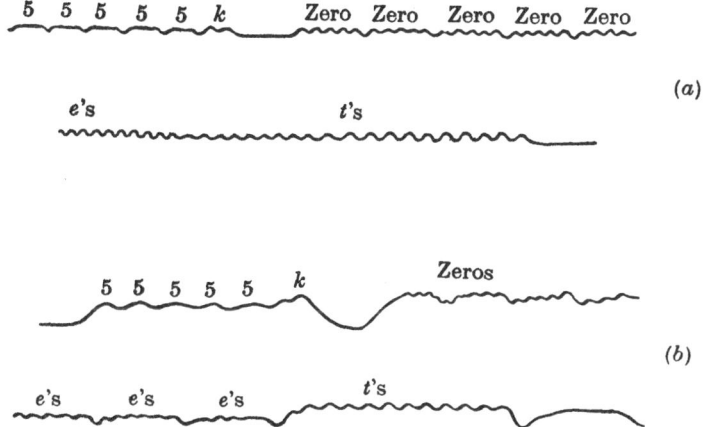

Fig. 79. (a) Signals received by siphon recorder using circuit of Fig. 78. (b) Signals received by siphon recorder using circuit of Fig. 78 but *without* shaping apparatus at PQ. The signals were transmitted in Morse code where 5 is five dots, k is dash dot dash, zero is five dashes, e is one dot, t is one dash. A dash is three times as long as a dot, the interval between consecutive dashes and/or dots in a letter is a dot, and that between letters is three dots (see [234] p. 207).

13·31. Loaded cables. We now turn our attention to cables having appreciable inductance. The central copper conductor is wrapped with thin wire or tape of magnetic material having a high initial permeability, of the order 4000. The nickel-iron alloys mumetal and permalloy are generally used for this purpose, and the inductance per unit length of the cable may be increased as much as 60-fold.

13·32. Sending current in very long loaded cable. From (12) § 13·21, when $E = E_0 H(t)$ is applied at $x = 0, t = 0$, it follows that
$$E_x \Rightarrow \varphi = E_0 e^{-\lambda x}, \tag{1}$$

so by (13) § 13·21,

$$I_x \rightleftharpoons \phi = E_0 \lambda \, e^{-\lambda x} / (\mathbf{L}p + \mathbf{R}) \qquad (2)$$

$$= E_0 \sqrt{\{(\mathbf{C}p + \mathbf{G}) / (\mathbf{L}p + \mathbf{R})\}} \, e^{-\lambda x}. \qquad (3)$$

At the head end of the cable $x = 0$, so that if we put $a = \mathbf{R}/2\mathbf{L}$, $b = \mathbf{G}/2\mathbf{C}$, then

$$\underset{x=0}{I} \rightleftharpoons \phi_0 = E_0 \sqrt{(\mathbf{C}/\mathbf{L})} \sqrt{\{(p + 2b)/(p + 2a)\}}. \qquad (4)$$

Writing K for the external multiplier,

$$\frac{\phi_0}{K} = \frac{(p + 2b)}{\sqrt{\{(p + 2a)(p + 2b)\}}} = \frac{(p + 2b)}{\sqrt{\{(p + \alpha)^2 - \beta^2\}}}, \qquad (5)$$

where $\alpha = (a + b)$, $\beta = (a - b)$. By (51) Appendix 10, with β for a,

$$I_0(\beta t) \rightleftharpoons p / \sqrt{(p^2 - \beta^2)}, \qquad (6)$$

so from (4) § 8·45

$$\int_0^t I_0(\beta t) \, dt \rightleftharpoons 1 / \sqrt{(p^2 - \beta^2)}. \qquad (7)$$

Applying (3) § 8·411 to (6), (7), with $-\alpha$ for a, the inversion of (5) is

$$\underset{x=0}{I} = K \left[e^{-\alpha t} I_0(\beta t) + 2b \int_0^t e^{-\alpha t} I_0(\beta t) \, dt \right]. \qquad (8)$$

Alternatively, using the Mellin theorem, (5) yields

$$\underset{x=0}{I} = \frac{E_0 \sqrt{(\mathbf{C}/\mathbf{L})}}{2\pi i} \left[\int_{Br_1} \frac{e^{zt} \, dz}{\sqrt{\{(z + \alpha)^2 - \beta^2\}}} \right.$$
$$\left. + 2b \int_{Br_1} \frac{e^{zt} \, dz}{z \sqrt{\{(z + \alpha)^2 - \beta^2\}}} \right]. \qquad (9)$$

Writing $z + \alpha = \zeta$, $dz = d\zeta$ in the first integral of (9), omitting the external factor and replacing ζ by z thereafter, we get

$$I_1 = \frac{e^{-\alpha t}}{2\pi i} \int_{Br_1} \frac{e^{zt} \, dz}{\sqrt{(z^2 - \beta^2)}} = e^{-\alpha t} I_0(\beta t), \qquad (10)$$

from (5) § 5·34. Owing to z in the denominator, the second integral in (9) is $2b$ times the integral from 0 to t of the first integral, so from (9) and (10) we obtain ultimately the head end or sending current

$$I \atop x=0 = E_0 \sqrt{(C/L)} \left[e^{-\alpha t} I_0(\beta t) + 2b \int_0^t e^{-\alpha t} I_0(\beta t)\, dt \right], \qquad (11)$$

this being the final form of solution. The integral in (11) can be evaluated only by computation using numerical data. Since $I_0(\beta t)$ is unity when $t = 0$, the current to the cable jumps to the value $E_0 \sqrt{(C/L)}$ at the transmitting end. When t is large enough [234] $I_0(\beta t) \sim e^{\beta t}/\sqrt{(2\pi\beta t)}$, so (11) may be written $(\alpha > \beta > 0)$

$$I \atop x=0 = E_0 \sqrt{(C/L)} \left[e^{(\beta-\alpha)t}/\sqrt{(2\pi\beta t)} \underset{t \to \infty}{} + 2b \int_0^\infty e^{-\alpha t} I_0(\beta t)\, dt \right] \quad (12)$$

$$= E_0 \sqrt{(C/L)}\, 2b/\sqrt{(\alpha^2 - \beta^2)} = E_0 \sqrt{(G/R)}, \qquad (13)$$

the integral being evaluated by aid of (51) Appendix 10. The same result can also be obtained more readily by making $| p | \ll 2a$, $2b$ in (4) (see §6·41 et seq.), which gives

$$I \atop x=0 \sim E_0 \sqrt{(C/L)} \sqrt{(b/a)}. \qquad (14)$$

Formula (13) represents the leakage current through the insulation, there being no normal conduction current.

13·33. Current at any point of cable. From §13·21,

$$\lambda = \sqrt{\{LC(p + R/L)(p + G/C)\}} = \frac{1}{v} \sqrt{\{(p + 2a)(p + 2b)\}},$$

where $v = 1/\sqrt{(LC)}$ is the velocity of propagation of current along the cable. For *any* value of $x \geq 0$, (3) §13·32 may be expressed in the form

$$\phi = E_0 \sqrt{(C/L)} \frac{(p + 2b)\exp\left[-\dfrac{x}{v} \sqrt{\{(p + 2a)(p + 2b)\}} \right]}{\sqrt{\{(p + 2a)(p + 2b)\}}}, \qquad (1)$$

where $(p + 2a)(p + 2b) = (p + \alpha)^2 - \beta^2$. The inverse of (1) may be obtained directly from (47) Appendix 10 and (3) §8·45. Alternatively the Mellin theorem may be applied.

Consider inversion of the first term in (1). Omitting the external factor, we have

$$I_1 = \frac{1}{2\pi i} \int_{Br_1} \frac{\exp\left[zt - \frac{x}{v}\sqrt{\{(z+\alpha)^2 - \beta^2\}}\right] dz}{\sqrt{\{(z+\alpha)^2 - \beta^2\}}}. \tag{2}$$

Writing $\zeta = (z+\alpha)$, $d\zeta = dz$, and replacing ζ by z thereafter, (2) becomes

$$I_1 = \frac{e^{-\alpha t}}{2\pi i} \int_{Br_1} \frac{\exp\left[zt - \frac{x}{v}\sqrt{(z^2 - \beta^2)}\right] dz}{\sqrt{(z^2 - \beta^2)}}. \tag{3}$$

Substituting $z = u/(t - x/v) + \beta^2(t - x/v)/4u$ in (3) and replacing u by z thereafter,* we obtain

$$I_1 = \frac{e^{-\alpha t}}{2\pi i} \int_{Br_1} \frac{e^{z + \beta^2(t^2 - x^2/v^2)/4z} dz}{z}, \tag{4}$$

the substitution being justifiable by § 10 Appendix 5.

By (7) § 5·33 $I_1 = e^{-\alpha t} I_0\{\beta \sqrt{(t^2 - x^2/v^2)}\},$ \hfill (5)

$t \geqslant x/v$, a restriction in accordance with § 4·41, the current being zero when $t < x/v$.

Now the second part of (1) is $2b/p$ times the first part, so by (5),

$$I_2 = 2b \int_{x/v}^{t} e^{-\alpha t} I_0\{\beta \sqrt{(t^2 - x^2/v^2)}\} dt, \tag{6}$$

the lower limit of integration being x/v, as in (5) § 8·45. Adding (5) and (6), and replacing the external factor from (1), the current at any point distant x from the transmitter is

$$I = E_0 \sqrt{(C/L)} \left[e^{-\alpha t} I_0\{\beta \sqrt{(t^2 - x^2/v^2)}\} \right.$$
$$\left. + 2b \int_{x/v}^{t} e^{-\alpha t} I_0\{\beta \sqrt{(t^2 - x^2/v^2)}\} dt \right], \tag{7}$$

$t \geqslant x/v$ [129].

The nature of the solution (7) indicates that there is no current at a point distant x from the transmitter until a time-

---

* This substitution may also be written
$$u = \tfrac{1}{2}(t - x/v) \{z + \sqrt{(z^2 - \beta^2)}\}.$$

interval $t_1 = x/v$ has elapsed after connecting the battery to the cable. The current rises precipitately to the value

$$E_0 \sqrt{(C/L)} \exp\left[-\frac{1}{2}\left(\frac{R}{L}+\frac{G}{C}\right)\frac{x}{v} \right],$$

i.e. it travels along the cable with a vertical wave front whose height steadily decreases with increase in x. The attenuation coefficient of the wave front is

$$e^{-\alpha x/v} = \exp\left[-\frac{1}{2}\left(\frac{R}{L}+\frac{G}{C}\right)x\sqrt{(LC)} \right],$$

where $x\sqrt{(LC)}$ is the time taken to reach the point x. As time progresses, the current assumes asymptotically a value which may be determined either by letting $t\to\infty$ in (7), or by taking $|p|\ll 2a$, $2b$ in (1),* and inverting. Using the latter procedure, we obtain

$$\underset{t\to\infty}{I} = E_0\sqrt{(C/L)}\sqrt{(b/a)}\,e^{-2x\sqrt{(ab)}/v} \tag{8}$$

$$= E_0\sqrt{(G/R)}\,e^{-x\sqrt{(GR)}}, \tag{9}$$

which is the leakage current through the insulation from the point x onwards. The relationship (7) between current and time is indicated in Fig. 80 for three cases. In (1), (2) I_x increases with increase in t, whereas in (3) I_x decreases (see [236], chapter 4).

13·34. P.d. at any point of cable. Taking the value of λ from § 13·33, then by (1) § 13·32 the p.d. at any point distant x from the transmitter is

$$\varphi = E_0 \exp\left[-\frac{x}{v}\sqrt{\{(p+2a)(p+2b)\}} \right]$$

$$= E_0 \exp\left[-\frac{x}{v}\sqrt{\{(p+\alpha)^2-p^2\}} \right]. \tag{1}$$

The inverse of (1) may be got immediately from (46) Appendix 10. Alternatively, the Mellin theorem may be used as shown below. The contour Br_2 is equivalent to Br_1, so we have

$$E = \frac{E_0}{2\pi i}\int_{Br_1} \frac{e^{zt-(x/v)\sqrt{\{(z+\alpha)^2-\beta^2\}}}\,dz}{z}. \tag{2}$$

---

* This is equivalent to the procedure in § 6·41 et seq.

This integral converges uniformly with respect to t in $0 < t_1 \leqslant t \leqslant t_2$, so it represents a continuous function of t in the interval. Differentiating under the sign with respect to t, we get

$$I_1 = \frac{\partial E}{\partial t} = \frac{E_0}{2\pi i} \int_{Br_2} e^{zt - (x/v)\sqrt{\{(z+\alpha)^2 - \beta^2\}}}\, dz, \qquad (3)$$

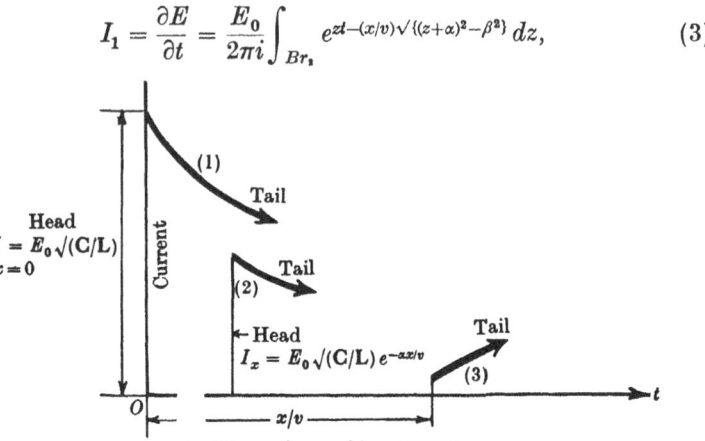

Fig. 80. Illustrating form of current-time curve at various x in a loaded cable.
(1) $x = 0$; the tail is asymptotic to $I = E_0 \sqrt{(G/R)}$.
(2) $x < 4 \sqrt{(LC)}/(CR - LG)$.
(3) $x > 4 \sqrt{(LC)}/(CR - LG)$. In (2), (3), the tail is asymptotic to
$$I = E_0 \sqrt{(G/R)}\, e^{-z\sqrt{(GR)}}.$$

this procedure being valid since (3) is uniformly convergent on Br_2. Integrating under the sign with respect to x and changing thereafter to an equivalent Br_1, yields

$$\int I_1\, dx = -\frac{E_0 v}{2\pi i} \int_{Br_1} \frac{e^{zt - (x/v)\sqrt{\{(z+\alpha)^2 - \beta^2\}}}\, dz}{\sqrt{\{(z+\alpha)^2 - \beta^2\}}} + \chi(t), \qquad (4)$$

where $\chi(t)$ is a function of t alone. Thus by (2), (5) § 13·33,

$$\int I_1\, dx = -E_0 v\, e^{-\alpha t} I_0\{\beta \sqrt{(t^2 - x^2/v^2)}\} + \chi(t). \qquad (5)$$

The integration is permissible in virtue of uniform convergence in $0 \leqslant x \leqslant x_1$, $0 < t_1 \leqslant t \leqslant t_2$. Differentiating (5) with respect to x, we obtain

$$I_1 = E_0 \frac{\beta x\, e^{-\alpha t}}{v} \frac{I_1\{\beta \sqrt{(t^2 - x^2/v^2)}\}}{\sqrt{(t^2 - x^2/v^2)}}. \qquad (6)$$

Then by (3) and (6), we get

$$E = E_0 \frac{\beta x}{v} \int_{x/v}^{t} \frac{e^{-\alpha t} I_1\{\beta \sqrt{(t^2 - x^2/v^2)}\}\, dt}{\sqrt{(t^2 - x^2/v^2)}} + \xi(x), \qquad (7)$$

where $\xi(x)$ is a function independent of t. To determine $\xi(x)$ we proceed as follows: When $t \to \infty$ the p.d. at x is, by (9) § 13·33,

$$E = E_0 - \mathbf{R} \int_0^x E_0 \sqrt{(G/R)}\, e^{-x\sqrt{(GR)}}\, dx = E_0 e^{-x\sqrt{(GR)}}. \qquad (8)$$

From (15) § 8·7 when $t \to \infty$, the value of the second member of (7), omitting E_0, is

$$\exp\left[-\frac{x}{v} \sqrt{(\alpha^2 - \beta^2)} \right] - \exp\left[-\frac{\alpha x}{v} \right]$$

$$= \exp[-x\sqrt{(GR)}] - \exp[-ax/v]. \qquad (9)$$

Now (8) gives the value of E in (7) when $t \to \infty$, so $\xi(x) = E_0 e^{-\alpha x/v}$, and for $t \geqslant \dfrac{x}{v}$

$$E = E_0\left[e^{-ax/v} + \frac{\beta x}{v} \int_{x/v}^{t} \frac{e^{-\alpha t} I_1\{\beta \sqrt{(t^2 - x^2/v^2)}\}\, dt}{\sqrt{(t^2 - x^2/v^2)}} \right]. \qquad (10)$$

The relationship between E and t is depicted in Fig. 81.

13·35. Impulsive p.d. applied to long loaded cable.
The problem is to determine the current at any point distant x from the transmitter, when the p.d. at $x = 0$, $t = 0$, is $SI(t)$, the cable being quiescent initially. By (7) § 13·33, the solution for an applied p.d. $H(t)$ may be written

$$\bar{I}_x = K\left[e^{-\alpha t} I_0(\beta u) + (\alpha - \beta) \int_{x/v}^{t} e^{-\alpha t} I_0(\beta u)\, dt \right], \qquad (1)$$

where $u = \sqrt{(t^2 - x^2/v^2)}$, $K = \sqrt{(C/L)}$, and $(\alpha - \beta) = 2b$. Applying (6) § 10·33 with $f(\beta u) = \bar{I}_x$, we get

$$\frac{d\bar{I}_x}{dt} = K\beta e^{-\alpha t}\left[\frac{t I_1(\beta u)}{u} - I_0(\beta u) \right], \qquad (2)$$

and for $t = x/v$, $\qquad\qquad \bar{I}_x = K e^{-\alpha x/v}. \qquad (3)$

Thus for $SI(t)$, the current is

$$I_x = SK \left\{ e^{-\alpha x/v} \underset{\substack{t=x/v \\ x \geqslant 0}}{I(t-x/v)} + \beta e^{-\alpha t} \left[\underset{\substack{t > x/v \\ x > 0}}{tI_1(\beta u)/u - I_0(\beta u)} \right] \right\}.$$

(4)

When $t = x/v$, the first term on the right-hand side of (4) represents a current impulse at any point x. It travels from $x = 0$ with velocity v, and is attenuated exponentially *en route*. The second

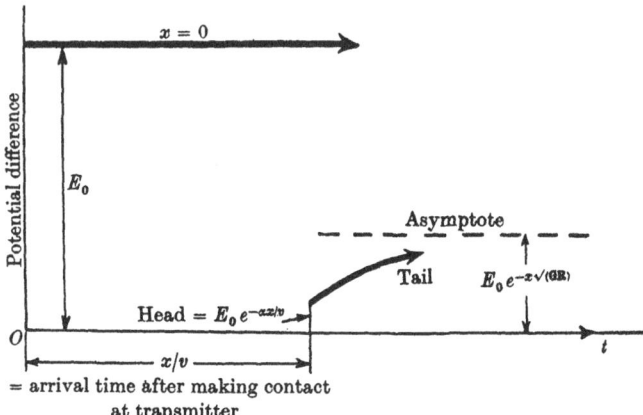

Fig. 81. Illustrating p.d.-time curve at $x = 0$, and $x > 2v/\beta$, in loaded cable.

and third terms represent the 'tail' of the disturbance for $t > x/v$. The electrical energy supplied instantaneously at any x when $t = x/v$, is dissipated gradually throughout the cable with increase in time, and $\to 0$ as $t \to +\infty$. At $t = (x/v + 0)$ the size of the tail is $SK\beta e^{-\alpha x/v}(\beta x/2v - 1)$. If then $\beta x/2v < 1$, current flows towards $x = 0$, whereas if $\beta x/2v > 1$, it flows towards the far end of the cable. The tail is zero at $t = (x/v + 0)$, when $\beta x/2v = 1$.

Taking $S = 5$ volt-seconds, $\mathbf{L} = 0\cdot1$ henry, $\mathbf{C} = 0\cdot4$ microfarad, $\mathbf{R} = 1\cdot8$ ohm, $\mathbf{G} = 10^{-6}$ ohm, the reader should plot current-time curves for $x = 500, 2000$ nautical miles.

13·36. Applied p.d. an arbitrary function of t, say, $\xi_2(t)$.
We suppose that $\xi_2(t)$ has a transform $\phi_2(p)$. This is written for E_0 in the transform solution. Inversion of the resulting expression gives the solution for initial quiescence. For non-quiescent

conditions, the procedure in § 13·13 may be used. Alternatively the product theorem in § 1, Appendix 8 may be applied as indicated in Prob. 93, p. 332. In using this method $\phi_2(p)$ need not be known (see § 13·522, also [236], pp. 60, 96).

13·41. Cable with terminal apparatus. The p.d. at any point in the cable (Fig. 82) is given by (10) § 13·21, while from (7) and (10) § 13·21

$$I \Rightarrow \phi = \frac{\lambda}{(\mathbf{L}p + \mathbf{R})} (A e^{-\lambda x} - B e^{\lambda x}) \tag{1}$$

$$= \frac{1}{Z_c} (A e^{-\lambda x} - B e^{\lambda x}), \tag{2}$$

where $Z_c = \sqrt{\{(\mathbf{L}p + \mathbf{R})/(\mathbf{C}p + \mathbf{G})\}}$ is the characteristic or surge impedance of the line. If there is apparatus at the sending and

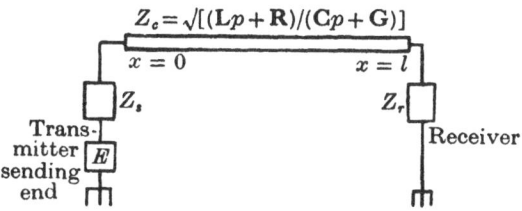

Fig. 82. Diagram for cable with terminal apparatus.

receiving ends of impedance Z_s, Z_r, respectively, with $E = E_0 H(t)$ the transform of the p.d. applied to the cable at $x = 0$ is by (10) § 13·21,

$$\varphi_0 = E_0 - \phi_0 Z_s = (A e^{-\lambda x} + B e^{\lambda x})_{x=0}, \tag{3}$$

or

$$E_0 - \phi_0 Z_s = A + B. \tag{4}$$

Since

$$\phi_0 = (A - B)/Z_c, \tag{5}$$

from (2) when $x = 0$, we have, by (4), (5),

$$A(1 + Z_s/Z_c) + B(1 - Z_s/Z_c) = E_0. \tag{6}$$

At the far end of the cable $x = l$, $\varphi_r = \phi_r Z_r$, and by this, (2) and (10) § 13·21, the p.d. is

$$E_r \Rightarrow \varphi_r = \frac{Z_r}{Z_c} (A e^{-\lambda l} - B e^{\lambda l}) = A e^{-\lambda l} + B e^{\lambda l}, \tag{7}$$

or

$$A e^{-\lambda l} (1 - Z_r/Z_c) + B e^{\lambda l} (1 + Z_r/Z_c) = 0. \tag{8}$$

From (6) and (8), we obtain

$$A = E_0(1+a_2)/[(1+a_1)(1+a_2)-(1-a_1)(1-a_2)e^{-2\lambda l}], \quad (9)$$

and

$$B = -E_0(1-a_2)e^{-2\lambda l}/[(1+a_1)(1+a_2)-(1-a_1)(1-a_2)e^{-2\lambda l}], \quad (10)$$

where $a_1 = Z_s/Z_c$, and $a_2 = Z_r/Z_c$. Inserting these values of A and B in (2), the current at any point in the cable is given by

$$I_x \Rightarrow \phi = \frac{E_0}{Z_s+Z_c}\left[\frac{e^{-\lambda x}+b_2 e^{-\lambda(2l-x)}}{1-b_1 b_2 e^{-2\lambda l}}\right], \quad (11)$$

where $b_1 = (1-a_1)/(1+a_1)$, and $b_2 = (1-a_2)/(1+a_2)$. If $Z_s = 0$ and $Z_r = Z_c$, then $a_1 = 0, a_2 = 1, b_1 = 1, b_2 = 0$; so (11) yields

$$I_x \Rightarrow \phi = \frac{E_0}{Z_c}e^{-\lambda x}, \quad (12)$$

which is identical with the formula (3) § 13·32 for the current in a line devoid of reflection.

13·42. Wave solution. Since $|b_1 b_2 e^{-2\lambda l}|$ can always be taken less than unity by suitable choice of the contour Br_1, thereby making the real part of z large enough (c in Fig. 22 (a)), (11) § 13·41 may be expanded as follows:

$$\phi = \frac{E_0}{Z_s+Z_c}[e^{-\lambda x}+b_2 e^{-\lambda(2l-x)}][1+b_1 b_2 e^{-2\lambda l}+b_1^2 b_2^2 e^{-4\lambda l}+\ldots]$$

$$(1)$$

$$= \frac{E_0}{Z_s+Z_c}[e^{-\lambda x}+b_2 e^{-\lambda(2l-x)}+b_1 b_2 e^{-\lambda(2l+x)}+b_1 b_2^2 e^{-\lambda(4l-x)}+\ldots].$$

$$(2)$$

The first term in (2) is the transform of the current wave in a very long reflectionless line; the second term represents a wave which has travelled to the receiver at $x = l$ (in the finite line considered) and is reflected back to x; the third represents a wave which has travelled to $x = l$, back to $x = 0$, and thence to x, and so on. Consequently (2) can be regarded—following Heaviside—as the wave solution of the transmission line. When inverting (2), Z_s, Z_r, Z_c, b_1, b_2 must be expressed as functions of z.

13·43. Application of (2) §13·42 to unloaded cable. As a simple illustration, we can consider the arrival current in an unloaded cable of length l connected direct to earth at its far end. Here $a_1 = a_2 = 0$, so $b_1 = b_2 = 1$, $Z_c = \sqrt{(R/Cp)}$, $\lambda = \sqrt{(CRp)}$. For quiescence initially, let the p.d. applied at $x = 0$, $t = 0$ be $E = E_0 H(t)$. Substituting these values in (2) §13·42, we obtain

$$\phi = 2E_0 \sqrt{(C/R)} \sqrt{p} \, (e^{-a\sqrt{p}} + e^{-3a\sqrt{p}} + e^{-5a\sqrt{p}} + \ldots), \qquad (1)$$

where $a = l\sqrt{(CR)}$. In inverting (1) we have to evaluate a series of integrals of the type given at (4) §5·222. Thus we find that the arrival current at the far end of the line is

$$I_l = 2E_0 \sqrt{(C/\pi Rt)} \, (e^{-a^2/4t} + e^{-9a^2/4t} + e^{-25a^2/4t} + \ldots), \qquad (2)$$

this being the wave solution of the problem. At (2) §13·25 we gave another solution of this problem in ascending powers of t, so (2) §13·25 and (2) above must be identical. So far as computation is concerned, (2) §13·25 is useful when t is moderately or very large, whereas (2) may be used when t is small.

13·44. Application of (2) §13·42 to loaded cable terminated by a resistance $\sqrt{(L/C)}$. In this case Z_c, the characteristic impedance at the receiving end, is replaced by $Z_r = \sqrt{(L/C)}$, so if $Z_s = 0$, $a_1 = 0$ and $b_1 = 1$. Thus (2) §13·42 becomes

$$\phi = \frac{E_0}{Z_c} \, [e^{-\lambda x} + b_2 e^{-\lambda(2l-x)} + b_2 e^{-\lambda(2l+x)} + b_2^2 e^{-\lambda(4l-x)} + \ldots], \qquad (1)$$

where $\qquad \lambda = \sqrt{\{(Lp + R)(Cp + G)\}}$,

and $\qquad b_2 = (1 - a_2)/(1 + a_2) = (Z_c - Z_r)/(Z_c + Z_r). \qquad (2)$

The leakance can usually be neglected, so we write $G = 0$, giving

$$\lambda = \sqrt{\{LCp(p + R/L)\}} = (1/v) \sqrt{\{(p+a)^2 - a^2\}}, \qquad (3)$$

and $\qquad Z_c = \sqrt{\left(\frac{Lp + R}{Cp}\right)} = \sqrt{\left(\frac{L}{C} \frac{p + 2a}{p}\right)}$

$$= \sqrt{\left(\frac{L}{C} \frac{(p+a)^2 - a^2}{p}\right)} \qquad \left(a = \frac{1}{2}\frac{R}{L}\right). \qquad (4)$$

Substituting for Z_c, Z_r in (2), we get

$$b_2 = \left[1 + \frac{p}{a} - \frac{1}{a} \sqrt{\{(p+a)^2 - a^2\}} \right].$$ (5)

We shall confine our attention to the receiving end of the cable where $x = l$. Since the time taken for a disturbance to travel from end to end of the cable is l/v, exclusion of all but the first two terms in (1) enables the current to be found during the interval $l/v < t < 3l/v$. At $t = l/v$ the head of the signal arrives at the receiving end. Reflection occurs, and when $l/v < t < 3l/v$ the current there is the sum of (a) that flowing from the transmitter, (b) that reflected from Z_r, represented, respectively, by the first two terms in (1).

The first term in (1) is

$$\phi_1 = \frac{E_0 \sqrt{(C/L)}\, p}{\sqrt{\{(p+a)^2 - a^2\}}} \exp\left[-\frac{x}{v} \sqrt{\{(p+a)^2 - a^2\}} \right].$$ (6)

By (1), (5) § 13·33 the inverse of (6) is

$$I_1 = E_0 \sqrt{(C/L)}\, e^{-at} I_0\{a \sqrt{(t^2 - x^2/v^2)}\}.$$ (7)

The second term in (1) is

$$I_2 \Rightarrow \phi_2 = \frac{E_0 \sqrt{(C/L)}\, p}{\sqrt{\{(p+a)^2 - a^2\}}} \exp\left[-\frac{2l - x}{v} \sqrt{\{(p+a)^2 - a^2\}} \right]$$
$$\times \left[1 + \frac{p}{a} - \frac{1}{a} \sqrt{\{(p+a)^2 - a^2\}} \right].$$ (8)

13·45. Inversion of (8) § 13·44. By (43) Appendix 10, with

$$y = \sqrt{(t^2 - k^2)}, \quad s = \sqrt{(p^2 - a^2)}, \quad k = (2l - x)/v,$$

$$I_0(ay) \Rightarrow (p/s)\, e^{-ks}.$$ (1)

Using (8) § 8·43

$$\frac{d}{dt}\{I_0(ay)\} = \frac{at I_1(ay)}{y} \Rightarrow \frac{p^2}{s} e^{-ks} - p\, e^{-kp},$$ (2)

so

$$\frac{t I_1(ay)}{y} \Rightarrow \frac{p}{a} \left(\frac{p}{s} e^{-ks} - e^{-kp} \right).$$ (3)

Further, by (12), § 8·7

$$-\frac{kI_1(ay)}{y} = \frac{p}{a}(e^{-kp} - e^{-ks}).\tag{4}$$

Adding (3), (4) leads to

$$(t-k)\frac{I_1(ay)}{y} = \frac{p}{s}e^{-ks}\left[\frac{p}{a} - \frac{s}{a}\right].\tag{5}$$

Applying (3) § 8·411 to (5), gives

$$e^{-at}(t-k)I_1\frac{\{a\sqrt{(t^2-k^2)}\}}{\sqrt{(t^2-k^2)}} = \frac{p\,e^{-k\sqrt{((p+a)^2-a^2)}}}{\sqrt{\{(p+a)^2-a^2\}}}$$

$$\times\left[1+\frac{p}{a} - \frac{\sqrt{\{(p+a)^2-a^2\}}}{a}\right],\tag{6}$$

the left-hand side being the inverse of (8) § 13·44 without $E_0\sqrt{(C/L)}$.

Thus from (6) and (8) § 13·44, we obtain

$$I_2 = E_0\sqrt{(C/L)}\,e^{-at}(t-k)\frac{I_1(ay)}{y}.\tag{7}$$

Writing $x = l$ in (7) § 13·44 and (7) above, by addition we get

$$I = E_0\sqrt{(C/L)}\,e^{-at}\left\{I_0(aw_1) + \left(t-\frac{l}{v}\right)\frac{I_1(aw_1)}{w_1}\right\},\tag{8}$$

where

$$w_1 = \sqrt{(t^2 - l^2/v^2)}.$$

Formula (8) holds for $l/v \leqslant t \leqslant 3l/v$, so when $t = 3l/v$, we have

$$I = E_0\sqrt{(C/L)}\,e^{-at}\left[I_0(\sqrt{8}\,al/v) + \frac{1}{\sqrt{2}}I_1(\sqrt{8}\,al/v)\right].\tag{9}$$

For a loaded cable we may take $a = 9\,\mathrm{sec.}^{-1}$, $l = 3500$ nautical miles, $v = 5000$ n.m. per sec. which gives $\sqrt{8}\,al/v = 17\cdot8$. With this value of the argument $I_0 \simeq I_1$, so the current at $x = l$ is about $1 + 1/\sqrt{2}$ times that in a line devoid of reflection.

13·46. Increased signal speed with loaded cable. The *initial* part of the current arrival curve in an *unloaded* cable rises slowly (Fig. 76). Only this part is important in rapid signalling, and consequently in an unloaded cable the speed is limited.

As shown in §§ 13·27, 13·28, 'shaping' offsets this defect up to a point, after which the signals tend to illegibility. The speed may be increased appreciably by using a loaded cable (§ 13·31). When **L, R, C, G** are *independent* of both current and frequency, the signal has a vertical *front* (Fig. 84, curve 3) throughout the cable. In practice, however, as shown in § 13·47, increase in **R** and decrease in **L** with rise in frequency cause the steepness of the signal profile to decrease with increase in the distance travelled along the cable. Nevertheless, the profile is much steeper than that in an unloaded cable of equal length. Hence a greater signalling speed may be attained in a loaded cable than in an unloaded one, before the signals at the receiver become unreadable. A speed ratio of five to one is practicable.

13·47. Practical data for loaded cable. We now describe experiments with a loaded cable 3467 nauts in length. The propagation time of the cable (length/velocity) was approximately 0·67 sec., this being the time interval between application of the battery at the sending end and arrival of the current at the receiver. To reduce reflection there, the cable was earthed through a resistance of value $\sqrt{(L/C)} = 500$ ohms. This is an approximation to the alternating current characteristic impedance $Z_c = \sqrt{\{(i\omega L + R)/(i\omega C + G)\}}$ when $\omega L \gg R$, $\omega C \gg G$. As shown in Fig. 83 (*a*), the p.d. across the resistance was applied to the input of a thermionic valve amplifier, thence to an oscillograph. Records of two tests when a single Morse dot was transmitted, are reproduced in Fig. 84, the circuit of Fig. 83 (*a*) being used, so that there was no shaping of the signal. The calculated value of the current for constant **L, R, C, G**, is shown by the heavy curve (3). The longer the duration of a dot, the closer the actual curve approaches the calculated one. Comparison with Fig. 76 shows the higher initial rate of rise in the loaded cable, but the rise is not precipitate as in theory. We demonstrated in § 10·52 that a Morse dot may be represented by a band frequency spectrum. Due to increase in **R** and decrease in **L**, the amplitudes of the higher frequency components are attenuated in comparison with the lower ones, so the initial rate of rise of the arrival current suffers reduction. The cable coefficients were almost

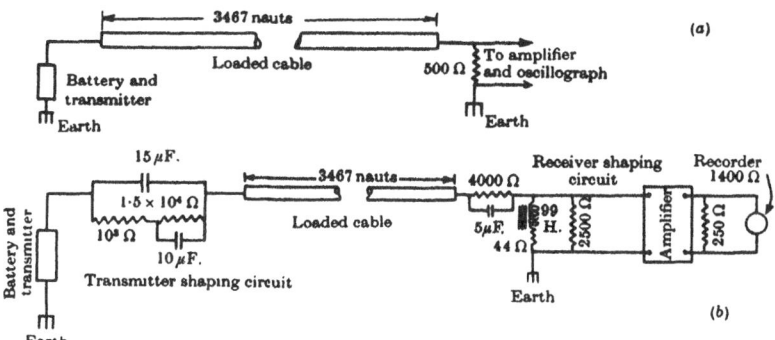

Fig. 83. (a) Test circuit for loaded cable, without shaping apparatus at receiver or transmitter. (b) Circuit for simplex (one way) working of loaded cable, with shaping apparatus at both ends.

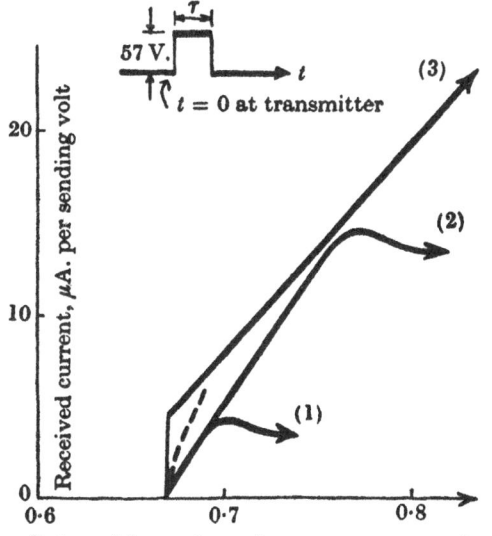

Interval from time of contact at transmitter (sec.)

Fig. 84. Received current in loaded cable using circuit of Fig. 83(a); (1) duration of contact at transmitter 0·023 sec.; (2) duration 0·1 sec.; (3) calculated current curve for constant L, R, C, G; broken line is initial part of current curve using shaping apparatus at the transmitter, showing enhanced rate of rise.

independent of *current*, up to that corresponding to a sending p.d. of 60 volts.

Using the circuit of Fig. 83 (*b*), a second series of tests was made to ascertain the performance of the cable during trans-

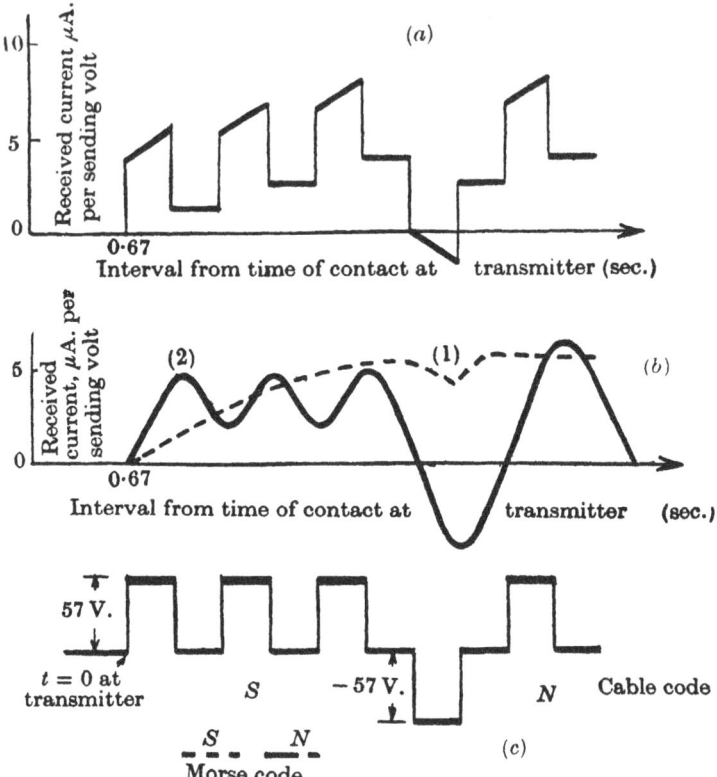

Fig. 85. (*a*) Calculated shape of received signal current in loaded cable with L, R, C, G constant, for circuit of Fig. 83(*a*). (*b*) (1) Signal SN using circuit of Fig. 83(*b*), but *without* shaping apparatus at transmitter. (2) As in (1), but *with* shaping apparatus at transmitter, showing enhanced initial rate of rise. (*c*) Signal SN (understand) in cable code and in Morse code.

mission of the code signal SN, which means 'understand'. At a speed of 600 letters per minute (120 words p.m.) the received signal was unreadable in absence of the shaping circuit at the *transmitter*, as illustrated in Fig. 85(*b*), curve 1. The effect of using the shaping circuit will be evident from curve 2. In Fig. 85 (b) dots are shown above the datum line and dashes below, the latter being obtained by reversing the battery connections at the transmitter, this being cable code practice.

A third series of tests, using signal shaping at both ends of the cable as in Fig. 83 (b), was made to determine the performance under working conditions. A message was transmitted consisting of letters, words, and numbers, arranged in a sequence guaranteed to reveal the existence of any source of signal irregularity in the complete system. Good signals were obtained up to a speed of 1200 letters per minute.

13·51. Electrical filters. An arrangement of electrical impedances known as a filter is shown schematically in Fig. 86 (a). The meshes $m = 0$ and $m = n$ have series impedances $\frac{1}{2}Z_1$, while the terminal impedance Z_r is such as to prevent reflexion from the far end of the filter. By applying Kirchhoff's laws, it can be shown that the currents in the $(m-1)$th, mth, and $(m+1)$th meshes, except the first and last, are related by the linear difference equation*

$$(Z_1 + 2Z_2)\,\phi_m - Z_2(\phi_{m+1} + \phi_{m-1}) = 0, \qquad (1)$$

with $I_m \Rightarrow \phi_m$.

Assume the solution, in absence of reflexion, to be

$$\phi_m = A\,e^{-m\lambda}, \qquad (2)$$

where λ is known as the propagation coefficient of the filter. After substituting into (1) and rearranging, we obtain

$$(1 + Z_1/2Z_2) = \tfrac{1}{2}(e^\lambda + e^{-\lambda}) = \cosh \lambda. \qquad (3)$$

Applying Kirchhoff's law to the section $m = 0$, with $E \Rightarrow \varphi$ we have

$$\varphi - \phi_0 Z_0 = \tfrac{1}{2}\phi_0 Z_1 + Z_2(\phi_0 - \phi_1). \qquad (4)$$

From (2), when $m = 0$ and 1, we get, respectively,

$$\phi_0 = A, \quad \phi_1 = A\,e^{-\lambda},$$

* The solution of difference equations using transform procedure is given in [225] . Z_0, Z_1, etc. are expressed in transform notation, i.e. as functions of p.

so (4) gives $\qquad A = \varphi/[\tfrac{1}{2}Z_1 + Z_0 + Z_2(1 - e^{-\lambda})]$

$$= \varphi/(Z_0 + Z_2 \sinh \lambda), \qquad (5)$$

by (3).

Whence from (2) and (5), the current in Z_1 in the mth mesh is

$$I_m \Rightarrow \phi_m = \frac{\varphi}{(Z_0 + Z_2 \sinh \lambda)(\sinh m\lambda + \cosh m\lambda)}. \qquad (6)$$

The current in the last mesh is given by

$$\phi_n(Z_r + \tfrac{1}{2}Z_1) = (\phi_{n-1} - \phi_n)Z_2,$$

or $\qquad\qquad \phi_n[\tfrac{1}{2}Z_1 + Z_2 + Z_r] = \phi_{n-1}Z_2. \qquad (7)$

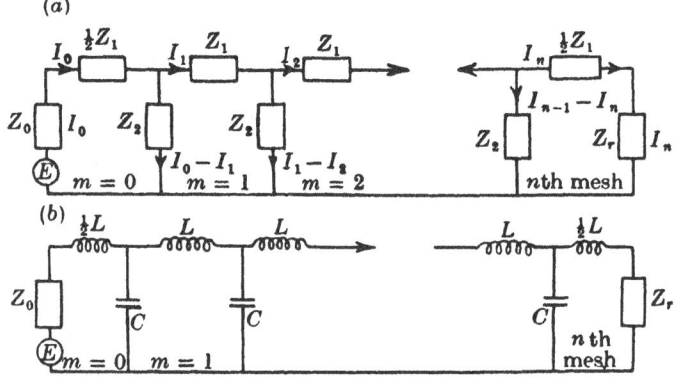

Fig. 86. (a) General diagram for electrical wave filter.
(b) Non-dissipative low-pass electrical filter.

Taking $\phi_n = A e^{-n\lambda}$, $\phi_{n-1} = A e^{-(n-1)\lambda}$ from (2) and substituting in (7), we find that

$$e^{-\lambda}[(Z_1/2Z_2) + 1 + (Z_r/Z_2)] = 1,$$

or by (3), $\qquad\qquad \cosh \lambda + (Z_r/Z_2) = e^{\lambda}, \qquad (8)$

so $\qquad\qquad\qquad Z_r = Z_2 \sinh \lambda \qquad\qquad (9)$

$$= \sqrt{(Z_1 Z_2 + \tfrac{1}{4}Z_1^2)}, \qquad (10)$$

by (3). The right-hand side of (10) is the iterative (repeated) impedance of the filter. If the termination Z_r has this value, there is no change of impedance, which implies absence of reflexion.

It is convenient to express (6) in an alternative form by writing $4\theta^2 = Z_1/Z_2$ in (3). This gives $\cosh \lambda = 1 + 2\theta^2$ and $\sinh \lambda = 2\sqrt{(\theta^2 + \theta^4)}$, so (6) becomes

$$I_m \Rightarrow \phi_m = \frac{\varphi}{\{Z_0 + \sqrt{(Z_1 Z_2)}\sqrt{(\theta^2 + 1)}\}\{\theta + \sqrt{(\theta^2 + 1)}\}^{2m}}. \quad (11)$$

The analysis herein applies also to mechanical and acoustical filters, provided L, C, R are replaced by their analogues. In an acoustical filter, the input would not take the form $H(t)$.

13·521. Low-pass filter. When p.d.s. of sine-wave form are applied at E, the filter passes currents of all frequencies below the cut-off point $\omega_c = 2/\sqrt{(LC)}$: above the cut-off point it passes but little.

In Fig. 86 (b), L and C are inductance and capacitance, respectively. For simplicity assume (i) $Z_0 = 0$; (ii) the resistance of L and the leakance of C are negligible; (iii) a battery of voltage E is applied to the input at $t = 0$. Then $Z_1 = Lp$, $Z_2 = 1/Cp$, so $\theta = \frac{1}{2}\sqrt{(Z_1/Z_2)} = \frac{1}{2}p\sqrt{(LC)}$, and $\sqrt{(Z_1 Z_2)} = \sqrt{(L/C)}$. Substituting these values in (11) § 13·51, we obtain

$$\phi_m = E\sqrt{(C/L)}/\sqrt{(p^2 a^2 + 1)}\{pa + \sqrt{(p^2 a^2 + 1)}\}^{2m}, \quad (1)$$

where $a = \frac{1}{2}\sqrt{(LC)}$.

Inverting (1) by the Mellin theorem or using formulae (4) § 5·34, (3) § 8·45, we have

$$I_m = \frac{E\sqrt{(C/L)}}{a}\int_0^t J_{2m}(t/a)\,dt, \quad (2)$$

since by § 8·421 t/a is written for t, when pa occurs instead of p. The cut-off frequency of the filter is $\omega_c = 2/\sqrt{(LC)} = 1/a$, so

$$I_m = E\sqrt{(C/L)}\,\omega_c\int_0^t J_{2m}(\omega_c t)\,dt = E\sqrt{(C/L)}\int_0^{\omega_c t} J_{2m}(y)\,dy. \quad (3)$$

When $t \to \infty$, the value of the first integral in (3) is $1/\omega_c$, so

$$I_m = E/\sqrt{(L/C)}, \quad (4)$$

a result also obtainable by making $|pa| \ll 1$ in (1), this being equivalent to the procedure in § 6·41 et seq. Formula (4) represents a unidirectional current, since $\sqrt{(L/C)}$ has the same dimensions as a resistance, namely, lt^{-1}.

13·522. Transition to uniform line. The product of inductance and capacitance up to the point x in a uniform line is $x^2 LC$. For the equivalent circuit with m meshes, each having inductance L and capacitance C, it is $m^2 LC = 4m^2 a^2$, where $2a = \sqrt{(LC)}$. For equality

$$4m^2 a^2 = x^2 LC, \tag{1}$$

so
$$a = x/2mv, \tag{2}$$

with $v = 1/\sqrt{(LC)}$. Since x, p, are finite, as $m \to \infty$, $a \to 0$, so $pa \to 0$. Thus in (1) § 13·521 with E_0 for E, if $C/L = \mathbf{C}/\mathbf{L}$, we get

$$\phi_m = E_0 \sqrt{(\mathbf{C}/\mathbf{L})}/(1+pa)^{2m} = E_0 \sqrt{(\mathbf{C}/\mathbf{L})}/(1+px/2mv)^{2m}, \tag{3}$$

and as $m \to \infty$ $\qquad I \Rrightarrow \phi = E_0 \sqrt{(\mathbf{C}/\mathbf{L})}\, e^{-px/v}. \tag{4}$

This is identical with (1) § 13·33 when $\mathbf{R} = \mathbf{G} = 0$, for the uniform loss-free line.

Potential difference a function of t. If the p.d. applied at $x = 0$ is $\xi_2(t) \Rrightarrow \phi_2(p)$, (4) becomes

$$I \Rrightarrow \phi = \sqrt{(\mathbf{C}/\mathbf{L})}\, \phi_2(p)\, e^{-px/v}, \tag{5}$$

so
$$I = \sqrt{(\mathbf{C}/\mathbf{L})}\, \xi_2(t - x/v). \tag{6}$$

By (1) § 13·34 with $\mathbf{R} = \mathbf{G} = 0$, we obtain

$$E = \xi_2(t - x/v). \tag{7}$$

Hence at any x, and $t > x/v$

$$E/I = \sqrt{(\mathbf{L}/\mathbf{C})}, \tag{8}$$

the characteristic impedance of the line. Moreover, the p.d. and current are transmitted without attenuation or distortion. If neither $\mathbf{R}$ nor $\mathbf{G}$ is zero but $\mathbf{R}/\mathbf{L} = \mathbf{G}/\mathbf{C}$, it is shown in [236], p. 97 that

$$I = \sqrt{(\mathbf{C}/\mathbf{L})}\, e^{-\alpha x/v}\, \xi_2(t - x/v), \tag{9}$$

and
$$E = e^{-\alpha x/v}\, \xi_2(t - x/v), \tag{10}$$

with $\alpha = [\mathbf{R}/\mathbf{L} + \mathbf{G}/\mathbf{C}]/2$. There is attenuation, but no change in signal shape, i.e. the condition $\mathbf{R}/\mathbf{L} = \mathbf{G}/\mathbf{C}$ corresponds to a distortionless line, and the ratio E/I is that in (8).

13·523. If in (11) § 13·51, $Z_0 = \theta \sqrt{(Z_1 Z_2)} = Z_1/2$, and $\varphi = E$, a constant, applied at $t = 0$, then

$$\phi_m = E/\sqrt{(Z_1 Z_2)}\, \{\theta + \sqrt{(\theta^2 + 1)}\}^{2m+1}. \tag{1}$$

Using the values of Z_1, Z_2 and θ from § 13·521,

$$\phi_m = \frac{E \sqrt{(C/L)}}{\{pa + \sqrt{(p^2 a^2 + 1)}\}^{2m+1}}. \tag{2}$$

By formula (55) Appendix 10

$$I_m = (2m+1) E \sqrt{\left(\frac{C}{L}\right)} \int_0^t \frac{J_{2m+1}(t/a) \, d(t/a)}{t/a} \tag{3}$$

$$= (2m+1) E \sqrt{\left(\frac{C}{L}\right)} \int_0^t \frac{J_{2m+1}\{2t/\sqrt{(LC)}\} \, dt}{t}$$

$$= E \sqrt{\left(\frac{C}{L}\right)} \left[\frac{2}{\sqrt{(LC)}} \int_0^t J_{2m}\left\{ \frac{2t}{\sqrt{(LC)}} \right\} dt - J_{2m+1}\left\{ \frac{2t}{\sqrt{(LC)}} \right\} \right], \tag{4}$$

by a recurrence formula [234]. (4) can be expressed in a form more convenient for computation. By writing $y = \omega_c t = 2t/\sqrt{(LC)}$, we get

$$I_m = E \sqrt{\left(\frac{C}{L}\right)} \left[\int_0^{\omega_c t} J_{2m}(y) \, dy - J_{2m+1}(y) \right]. \tag{5}$$

Inverting (2) by the Mellin theorem and making $za \ll 1$, or by writing $p = 0$ in (2) (see § 6·41 et seq.), we find that when $t \to \infty$,

$$I_m \to E \sqrt{(C/L)}, \tag{6}$$

as at (4) § 13·521. This is to be expected, since Z_0 is a pure inductance and there is no p.d. across it due to a constant unidirectional current. When the applied p.d. E is either sinusoidal or a complex wave due to speech or music, each change in wave form is accompanied by a transient in the filter.

13·53. High-pass filter. This filter (Fig. 87 (a)) passes sinusoidal currents of all frequencies above the cut-off point $\omega_c = 1/2\sqrt{(LC)}$. The filter may be formed by interchanging the corresponding elements in the low-pass filter Fig. 86 (b), except that the capacitances in the zeroth and nth meshes are doubled to comply with Fig. 86 (a), where the impedances are $\frac{1}{2}Z_1$. Taking $Z_0 = 0$, $Z_1 = 1/pC$, $Z_2 = pL$, $\theta = 1/2p\sqrt{(LC)} = b/p$, where $b = 1/2\sqrt{(LC)}$; also $\sqrt{(Z_1 Z_2)} = \sqrt{(L/C)}$, and substituting these values in (11) § 13·51, we get for an applied battery voltage E,

$$\phi_m = E \sqrt{(C/L)}/\sqrt{\{(b/p)^2 + 1\}} \, [(b/p) + \sqrt{\{(b/p)^2 + 1\}}]^{2m}. \tag{1}$$

This transform solution is identical with that of (1) § 13·521 provided b is written for a and $1/p$ for p. Now it can be shown that*

$$\phi(1/p) \leftrightharpoons -\left[f(u) J_0\{2\sqrt{(ut)}\} \right]_0^\infty + \int_0^\infty J_0\{2\sqrt{(ut)}\} f'(u)\,du, \quad (2)$$

where $f(u)$ is the inverse of $\phi(p)$, i.e. that of (1) § 13·521. Thus from (3) § 13·521, writing τ for t, we have

$$f(u) = f(\omega_c \tau) = E\sqrt{(C/L)} \int_0^\tau J_{2m}(\omega_c \tau)\,d(\omega_c \tau). \quad (3)$$

When $\tau = 0$, $u = \omega_c \tau = 0$ and $f(0) = 0$; while when $\tau \to \infty$, $f(u) = E\sqrt{(C/L)}$ and $J_0\{2\sqrt{(ut)}\} = 0$. Consequently the bracketed quantity in (2) is zero and

$$\phi(1/p) \leftrightharpoons \int_0^\infty J_0\{2\sqrt{(ut)}\} f'(\omega_c \tau)\,d(\omega_c \tau). \quad (4)$$

From (3) we get

$$\frac{d}{du} f(u) = E\sqrt{(C/L)}\, J_{2m}(\omega_c \tau). \quad (5)$$

Substituting from (5) into (4), the current in the mth mesh of Fig. 87 (a) at any time $t > 0$ is

$$I_m = E\sqrt{(C/L)} \int_0^\infty J_0\{2\sqrt{(\omega_c \tau t)}\} J_{2m}(\omega_c \tau)\,d(\omega_c \tau) \leftrightharpoons \phi(1/p). \quad (6)$$

To obtain I_m when $t \to 0$, as in § 6·14 we take c in Fig. 22 to be large but finite, so that on Br_1, $|z| \gg b$. Then from (1) with $|p| \gg b$, we get

$$I_m \leftrightharpoons \phi_m = E_0 \Big/ \sqrt{\left(\frac{L}{C}\right)},$$

a value attained immediately the circuit is completed, since all the uncharged condensers behave as short-circuits. $\sqrt{(L/C)}$ has the dimensions of a resistance as in the low-pass filter, being equal to $\sqrt{(Z_1 Z_2)}$.

* In reference [162] it is shown that

$$\int_0^\infty \sqrt{(t/u)}\, J_1\{2\sqrt{(ut)}\} f(u)\,du \leftrightharpoons \phi(1/p).$$

The right-hand side of (2) is obtained on integrating by parts using

$$d\{J_0(v)\}/dv = -J_1(v).$$

13·541. Artificial line. A network akin to that employed for balancing unloaded submarine telegraph cables used in duplex working is illustrated in Fig. 87 (b). Here $Z_0 = 0$, $Z_1 = R, Z_2 = 1/pC$, giving $\theta = \frac{1}{2}\sqrt{(pCR)}$ and $\sqrt{(Z_1 Z_2)} = \sqrt{(R/pC)}$.

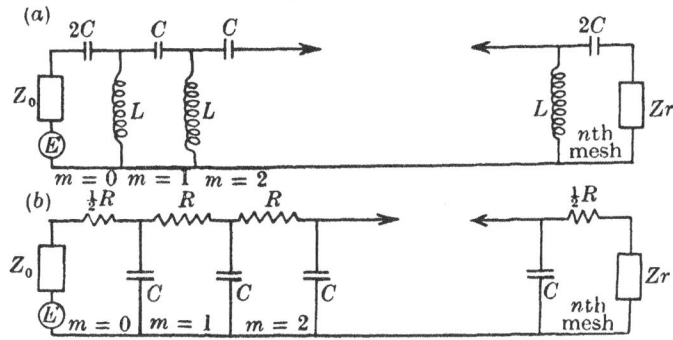

Fig. 87. (a) Non-dissipative high-pass electrical filter. (b) Electrical artificial line for non-inductive cable.

Substituting the values of θ and $\sqrt{(Z_1 Z_2)}$ in (11) § 13·51, the transform of the current in the mth mesh, for an applied battery voltage E, is

$$\phi_m = E\sqrt{(C/R)}\, b^{m+\frac{1}{2}}\sqrt{p}/\sqrt{(p+b)}\{\sqrt{p}+\sqrt{(p+b)}\}^{2m}, \qquad (1)$$

where $b = 4/CR$. Multiplying above and below by $\sqrt{p}$ and squaring the bracketed quantity in the denominator of (1), we obtain

$$\phi_m = \frac{E\sqrt{(C/R)}\, b^{m+\frac{1}{2}}\, p}{\sqrt{(p^2+pb)}\,[(2p+b)+2\sqrt{(p^2+pb)}]^m} \qquad (2)$$

$$= \frac{(2E/R)(b/2)^m\, p}{\sqrt{\{(p+b/2)^2 - b^2/4\}}\,[(p+b/2)+\sqrt{\{(p+b/2)^2 - b^2/4\}}]^m}. \qquad (3)$$

Thus
$$I_m = 2(E/R)\, e^{-2t/CR}\, I_m(2t/CR), \qquad (4)$$

by (5) § 5·34 and (1) § 8·423, the last factor in (4) being a modified Bessel function of the first kind.

When $t = 0$, $I_m = 0$ $(m > 0)$, and when t is large enough for the asymptotic formula for the Bessel function [234] to be used, we have
$$I_m(2t/CR) \sim \frac{1}{2}e^{+2t/CR}\sqrt{(CR/\pi t)}, \qquad (5)$$

so from (4), (5)
$$I_m \sim E\sqrt{(C/\pi Rt)}, \qquad (6)$$

which is identical in type with the formula for the input current to a very long unloaded cable as given at (7) § 13·23. This formula can be obtained also from (1) by neglecting p in the denominator in comparison with b, and inverting, $\sqrt{p}$ in the numerator being retained (see § 6·41 et seq.).

13·542. Transition to uniform line. Proceeding as in § 13·522, we have

$$4m^2/b = x^2 CR, \tag{1}$$

so

$$\sqrt{b} = 2m/x\sqrt{(CR)}. \tag{2}$$

Now with E_0 for E, (1) § 13·541 may be written

$$\phi_m = E_0\sqrt{(C/R)}\,\sqrt{p}/\sqrt{(1+p/b)}\,[\sqrt{(p/b)} + \sqrt{(1+p/b)}]^{2m}, \tag{3}$$

and since x, p are finite, as $m \to +\infty$, $b \to +\infty$, so $p/b \to 0$. Thus with $C/R = \mathbf{C}/\mathbf{R}$, we get

$$\lim_{m\to\infty} \phi_m = \lim_{m\to\infty} E_0\sqrt{(\mathbf{C}/\mathbf{R})}\sqrt{p}/[1+\sqrt{(p/b)}]^{2m} = E_0\sqrt{(\mathbf{C}/\mathbf{R})}\,\sqrt{p}\,e^{-x\sqrt{(\mathbf{CR}\,p)}}, \tag{4}$$

so

$$\phi \mathbf{\subset} I = E_0\sqrt{(\mathbf{C}/\pi\mathbf{R}t)}\,e^{-x\,\mathbf{CR}/4t}, \tag{5}$$

by (34) Appendix 10, which is identical with (6) § 13·23.

13·551. Variation in Z_r with ω. (10) § 13·51 may be written

$$Z_r = \sqrt{(Z_1 Z_2)}\sqrt{(1+\theta^2)}, \tag{1}$$

where $\theta = \frac{1}{2}\sqrt{(Z_1/Z_2)}$. Substituting the values of Z_1, Z_2 for a low-pass filter from § 13·521, we get

$$Z_r(p) = \sqrt{(L/C)}\sqrt{[1+(p/\omega_c)^2]}. \tag{2}$$

For sinusoidal current, we replace p by $i\omega$ and (2) becomes

$$Z_r(i\omega) = R_0\sqrt{[1-(\omega/\omega_c)^2]} \qquad (0 \leqslant \omega < \omega_c) \tag{3}$$

$$= iR_0\sqrt{[(\omega/\omega_c)^2 - 1]} \qquad (\omega_c < \omega < \infty), \tag{4}$$

with $R_0 = \sqrt{(L/C)}$. (3), (4) show that in the 'pass' range, as ω increases from 0 to ω_c, Z_r is real and decreases from R_0 to 0. In the 'stop' range, where $\omega > \omega_c$, Z_r is imaginary and increases with increase in ω. When loss occurs Z_r is complex (see § 13·552).

Our discussion of filters is based upon absence of reflexion (see § 13·51), so we have assumed tacitly that Z_r takes its appropriate value automatically. In a practical (dissipative) filter, however, this is not possible, and reflexion occurs throughout the frequency range, although it is not serious usually.

13·552. Numerical example of dissipative low-pass filter.
Referring to Fig. 88, we consider sinusoidal current and use the values
$L = 0\cdot1$ H, $R = 10\,\Omega$, $C = 0\cdot1\,\mu$F. Then with

$$Z_0 = 0, \quad Z_1 = R + i\omega L, \quad Z_2 = 1/i\omega C,$$

we get $\theta^2 = (R + i\omega L)\,i\omega C/4$, so by (1) § 13·551

$$Z_r = \frac{1}{2}\sqrt{\left(\frac{R + i\omega L}{i\omega C}\right)}\sqrt{\{(4 - \omega^2 LC) + i\omega CR\}} \tag{1}$$

$$= \left(\frac{R^2 + \omega^2 L^2}{\omega^2 C^2}\right)^{\frac{1}{4}}\{[1 - (\omega/\omega_c)^2]^2 + \omega^2 C^2 R^2/16\}^{\frac{1}{4}}\,e^{i(\theta_1 + \theta_2 - \frac{1}{2}\pi)/2}, \tag{2}$$

where $\theta_1 = \tan^{-1}\{\omega CR/4[1 - (\omega/\omega_c)^2]\}$, and $\theta_2 = \tan^{-1}(\omega L/R)$. Then
$\omega_c = 2/\sqrt{(LC)} = 2 \times 10^4$, i.e. 3183 cycles per second. Consider first the
range above 200 c.p.s., where $\omega^2 L^2 > 1\cdot4 \times 10^4$, $R^2 = 100$, and for $\omega \geqslant 1200$
the first factor in (2) is nearly $\sqrt{(L/C)}$. Also when $\omega = 1200$, we have

$$\omega/\omega_c = 0\cdot06, \quad \omega^2 C^2 R^2/16 = 8\cdot75 \times 10^{-8},$$

$$\theta_1 \simeq 0, \quad \theta_2 \simeq \tfrac{1}{2}\pi \quad \text{giving} \quad (\theta_1 + \theta_2 - \tfrac{1}{2}\pi) \simeq 0.$$

Hence the effect of R is negligible, and

$$Z_r \simeq \sqrt{(L/C)} = 1000\,\text{ohms}. \tag{3}$$

When $\quad \omega = \omega_c, \quad (\omega_c^2 C^2 R^2)^{\frac{1}{4}}/2 = (200)^{-\frac{1}{4}}, \quad \theta_1 = \tfrac{1}{2}\pi, \quad \theta_2 \simeq \tfrac{1}{2}\pi,$

so $\qquad\qquad\qquad (\theta_1 + \theta_2 - \tfrac{1}{2}\pi)/2 \simeq \tfrac{1}{4}\pi.$

Using these in (2), we obtain

$$Z_r \simeq (200)^{-\frac{1}{4}}\sqrt{(L/C)}\,e^{\frac{1}{4}i\pi} = \{(1 + i)/20\}\sqrt{(L/C)} \tag{4}$$

$$= 50(1 + i)\,\text{ohms}, \tag{5}$$

as compared with zero for $R = 0$.

With ω small and $R > 0$, $\theta_1 \simeq 0$, $\theta_2 \simeq 0$ and

$$Z_r \simeq \sqrt{(R/\omega C)}\,e^{-\frac{1}{4}i\pi} = \sqrt{(R/\omega C)}\,(1 - i)/\sqrt{2}, \tag{6}$$

so that both the real and imaginary parts may be large.

When $\qquad\qquad \omega^2 = 2/LC, \quad \omega = \sqrt{2} \times 10^4, \quad (\omega/\omega_c)^2 = \tfrac{1}{2},$

$$\theta_1 = \tan^{-1}(\tfrac{1}{2}\omega CR), \quad \theta_2 = \tan^{-1}(2/\omega CR) = \cot^{-1}(\tfrac{1}{2}\omega CR),$$

so $\qquad\qquad\qquad (\theta_1 + \theta_2) = \tfrac{1}{2}\pi,$

and $\qquad\qquad Z_r \simeq \sqrt{(L/C)}/\sqrt{2} = 707\,\text{ohms}, \tag{7}$

at 2250 c.p.s. These data illustrate the variation in Z_r with $\omega \leqslant \omega_c$. As an
exercise the reader may consider the variation in Z_r when $\omega > \omega_c$. In the

above filter, if there is no reflection, the current in the mth inductance corresponding to an applied p.d. $E_0 \sin \omega_c t$ and $Z_0 = 0$ is

$$I_m \simeq \frac{\sqrt{2}\,E_0 \sin{(\omega_c t - \frac{1}{4}\pi - 2m\alpha)}}{\sqrt{R\,(L/C)^{\frac{1}{2}}}\,\{1 + \sqrt{R\,(C/L)^{\frac{1}{2}}}\}^m}, \qquad (8)$$

where $\alpha \simeq 87\cdot 25°$. Moreover, when $\omega = \omega_c$, the currents in successive inductances are almost in phase opposition.

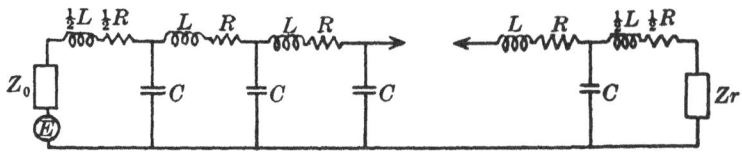

Fig. 88.   Dissipative low-pass filter.

13·553. $E_0 H(t)$ applied to input of dissipative low-pass filter. Referring to Fig. 88, $Z_1 = pL + R$, $Z_2 = 1/pC$, so

$$\sqrt{(Z_1 Z_2)} = \sqrt{\{(pL + R)/pC\}} = \sqrt{(L/C)}\,\sqrt{\{(p + 2\kappa)/p\}},$$

with $2\kappa = R/L$. Also

$$\theta = \sqrt{(Z_1/Z_2)/2} = \tfrac{1}{2}\sqrt{(LC)}\,\sqrt{\{(p + 2\kappa)\,p\}} = a\,\sqrt{\{(p + \kappa)^2 - \kappa^2\}},$$

where $a = \tfrac{1}{2}\sqrt{(LC)}$. Substituting into (11) § 13·51, and taking $Z_0 = 0$, we get

$$\phi_m = \frac{E_0\,\sqrt{(C/L)}}{a^{2m+1}}\,\sqrt{\left(\frac{p}{p+2\kappa}\right)}\,\sqrt{\left(\frac{1}{(p+\kappa)^2 - \kappa^2 + 1/a^2}\right)}$$

$$\times \left(\frac{1}{\sqrt{\{[p+\kappa]^2 - \kappa^2\}} + \sqrt{\{[p+\kappa]^2 - \kappa^2 + 1/a^2\}}}\right)^{2m}. \qquad (1)$$

Applying (3) § 8·411 to (1) gives

$$e^{\kappa t} I_m = \frac{E_0\,\sqrt{(C/L)}}{a}\,\frac{a^{-2m}p/\sqrt{(p^2 - \kappa^2)}}{\sqrt{(p^2 - \kappa^2 - 1/a^2)}\,\{\sqrt{(p^2 - \kappa^2)} + \sqrt{(p^2 - \kappa^2 + 1/a^2)}\}^{2m}}. \qquad (2)$$

In [236], p. 46, it is shown that if $f(t) \Rightarrow \phi(p)$, then

$$\int_0^t I_0\{\kappa\,\sqrt{(t^2 - x^2)}\}\,f(x)\,dx \Rightarrow \frac{p}{p^2 - \kappa^2}\,\phi\{\sqrt{(p^2 - \kappa^2)}\} = \psi(p), \quad \text{say}. \qquad (3)$$

Writing $\sqrt{(p^2 + \kappa^2)}$ for p in (3) gives

$$\phi(p) = \frac{p^2}{\sqrt{(p^2 + \kappa^2)}}\,\psi\{\sqrt{(p^2 + \kappa^2)}\}. \qquad (4)$$

Applying (4) to the second fraction in (2), i.e. writing $\sqrt{(p^2 + \kappa^2)}$ for p and multiplying the result by $p^2/\sqrt{(p^2 + \kappa^2)}$, yields

$$\phi(p) = \frac{a^{-2m}p}{\sqrt{(p^2 + 1/a^2)}\{p + \sqrt{(p^2 + 1/a^2)}\}^{2m}} \tag{5}$$

$$\subset J_{2m}(t/a), \tag{6}$$

by (50) Appendix 10, with $\nu = 2m$.

Hence by (2), (3), (5), (6), for quiescence initially, we obtain the current in the mth inductance,

$$I_m = E_0\sqrt{(C/L)}\, e^{-\kappa t}\int_0^t I_0\{\kappa\sqrt{(t^2 - x^2)}\}\, J_{2m}(x/a)\, d(x/a). \tag{7}$$

Writing $x/a = y$, with $1/a = \omega_c$, (7) becomes

$$I_m = E_0\sqrt{(C/L)}\, e^{-\kappa t}\int_0^{\omega_c t} I_0\{\kappa\sqrt{(t^2 - y^2/\omega_c^2)}\}\, J_{2m}(y)\, dy. \tag{8}$$

Attenuation of currents of frequency $\omega > \omega_c$ is relatively high, provided the loss in the meshes is small. When $R = 0$, $\kappa = 0$, and (8) degenerates to (3) § 13·521, since $I_0(0) = 1$.

13·61. Sinusoidal input p.d. Hitherto we have considered the response of a filter to an applied p.d. of the unit function type. In telephone engineering and radio communication, filters are subject to complicated wave forms. These can be analysed into infinite frequency spectra in the case of transients, and into Fourier series for continuous or repeated wave forms (see §§ 10·51–10·56). We shall now determine the response of a low-pass filter to a suddenly applied sinusoidal input p.d. of any frequency. Since the system is assumed to be linear, the influence of two or more sine waves may be ascertained by obtaining the separate solutions and superimposing them. To avoid complication, we shall consider the loss-free filter of Fig. 86 (b) with $Z_0 = 0$. The transform solution is obtained on replacing E in (1) § 13·521 by the transform of the applied p.d. $E_0 \sin \omega t$. Thus, for initial quiescence, the transform of the current in the mth inductance is given by

$$\phi_m = \frac{E_0\sqrt{(C/L)}\,\omega p}{(p^2 + \omega^2)\sqrt{(p^2 a^2 + 1)}\{pa + \sqrt{(p^2 a^2 + 1)}\}^{2m}}. \tag{1}$$

13·621. Steady state solution. The complex integral corresponding to (1) § 13·61 is, with $a = 1/\omega_c$,

$$I_m = \frac{E_0 \sqrt{(C/L)} \, \omega \omega_c^{2m+1}}{2\pi i} \int_{Br_1} \frac{e^{zt} \, dz}{(z^2 + \omega^2) \sqrt{(z^2 + \omega_c^2)} \{z + \sqrt{(z^2 + \omega_c^2)}\}^{2m}}. \tag{1}$$

Suitable contours equivalent to Br_1 are depicted in Figs. 89, 90. To find the sinusoidal current in the mth mesh, we evaluate (1) at the two poles $z = \pm i\omega$, i.e. round the circles in Figs. 89, 90. For the pole at $z = i\omega$, if the external factor in (1) be omitted, we get the contribution

$$e^{i\omega t}/2i\omega \sqrt{(\omega_c^2 - \omega^2)} \{\sqrt{(\omega_c^2 - \omega^2)} + i\omega\}^{2m}. \quad (\omega < \omega_c) \tag{2}$$

Writing $\sqrt{(\omega_c^2 - \omega^2)} = x$, and $\omega = y$, we have $r = \sqrt{(x^2 + y^2)} = \omega_c$, $\theta = \tan^{-1}(y/x) = \tan^{-1}\omega/\sqrt{(\omega_c^2 - \omega^2)} = \sin^{-1}\omega/\omega_c$, so

$$\{\sqrt{(\omega_c^2 - \omega^2)} + i\omega\}^{2m} = r^{2m} e^{2im\theta} = \omega_c^{2m} e^{2im\theta}. \tag{3}$$

Substituting from (3) into (2), the contribution from the pole at $z = i\omega$ is
$$e^{i(\omega t - 2m\theta)}/2i\omega\omega_c^{2m} \sqrt{(\omega_c^2 - \omega^2)}. \tag{4}$$

Changing i to $-i$ in (4), the contribution from the pole at $z = -i\omega$ is
$$-e^{-i(\omega t - 2m\theta)}/2i\omega\omega_c^{2m} \sqrt{(\omega_c^2 - \omega^2)}. \tag{5}$$

Adding (4), (5) and introducing the external factor from (1), the steady current in the mth inductance is

$$I_{m\,(\text{steady})} = \frac{E_0 \sqrt{(C/L)} \, \omega_c \sin(\omega t - 2m\theta)}{\sqrt{(\omega_c^2 - \omega^2)}} \tag{6}$$

$$= \frac{E_0 \sqrt{(C/L)}}{\sqrt{[1 - (\omega/\omega_c)^2]}} \sin(\omega t - 2m\theta), \tag{7}$$

provided $\omega_c > \omega$. From (7), as $\omega \to \omega_c$, the current increases very rapidly.

When $\omega > \omega_c$, corresponding to (2) we have

$$\frac{(-1)^{m+1} e^{i\omega t}}{2\omega \sqrt{(\omega^2 - \omega_c^2)} [\sqrt{(\omega^2 - \omega_c^2)} + \omega]^{2m}}$$
$$= \frac{(-1)^{m+1} e^{i\omega t}}{2\omega\omega_c^{2m+1} \sqrt{[(\omega/\omega_c)^2 - 1]} \{(\omega/\omega_c) + \sqrt{[(\omega/\omega_c)^2 - 1]}\}^{2m}}. \tag{8}$$

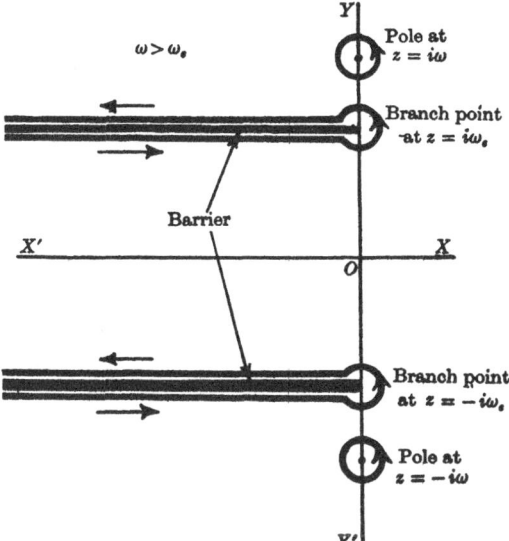

Fig. 89.

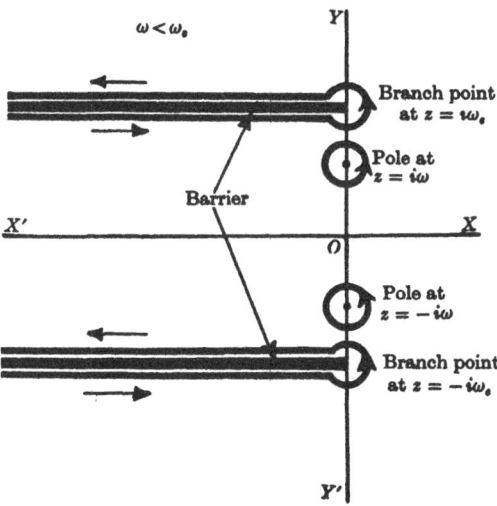

Fig. 90.

This is the contribution to (1) from the pole at $z = i\omega$. Changing i to $-i$ in (8) gives the contribution from the pole at $z = -i\omega$. Adding these and introducing the external factor from (1), yields

$$I_{m \text{ (steady)}} = \frac{(-1)^{m+1} E_0 \sqrt{(C/L)} \cos \omega t}{\sqrt{[(\omega/\omega_c)^2 - 1]} \{(\omega/\omega_c) + \sqrt{[(\omega/\omega_c)^2 - 1]}\}^{2m}}, \quad (9)$$

the factor $(-1)^{m+1}$ signifying that the currents in consecutive meshes are in phase opposition. As before, the current increases rapidly when $\omega \to \omega_c$, but it is attenuated considerably as ω increases beyond ω_c. If $\omega = \omega_c$, (1) cannot be evaluated as shown above, since there are no poles, the singularities then being branch points at $z = \pm i\omega_c$ (see § 2·68). An asymptotic formula is derived in § 13·625.

13·622. Transient state solution. For this we have to integrate (1) § 13·621 round the branch points $z = \pm i\omega_c$. Two different contours are illustrated in Fig. 91 (a), (b). Integration round the latter would be a difficult matter. Moreover, the former will be used, since an asymptotic formula may be obtained which enables the physical behaviour of the filter to be understood when t is large (in a relative sense).

Referring to Fig. 91 (a), we move the origin to the branch point $z = i\omega_c$ by writing $z = (\zeta + i\omega_c)$, and (1) § 13·621 becomes

$$I_{m_1} = E_0 \sqrt{(C/L)} \, \omega \omega_c^{2m+1} e^{i\omega_c t} \times$$

$$\frac{1}{2\pi i} \int_{Br_1} \frac{e^{\zeta t} d\zeta}{(\zeta^2 + 2i\omega_c \zeta + \omega^2 - \omega_c^2) \sqrt{(\zeta^2 + 2i\omega_c \zeta)} \{(\zeta + i\omega_c) + \sqrt{(\zeta^2 + 2i\omega_c \zeta)}\}^{2m}}. \quad (1)$$

13·623. Approximation to reciprocal of denominator of (1) § 13·622 taking $|\zeta|$ small.

1st bracket gives
$$-(\omega_c^2 - \omega^2)[1 - 2i\omega_c \zeta/(\omega_c^2 - \omega^2)]; \quad (1)$$

2nd bracket gives
$$\sqrt{(2i\omega_c \zeta)} \sqrt{(1 + \zeta/2i\omega_c)}; \quad (2)$$

3rd bracket gives
$$(-1)^m \omega_c^{2m}[1 + \sqrt{(2\zeta/i\omega_c)} + (\zeta/i\omega_c)]^{2m}. \quad (3)$$

Using the binomial theorem, with $|\zeta|$ small, we obtain the following approximations for the reciprocals of (1)–(3):

(i)
$$-[1 + 2i\omega_c \zeta/(\omega_c^2 - \omega^2)]/(\omega_c^2 - \omega^2); \quad (4)$$

(ii)
$$e^{-\frac{1}{4}i\pi}(1 + i\zeta/4\omega_c)/\sqrt{(2\omega_c \zeta)}; \quad (5)$$

(iii)
$$\{(-1)^m/\omega_c^{2m}\}[1 - 2^{\frac{1}{2}}m\sqrt{\zeta}/\sqrt{(i\omega_c)} - 4m^2 i\zeta/\omega_c]. \quad (6)$$

The product of (4), (5) and (6) is

$$\frac{(-1)^{m+1}e^{-\frac{1}{4}i\pi}}{\sqrt{2}\,(\omega_c^2-\omega^2)\,\omega_c^{2m+\frac{1}{2}}\,\sqrt{\zeta}}\left[1+\frac{2i\omega_c\zeta}{(\omega_c^2-\omega^2)}\right]\left[1+\frac{i\zeta}{4\omega_c}\right]\left[1-\frac{2^{\frac{1}{2}}m\sqrt{\zeta}}{\sqrt{(i\omega_c)}}-\frac{4m^2i\zeta}{\omega_c}\right] \quad (7)$$

$$\simeq\frac{(-1)^{m+1}e^{-\frac{1}{4}i\pi}}{\sqrt{2}\,(\omega_c^2-\omega^2)\,\omega_c^{2m+\frac{1}{2}}\,\sqrt{\zeta}}\left[1-\frac{2^{\frac{1}{2}}m\sqrt{\zeta}}{\sqrt{(i\omega_c)}}+i\zeta\left\{\frac{2\omega_c}{(\omega_c^2-\omega^2)}+\frac{(\frac{1}{4}-4m^2)}{\omega_c}\right\}\right] \quad (8)$$

$$=\frac{(-1)^{m+1}e^{-\frac{1}{4}i\pi}}{\sqrt{2}\,(\omega_c^2-\omega^2)\,\omega_c^{2m+\frac{1}{2}}}\left[\frac{1}{\sqrt{\zeta}}-\frac{2^{\frac{1}{2}}m}{\sqrt{(i\omega_c)}}+i\sqrt{\zeta}\,(4m^2/\omega_c)\left\{\frac{(9\omega_c^2-\omega^2)}{16m^2(\omega_c^2-\omega^2)}-1\right\}\right]. \quad (9)$$

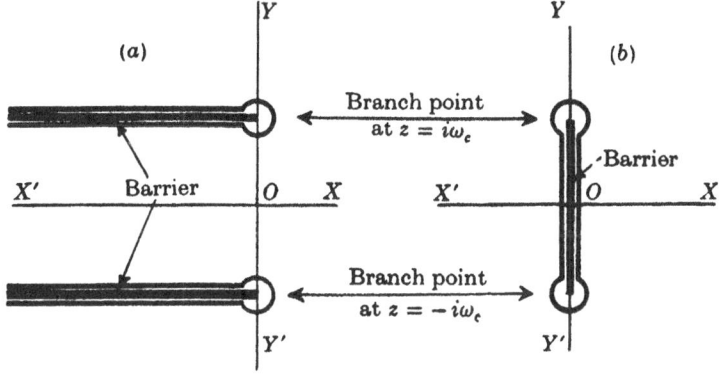

Fig. 91. (b) applies when $\omega>\omega_c$.

13·624. Evaluation of (1) §13·622. Substituting from (9) §13·623 into (1) §13·622, we obtain for the branch point at $z=i\omega_c$

$$I_{m_1}\sim\frac{(-1)^{m+1}E_0\sqrt{(C/L)}\,\omega\sqrt{\omega_c}\,e^{i(\omega_ct-\frac{1}{4}\pi)}}{\sqrt{2}\,(\omega_c^2-\omega^2)\,2\pi i}$$

$$\times\int_{Br_1}e^{\zeta t}\,d\zeta\left[\frac{1}{\sqrt{\zeta}}-\frac{2^{\frac{1}{2}}m}{\sqrt{(i\omega_c)}}+i\sqrt{\zeta}\,\frac{4m^2}{\omega_c}\left\{\frac{(9\omega_c^2-\omega^2)}{16m^2(\omega_c^2-\omega^2)}-1\right\}\right] \quad (1)$$

$$=(-1)^{m+1}\frac{E_0\sqrt{(C/2\pi tL)}\,\omega\sqrt{\omega_c}\,e^{i(\omega_ct-\frac{1}{4}\pi)}}{(\omega_c^2-\omega^2)}\left[1-i\frac{2m^2}{\omega_ct}\left\{\frac{(9\omega_c^2-\omega^2)}{16m^2(\omega_c^2-\omega^2)}-1\right\}\right], \quad (2)$$

provided $\omega\neq\omega_c$. The contribution associated with the branch point at $z=-i\omega_c$ is found from (2) by writing $-i$ for i, which gives

$$I_{m_1}\sim(-1)^{m+1}\frac{E_0\sqrt{(C/2\pi tL)}\,\omega\sqrt{\omega_c}\,e^{-i(\omega_ct-\frac{1}{4}\pi)}}{(\omega_c^2-\omega^2)}\left[1+i\frac{2m^2}{\omega_ct}\left\{\frac{(9\omega_c^2-\omega^2)}{16m^2(\omega_c^2-\omega^2)}-1\right\}\right]. \quad (3)$$

Adding (2), (3), the transient in the mth mesh for large values of t, is

$$I_{m\text{(transient)}} \sim (-1)^{m+1} \frac{E_0 \sqrt{(2C/\pi t L)}\, \omega \sqrt{\omega_c}}{(\omega_c^2 - \omega^2)}$$

$$\times \left[\cos(\omega_c t - \tfrac{1}{4}\pi) + \frac{2m^2}{\omega_c t} \left\{ \frac{(9\omega_c^2 - \omega^2)}{16m^2(\omega_c^2 - \omega^2)} - 1 \right\} \sin(\omega_c t - \tfrac{1}{4}\pi) \right], \quad (4)$$

provided $\omega \neq \omega_c$. When t is so *large* that the factor preceding the sine term $\ll 1$, (4) reduces to

$$I_{m\text{(transient)}} \sim (-1)^{m+1} \frac{E_0 \sqrt{(C/L)}\, \omega \omega_c}{(\omega_c^2 - \omega^2)} [\sqrt{(2/\pi \omega_c t)} \cos(\omega_c t - \tfrac{1}{4}\pi)] \quad (5)$$

$$\simeq (-1)^{m+1} \frac{E_0 \sqrt{(C/L)}\, \omega \omega_c}{(\omega_c^2 - \omega^2)} J_0(\omega_c t) \quad (\omega \neq \omega_c). \quad (6)$$

The total current is the sum of the transient and steady currents, the latter being represented by either (7) or (9) §13·621, according as $\omega_c >$ or $< \omega$.

13·625. Evaluation of (1) §13·621 when $\omega = \omega_c$. The integral in question is

$$I_m = \frac{E_0 \sqrt{(C/L)}\, \omega_c^{2m+2}}{2\pi i} \int_{Br_1} \frac{e^{zt}\, dz}{(z^2 + \omega_c^2)^{\frac{1}{4}} \{z + \sqrt{(z^2 + \omega_c^2)}\}^{2m}}. \quad (1)$$

The integrand has branch points at $z = \pm i\omega_c$. Moving the origin to $z = i\omega_c$, by (1) §13·622 with $\omega = \omega_c$, we get

$$I_{m_1} = \frac{E_0 \sqrt{(C/L)}\, \omega_c^{2m+2} e^{i\omega_c t}}{2\pi i} \int_{Br_1} \frac{e^{\zeta t}\, d\zeta}{(\zeta^2 + 2i\omega_c \zeta)^{\frac{1}{4}} [(\zeta + i\omega_c) + \sqrt{(\zeta^2 + 2i\omega_c \zeta)}]^{2m}}. \quad (2)$$

Taking $|\zeta|$ small, the reciprocal of the first bracket in the denominator is

$$(2i\omega_c \zeta)^{-\frac{1}{4}} (1 + \zeta/2i\omega_c)^{-\frac{1}{4}} \simeq (2\omega_c \zeta)^{-\frac{1}{4}} e^{-\frac{1}{8}\pi i} (1 + 3i\zeta/4\omega_c). \quad (3)$$

For the second bracket we use the right-hand side of (6) §13·623. The product of this and (3) is

$$\frac{(-1)^{m+1} e^{\frac{1}{8}\pi i}}{2^{\frac{1}{4}} \omega_c^{2m+3} \zeta^{\frac{1}{4}}} \left[1 - \frac{2^{\frac{3}{4}} m \sqrt{\zeta}}{\sqrt{(i\omega_c)}} - \frac{i\zeta}{4\omega_c} (16m^2 - 3) \right]. \quad (4)$$

Thus by (2) and (4)

$$I_{m_1} \sim (-1)^{m+1} \frac{E_0 \sqrt{(C/L)} \sqrt{\omega_c}\, e^{i(\omega_c t + \frac{1}{8}\pi)}}{2^{\frac{1}{4}} 2\pi i}$$

$$\times \int_{Br_1} e^{\zeta t}\, d\zeta \left[\frac{1}{\zeta^{\frac{1}{4}}} - \frac{2^{\frac{3}{4}} m\, e^{-\frac{1}{4}i\pi}}{\sqrt{\omega_c}\, \zeta} - \frac{i}{\sqrt{\zeta}} \left(\frac{16m^2 - 3}{4\omega_c} \right) \right] \quad (5)$$

$$= (-1)^{m+1} \frac{E_0 \sqrt{(C/L)} \sqrt{\omega_c}\, e^{i(\omega_c t + \frac{1}{8}\pi)}}{2^{\frac{1}{4}}}$$

$$\times [2\sqrt{(t/\pi)} - (2^{\frac{3}{4}} m\, e^{-\frac{1}{4}i\pi}/\sqrt{\omega_c}) - i(16m^2 - 3)/4\sqrt{(\pi t)}\, \omega_c]. \quad (6)$$

The contribution to (1) from the branch point at $z = -i\omega_c$ is obtained from (6) by writing $-i$ for i. Adding the two contributions, we find that

$$I_m \sim (-1)^{m+1} E_0 \sqrt{(2C/\pi L)} \Bigg[\sqrt{(\omega_c t)} \cos(\omega_c t + \tfrac{1}{4}\pi) - \sqrt{(2\pi)} \, m \cos \omega_c t$$
$$+ \frac{(16m^2 - 3)}{8\sqrt{(\omega_c t)}} \sin(\omega_c t + \tfrac{1}{4}\pi) \Bigg], \quad (7)$$

which is the asymptotic formula for the *total* current in the mth inductance.

From the first term of (7) it appears that the current increases in proportion to $\sqrt{t}$, while the second indicates a cosine wave of constant amplitude. By hypothesis the filter and the supply source are devoid of inherent loss, so the current increases with lapse of time. In practice the loss in the source and in the inductance arms would prevent this occurring, while the steady alternating current would suffer reduction. Although (7) departs from the result which would be obtained by inclusion of resistance, it enables us to understand that in practice the transient due to an applied sine-wave voltage of cut-off frequency will be relatively large. This conclusion applies to all practical selective systems. Whereas they perform the function of being selective during the steady state, they fail singularly in the case of certain transients.

13·626. Total current when t is small. The required approximation is found by applying § 6·31 to (1) § 13·621. Thus, making $|z| \gg \omega$ and ω_c, both of which are finite, we obtain

$$I_{m(\text{total})} \simeq \frac{E_0 \sqrt{(C/L)} \, \omega \omega_c^{2m+1}}{2^{2m} 2\pi i} \int_{Br_1} \frac{e^{zt} dz}{z^{2m+3}} \quad (1)$$

$$= \frac{E_0 \sqrt{(C/L)} \, \omega \omega_c^{2m+1} t^{2m+2}}{2^{2m}(2m+2)!}. \quad (2)$$

The rate of rise of current is dI_m/dt and varies as t^{2m+1}. Thus the further the mesh from the zeroth section, the slower is the growth of the current initially. This is due to the charging currents to the capacitances. That in the zeroth mesh takes the largest initial charge, the remainder being charged more slowly as m increases, since each robs those succeeding it, until the steady state is attained.

13·7. Sine-wave p.d. applied to dissipative filter. We consider the filter of Fig. 88, but with a resistance $1/G$ across each C, such that $G/C = R/L$,* the p.d. being $E = E_0 \sin \omega t$. The analysis is similar to that in previous sections, and is left as an exercise for the reader. Salient results are as follows:

1. *Steady state solution.*

$$I_{m(\text{steady})} = \{ E_0 \sqrt{(C/L)} \, \omega_c^{2m+1}/r_1 r^{2m} \} \sin(\omega t - \theta_1 - 2m\theta), \quad (1)$$

* This relationship corresponds to that for a distortionless cable as in § 13·522 and [236], p. 88.

where

$$\alpha = G/C = R/L, \quad \sqrt{\{\omega_c^2 + (\alpha + i\omega)^2\}} = r_1 e^{i\theta_1} \quad (\alpha + i\omega) = r_2 e^{i\theta_2},$$

$$r_1 e^{i\theta_1} (r_1 e^{i\theta_1} + r_2 e^{i\theta_2})^{2m} = r_1 r^{2m} e^{i(\theta_1 + 2m\theta)}, \quad r_1^2 = \sqrt{\{(\omega_c^2 - \omega^2 + \alpha^2)^2 + 4\alpha^2\omega^2\}},$$

$$r_2 = (\omega^2 + \alpha^2)^{\frac{1}{2}}, \quad r^2 = r_1^2 + r_2^2 + 2r_1 r_2 \cos(\theta_1 - \theta_2),$$

$$\theta_1 = \tfrac{1}{2} \tan^{-1}[2\alpha\omega/(\omega_c^2 - \omega^2 + \alpha^2)], \quad \theta_2 = \tan^{-1}(\omega/\alpha),$$

$$\theta = \tan^{-1}[(r_1 \sin\theta_1 + r_2 \sin\theta_2)/(r_1 \cos\theta_1 + r_2 \cos\theta_2)].$$

If $\alpha = 0$, and $\omega_c > \omega$, (1) degenerates to (7) § 13·621. With $\omega = \omega_c \gg \alpha > 0$, (1) yields

$$I_{m(\text{steady})} \simeq (-1)^m \{(E_0/R)/[1 + 2(\sqrt{\gamma} + \gamma)]^m\} \sin(\omega_c t - \tfrac{1}{4}\pi), \qquad (2)$$

where $\gamma = \alpha/\omega_c = R(C/L)^{\frac{1}{2}}/2$. If R is small, (2) indicates that the current at the cut-off frequency is finite, but large, whereas in the absence of loss it tends to infinity as $\omega \to \omega_c$ (see § 13·621).

2. *Transient solution.* If $\alpha > 0$, ω is finite and t is large enough,

$$I_{m(\text{transient})} \sim (-1)^m \frac{E_0 \sqrt{(2C/\pi Lt)}\, \omega \sqrt{\omega_c}\, e^{-\alpha t}}{(c^2 + d^2)} \left[\left\{ c - \left(\frac{ce - fd}{2t} \right) \right\} \cos(\omega_c t - \tfrac{1}{4}\pi) \right.$$

$$\left. + \left\{ d - \left(\frac{cf + de}{2t} \right) \right\} \sin(\omega_c t - \tfrac{1}{4}\pi) \right], \qquad (3)$$

with $c = \omega^2 - \omega_c^2 + \alpha^2, \quad d = -2\alpha\omega_c, \quad e = (2\alpha c - d\omega_c)/(c^2 + d^2),$

$$f = \{2(\alpha d + c\omega_c)/(c^2 + d^2)\} + (16m^2 - 1)/4\omega_c.$$

If $\alpha = d = e = 0$, (3) reduces to (4) § 13·624. When $t \gg (ce - fd)$ and $(cf + de)$, (1) yields

$$I_{m(\text{transient})} \sim (-1)^m \frac{E_0 \sqrt{(2C/\pi Lt)}\, \omega \sqrt{\omega_c}\, e^{-\alpha t}}{\sqrt{(c^2 + d^2)}} \cos[\omega_c t - \tfrac{1}{4}\pi - \tan^{-1}(d/c)].$$

$$(4)$$

The ratio of the current amplitude (variable) in (4) to that in the loss-free filter, where $R = G = 0$, as given by (5) § 13·624, is

$$e^{-\alpha t}/\sqrt{\{[1 - \gamma^2/\{1 - (\omega/\omega_c)^2\}]^2 + 4\gamma^2/[1 - (\omega/\omega_c)^2]^2\}} \quad (\omega \neq \omega_c). \qquad (5)$$

For assigned values of α, t, the nearer ω is to ω_c, the smaller is the ratio in (5). One effect of loss is to introduce the damping factor $e^{-\alpha t}$ into the expression for the transient current.

XIV

SOLENOID WITH METAL CORE: CONDENSER MICROPHONE: LOUD SPEAKER HORN

14·11. Inductance of solenoid with solid cylindrical core. The solution of the problem of eddy current distribution in a solid circular metal cylinder, within a long solenoid carrying alternating current of sine-wave form, is well known [234]. In the present case we shall find the inductance when a voltage E is applied to the solenoid at $t = 0$, by a battery or a D.C. generator. The analysis is valid also for a toroidal coil having a solid metal core of circular cross-section, provided the mean radius of the toroid is large compared with that of the winding section. It will be assumed that the magnetic flux is mainly that within a core of constant magnetic permeability. Owing to its acting as a form of short-circuited one-turn secondary winding, eddy currents are induced in the core. These currents retard the growth of flux, but they enhance the rate of rise of current in the solenoid owing to reduction in inductance.

14·12. The arrangement of the electrical circuit is shown in Fig. 92 (a). We use the following symbols: E = applied voltage; R = resistance of circuit; L = inductance due to magnetic flux in core, this being variable owing to eddy currents; I = current in circuit. To determine the value of L we must begin by considering the effect of the eddy currents induced in the metal core on H the magnetizing force due to the solenoidal current I. To make the analysis tractable, we shall assume that R is large enough for the current to attain a sensibly constant value soon after the switch is closed. By differentiating (35) in [234], p. 143, with respect to r, the D.E. for the magnetizing force at any radius r of the cylindrical core is

$$\frac{\partial^2 H}{\partial r^2} + \frac{1}{r}\frac{\partial H}{\partial r} - \frac{4\pi\mu}{\rho}\frac{\partial H}{\partial t} = 0, \tag{1}$$

where μ = permeability of core; ρ = specific resistance of core.

Writing k^2 for $-\dfrac{4\pi\mu}{\rho}\dfrac{\partial}{\partial t}$, and ϕ for H, i.e. $H \Rightarrow \phi$, since $H = 0$ when $t = 0$, we obtain

$$\frac{d^2\phi}{dr^2}+\frac{1}{r}\frac{d\phi}{dr}+k^2\phi = 0, \qquad (2)$$

whose solution appropriate to the present problem is

$$\phi = AJ_0(kr). \qquad (3)$$

(a)   Long solenoid   **(b)**

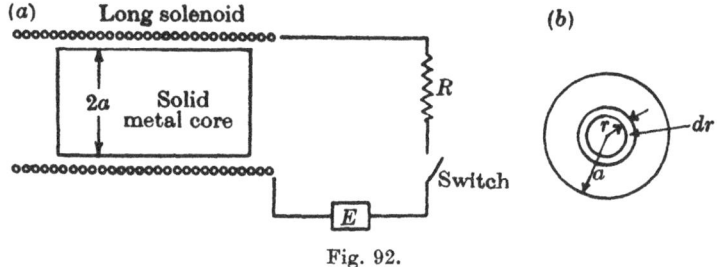

Fig. 92.

At the surface of the core of radius a the magnetizing force $H_0 = 4\pi n I/l$, n being the number of turns uniformly spaced on the solenoid and l its axial length. Since $H \Rightarrow \phi$ and I is constant, we have $\phi_0 = 4\pi n I/l$. Using this condition in (3) gives

$$\phi = \frac{4\pi n I}{l}\frac{J_0(kr)}{J_0(ka)}. \qquad (4)$$

Taking the inductance to be the product of the terms n and the total core flux per unit current in the solenoid, with I constant,

$$L = \frac{2\pi n\mu}{I}\int_0^a rH\,dr. \qquad (5)$$

Since L is a function of t, if we write $L \Rightarrow \phi_1$, (4) and (5) yield

$$\phi_1 = \frac{8\pi^2 n^2\mu}{lJ_0(ka)}\int_0^a rJ_0(kr)\,dr \qquad (6)$$

$$= \frac{BJ_1(ka)}{kaJ_0(ka)}, \qquad (7)$$

where $B = 8\pi^2 n^2 a^2 \mu / l$. Inverting (7) by the Mellin theorem,

$$L = \frac{B/ca}{2\pi i} \int_{Br_1} \frac{e^{zt} J_1(ca \sqrt{z}) \, dz}{z^{\frac{1}{2}} J_0(ca \sqrt{z})}, \tag{8}$$

with $c = i \sqrt{(4\pi\mu/\rho)}$. The singularities of the integrand are a simple pole at the origin, and simple poles arising from the zeros of $J_0(ca \sqrt{z})$, which have a limit point at infinity. The first singularity may be identified by making z small. Then

$$J_1(ca \sqrt{z}) \simeq \tfrac{1}{2} ca \sqrt{z}, \quad J_0 \simeq 1,$$

so the integrand tends to $ca \, e^{zt}/2z$ as $z \to 0$.

Writing $ca \sqrt{z} = \zeta$,

$$J_1(ca \sqrt{z}) = \tfrac{1}{2}\zeta \left[1 - \frac{(\tfrac{1}{2}\zeta)^2}{2!} + \frac{(\tfrac{1}{2}\zeta)^4}{2! \, 3!} - \cdots \right]. \tag{9}$$

Substituting from (9) into (8),

$$L = \frac{\tfrac{1}{2}B}{2\pi i} \int_{Br_1} e^{zt} \left[1 - \frac{(\tfrac{1}{2}\zeta)^2}{2!} + \frac{(\tfrac{1}{2}\zeta)^4}{2! \, 3!} - \cdots \right] \frac{dz}{z J_0(\zeta)}, \tag{10}$$

and since $J_0(\zeta) \to 1$ as $z \to 0$, the contribution from the pole at $z = 0$ is $\tfrac{1}{2}B$. Using (8), the contribution from the poles due to $1/J_0(\zeta)$ is

$$\frac{B}{ca} \sum_{n=1}^{\infty} \left[\frac{e^{zt} J_1(\zeta)}{z^{\frac{1}{2}} \dfrac{d}{dz} J_0(\zeta)} \right]_{\substack{\zeta = ca\sqrt{z} = \alpha_n \\ z = \alpha_n^2/c^2 a^2}}, \tag{11}$$

α_n being the nth positive root of $J_0(\zeta) = 0$. Since

$$\frac{d}{dz} J_0(\zeta) = J_0'(\zeta) \frac{d\zeta}{dz} = -J_1(\zeta) \frac{d\zeta}{dz},$$

(11) may be written

$$-\frac{2B}{c^2 a^2} \sum_{n=1}^{\infty} \left[\frac{e^{zt} J_1(\zeta)}{z J_1(\zeta)} \right]_{z = \alpha_n^2/c^2 a^2} = -2B \sum_{n=1}^{\infty} \frac{e^{\alpha_n^2 t/c^2 a^2}}{\alpha_n^2}. \tag{12}$$

Adding $\tfrac{1}{2}B$ to (12), the inductance of the solenoid at any time $t \geqslant 0$ is

$$L = \frac{4\pi^2 n^2 a^2 \mu}{l} \left[1 - 4 \sum_{n=1}^{\infty} \frac{e^{-\alpha_n^2 \rho t / 4\pi \mu a^2}}{\alpha_n^2} \right]. \tag{13}$$

When t is large and the eddy currents in the metal core are negligible, (13) degenerates to the well-known formula

$$L \sim 4\pi^2 n^2 a^2 \mu / l. \tag{14}$$

14·2. Inductance when t is small. We take (8) §14·12 and make z large, in accordance with §6·31. Using the asymptotic formulae for the Bessel functions (see [234]),

$$L \simeq \frac{(B/ca)}{2\pi i} \int_{Br_1} \frac{e^{zt} \cos (\zeta - \tfrac{3}{4}\pi)\, dz}{z^{\tfrac{3}{2}} \cos (\zeta - \tfrac{1}{4}\pi)}. \tag{1}$$

Now $\cos (\zeta - \tfrac{3}{4}\pi) = \tfrac{1}{2}[e^{i(\zeta - \tfrac{3}{4}\pi)} + e^{-i(\zeta - \tfrac{3}{4}\pi)}]$, so writing $\zeta = is \sqrt{z}$, where $s = a \sqrt{(4\pi\mu/\rho)}$, being real > 0,

$$\cos (is \sqrt{z} - \tfrac{3}{4}\pi) = \tfrac{1}{2}[\exp (-s \sqrt{z} - \tfrac{3}{4}\pi i) + \exp (s \sqrt{z} + \tfrac{3}{4}\pi i)] \tag{2}$$

$$\sim \tfrac{1}{2} \exp (s \sqrt{z} + \tfrac{3}{4}\pi i), \tag{3}$$

when z is large. Similarly,

$$\cos (is \sqrt{z} - \tfrac{1}{4}\pi) \sim \tfrac{1}{2} \exp (s \sqrt{z} + \tfrac{1}{4}\pi i). \tag{4}$$

Substituting from (3), (4) into (1), we find that when t is small

$$L \simeq \frac{iB/ca}{2\pi i} \int_{Br_1} \frac{e^{zt}\, dz}{z^{\tfrac{3}{2}}} \tag{5}$$

$$= 8\pi n^2 a \sqrt{(\mu\rho t)}/l. \tag{6}$$

Formula (6) shows that the inductance is zero at $t = 0$, this being due to lack of flux in the metal core. Thus from (13) §14·12 when $t = 0$, by virtue of uniform convergence we have

$$4 \sum_{n=1}^{\infty} 1/\alpha_n^2 = 1. \tag{7}$$

From (13) §14·12, we see also that the greater the resistivity and the smaller the radius of the metal core, the more rapidly does L attain the value given by (14) §14·12. If ρ were very small and a very large, the exponential indices in (13) §14·12 would be comparatively small unless t were large. Under these conditions the eddy currents in the core would persist for an appreciable time, thereby retarding the rise in L to the value at (14) §14·12.

14·3. Magnetizing force at any radius r.

Inverting (4) § 14·12,

$$H = \frac{4\pi nl}{l} \frac{1}{2\pi i} \int_{Br_1} \frac{e^{zt} J_0(cr\sqrt{z})\, dz}{zJ_0(ca\sqrt{z})}. \tag{1}$$

Both Bessel functions in (1) $\to 1$ as $z \to 0$, so the contribution from the pole at $z = 0$ is $4\pi nI/l$, this being the magnetizing force at the surface of the core. By § 14·12 the contribution to (1) from the singularities of $1/J_0(ca\sqrt{z})$ is

$$\frac{-8\pi nI}{cal} \left[\sum_{n=1}^{\infty} \frac{e^{zt} J_0(cr\sqrt{z})}{\sqrt{z}\, J_1(ca\sqrt{z})} \right]_{\substack{ca\sqrt{z}=\alpha_n \\ z=\alpha_n^2/c^2a^2}} \tag{2}$$

$$= -(8\pi nI/l) \sum_{n=1}^{\infty} \frac{\exp(\alpha_n^2 t/c^2 a^2)\, J_0(\alpha_n r/a)}{\alpha_n J_1(\alpha_n)}. \tag{3}$$

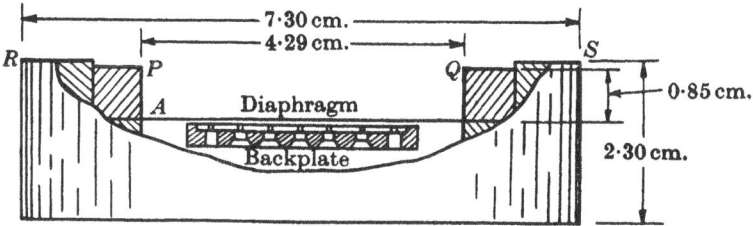

Fig. 93. Schematic diagram of condenser microphone.

Adding the two contributions, the magnetizing force at any radius r when $t > 0$ is

$$H = (4\pi nI/l) \left[1 - 2 \sum_{n=1}^{\infty} \frac{\exp(-\alpha_a^2 \rho t/4\pi\mu a^2)\, J_0(\alpha_n r/a)}{\alpha_n J_1(\alpha_n)} \right]. \tag{4}$$

14·41. Condenser microphone.

This instrument, shown schematically in Fig. 93, has a very thin aluminium-alloy diaphragm (membrane) which is subjected to a large radial tension. The diaphragm motion is rendered aperiodic by a special construction at the back, which introduces damping. For our present purpose the diaphragm can be regarded as a membrane, the mechanical resistance being proportional to the axial velocity, while secondary effects due to sound radiation from the diaphragm, by virtue of its vibration, are to be neglected. The problem is to discuss the motion when (a) a constant unidirectional air-pressure is applied over and above that due to the atmosphere, (b) the excess pressure has a sine-wave form (excess pressure is on one side of membrane only). The analysis below is approximate but instructive.

14·42. Unidirectional pressure of form $H(t)$. From [234], p. 12, if the displacement is small, the forces acting on the membrane are

$$\text{elastic} + \text{inertive} + \text{resistive} = \text{driving},$$

so $-2\pi x\, \partial x \tau \left(\dfrac{\partial^2 \xi}{\partial x^2} + \dfrac{1}{x}\dfrac{\partial \xi}{\partial x}\right) + 2\pi x\, \partial x\, \mathbf{m}\, \dfrac{\partial^2 \xi}{\partial t^2} + 2\pi x\, \partial x\, \mathbf{r}\, \dfrac{\partial \xi}{\partial t} = 2\pi x\, \partial x \mathbf{f},$

$$\tag{1}$$

or $\dfrac{\partial^2 \xi}{\partial x^2} + \dfrac{1}{x}\dfrac{\partial \xi}{\partial x} - \left(\dfrac{\mathbf{m}}{\tau}\dfrac{\partial^2 \xi}{\partial t^2} + \dfrac{\mathbf{r}}{\tau}\dfrac{\partial \xi}{\partial t}\right) = -\mathbf{f}/\tau,$ (2)

where τ = radial tension per unit arc length, $\mathbf{m}$ = mass per unit area, $\mathbf{f}$ = force per unit area, i.e. pressure, and $\mathbf{r}$ = resistance to motion per unit area, per unit velocity. Writing ϕ for ξ, i.e. $\xi(x,t) \Rightarrow \phi(x,p)$, and p for $\partial/\partial t$ in (2), the transform equation for initial quiescence is

$$\frac{d^2\phi}{dx^2} + \frac{1}{x}\frac{d\phi}{dx} - (\mathbf{m}p^2 + \mathbf{r}p)\,\phi/\tau = -\mathbf{f}/\tau, \tag{3}$$

of which the formal solution is

$$\phi = A I_0(\lambda x) + B K_0(\lambda x) + \mathbf{f}/\lambda^2\tau, \tag{4}$$

where I_0 and K_0 are modified Bessel functions [234], and

$$\lambda^2 = \left[\frac{1}{\tau}(\mathbf{m}p^2 + \mathbf{r}p)\right].$$

When $x = 0$, $K_0(\lambda x)$ is infinite, but the central displacement is finite, so $B = 0$, and the appropriate solution is

$$\phi = A I_0(\lambda x) + \mathbf{f}/\lambda^2\tau. \tag{5}$$

At the edge of the membrane $x = a$ and $\xi = 0$, and, therefore, $\phi = 0$, so from (5)

$$A = -\mathbf{f}/\lambda^2\tau I_0(\lambda a). \tag{6}$$

Substituting for A from (6) into (5), we obtain

$$\phi = (\mathbf{f}/\lambda^2 \boldsymbol{\tau}) \left[1 - I_0(\lambda x)/I_0(\lambda a)\right]. \tag{7}$$

Using the Mellin theorem, the membrane displacement is given by

$$\xi = \frac{\mathbf{f}/\boldsymbol{\tau}}{2\pi i} \int_{Br_1} e^{zt} \left[1 - \frac{I_0(\lambda x)}{I_0(\lambda a)}\right] \frac{dz}{z\lambda^2} \tag{8}$$

$$= \frac{\mathbf{f}/\mathbf{m}}{2\pi i} \int_{Br_1} e^{zt} \left\{1 - \frac{I_0[\sqrt{(\alpha z^2 + \beta z)]}}{I_0[\sqrt{(\alpha_1 z^2 + \beta_1 z)]}}\right\} \frac{dz}{z^2(z + \mathbf{r}/\mathbf{m})}, \tag{9}$$

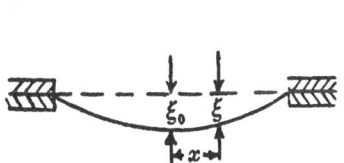

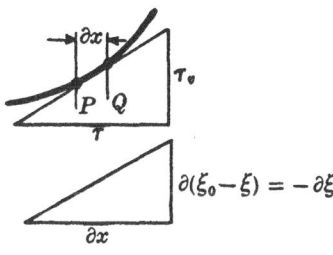

Fig. 94 (a). Illustrating membrane of condenser microphone.

Fig. 94 (b). $\boldsymbol{\tau}$ = radial tension per unit length.

where $\alpha = \mathbf{m}x^2/\boldsymbol{\tau}$, $\beta = \mathbf{r}x^2/\boldsymbol{\tau}$, $\alpha_1 = \mathbf{m}a^2/\boldsymbol{\tau}$, $\beta_1 = \mathbf{r}a^2/\boldsymbol{\tau}$. Since the Bessel function $I_0(y)$ is an even function, the integrand of (9) has no branch point. The singularities are poles and a limit point at infinity. We have to find the sum of the residues. Expanding the numerator of the bracketed quantity, we get

$$\frac{e^{zt}}{z^2(z + \mathbf{r}/\mathbf{m}) I_0[\sqrt{(\alpha_1 z^2 + \beta_1 z)]}} \left\{ \frac{(\beta_1 z + \alpha_1 z^2) - (\beta z + \alpha z^2)}{2^2} \right.$$

$$\left. + \frac{(\beta_1 z + \alpha_1 z^2)^2 - (\beta z + \alpha z^2)^2}{2^2 4^2} + \dots \right\} \tag{10}$$

$$= \frac{e^{zt}}{z(z + \mathbf{r}/\mathbf{m}) I_0[\sqrt{(\alpha_1 z^2 + \beta_1 z)]}} \left\{ \frac{\beta_1 - \beta}{4} + \frac{(\alpha_1 - \alpha) z}{4} \right.$$

$$\left. + \text{terms in } z, z^2, \dots \right\}. \tag{11}$$

When $z = 0$, the Bessel function in the denominator of (11) is unity and a pole occurs, the residue being $m(\beta_1 - \beta)/4r$. Thus by (9) the contribution from this pole is

$$\xi_1 = f(\beta_1 - \beta)/4r = f(a^2 - x^2)/4\tau, \qquad (12)$$

which represents the unidirectional displacement of the membrane due to the static excess air-pressure f. There is no pole at $z = -r/m$.

We now come to the evaluation of (9) at the singularities arising from the positive zeros of $I_0 \sqrt{(a_1 z^2 + \beta_1 z)}$, i.e. of

$$J_0[i \sqrt{(\alpha_1 z^2 + \beta_1 z)}].$$

If the zeros of $J_0(ui)$, i.e. the roots of $J_0(ui) = 0$, are $\pm k_n$, where $n = 0, 1, 2, \ldots$, then

$$i \sqrt{(\alpha_1 z^2 + \beta_1 z)} = \pm k_n, \qquad (13)$$

or

$$\alpha_1 z^2 + \beta_1 z + k_n^2 = 0, \qquad (14)$$

which gives

$$\left.\begin{matrix} z_1 \\ z_2 \end{matrix}\right\} = -\frac{\beta_1}{2\alpha_1} \pm \sqrt{\{(\beta_1^2/4\alpha_1^2) - (k_n^2/\alpha_1)\}}. \qquad (15)$$

The values of k_n correspond to the vibrational modes of the membrane. If $k_n^2 < \beta_1^2/4\alpha_1$, the motion is aperiodic, but when $k_n^2 > \beta_1^2/4\alpha_1$, the third member of (15) is imaginary, and each value of k_n corresponds to a damped oscillation. In practice, however, r increases with increase in k_n, and the influence of the higher modes is negligible. By (9) the contribution from the singularities at z_1, z_2 is

$$\xi_2 = (f/m) \sum_{n=0}^{\infty} \left[\frac{e^{zt}}{z^2(z + r/m)} \right.$$

$$\left. \times \left\{ \frac{J_0[i \sqrt{(\alpha_1 z^2 + \beta_1 z)}] - J_0[i \sqrt{(\alpha z^2 + \beta z)}]}{d\{J_0[i \sqrt{(\alpha_1 z^2 + \beta_1 z)}]\}/dz} \right\} \right]_{z=z_1}^{z=z_2}. \qquad (16)$$

Now

$$\frac{dJ_0}{dz} = \frac{dJ_0[i \sqrt{(\alpha_1 z^2 + \beta_1 z)}]}{d[i \sqrt{(\alpha_1 z^2 + \beta_1 z)}]} \cdot \frac{d[i \sqrt{(\alpha_1 z^2 + \beta_1 z)}]}{dz}$$

$$= J_1[i \sqrt{(\alpha_1 z^2 + \beta_1 z)}] \cdot (2\alpha_1 z + \beta_1)/2i \sqrt{(\alpha_1 z^2 + \beta_1 z)}. \qquad (17)$$

Substituting from (17) into (16) and putting $z = z_1,\ z_2$, we obtain

$$\xi_2 = -(f\tau/a^2 m^2)\sum_{n=0}^{\infty}\left\{\left[\frac{e^{z_1 t}}{z_1^2(z_1 + r/m)(z_1 + r/2m)}\right.\right.$$

$$\left.\left. + \frac{e^{z_2 t}}{z_2^2(z_2 + r/m)(z_2 + r/2m)}\right]\frac{k_n J_0(k_n x/a)}{J_1(k_n)}\right\}. \qquad (18)$$

The diaphragm displacement at any time t is

$$\xi = \xi_1 + \xi_2, \qquad (19)$$

$\xi_1,\ \xi_2$ being obtained from (12) and (18). The displacement represented by ξ_2 will decrease steadily with increase in t, since the real parts of $z_1,\ z_2$ are negative. Ultimately the membrane takes up a displacement given by (12), so its cross-section by an axial plane is parabolic. (18) represents an infinite series of exponential terms, corresponding to aperiodic and oscillatory motions arising from the modes of the membrane. When z_1 and z_2 are complex, (18) will represent the damped oscillations due to the vibrational modes.

14·43. Sinusoidal sound pressure. When the sound pressure on the membrane is sinusoidal, f in (7) § 14·42 is replaced by the transform of $f \sin \omega t$, so (9) § 14·42 becomes

$$\xi = \frac{f\omega/m}{2\pi i}\int_{Br_1}\left\{1 - \frac{I_0[\sqrt{(\alpha z^2 + \beta z)}]}{I_0[\sqrt{(\alpha_1 z^2 + \beta_1 z)}]}\right\}\frac{e^{zt}\,dz}{z(z + r/m)(z^2 + \omega^2)}. \qquad (1)$$

By expanding as at (10) and (11) § 14·42, we find that (1) has no pole at $z = 0$, so the membrane is not permanently displaced. The singularities occur at $z = \pm i\omega$, and at the zeros of

$$I_0[\sqrt{(\alpha_1 z^2 + \beta_1 z)}],$$

as before.

There is no pole at $z = -r/m$, but those at $z = \pm i\omega$ contribute the oscillating displacement of the membrane caused by the sinusoidal sound pressure. As before, the singularities of I_0 contribute an infinite number of terms representing aperiodic and oscillatory motions relating to vibrational modes of the membrane. The presence of these terms indicates that a phase change occurs at the commencement of the diaphragm motion, i.e. it is out of phase with the applied sound pressure.

14·44. Initial form of membrane displacement.

When t is small enough, we can ascertain the nature of the displacement by making z in (1) § 14·43 large, as in §§ 4·61, 6·31. We neglect, therefore, $r/\mathbf{m}$, ω^2, terms in β, β_1, and replace the Bessel functions by their asymptotic values [234], e.g.

$$I_0(u) \sim e^u/\sqrt{(2\pi u)}.$$

Then (1) § 14·43 becomes

$$\xi \simeq \frac{\mathbf{f}\omega/\mathbf{m}}{2\pi i} \int_{Br_1} \{1 - (\alpha_1/\alpha)^{\frac14} \exp\left[(\sqrt{\alpha} - \sqrt{\alpha_1})\,z\right]\} \frac{e^{zt}\,dz}{z^4} \tag{1}$$

$$\simeq \frac{\mathbf{f}\omega/\mathbf{m}}{3!} \{t^3 - \sqrt{(a/x)}\,[t - \sqrt{(\mathbf{m}/\tau)}\,(a-x)]^3\}, \tag{2}$$

provided ω is not too large, $x > 0$ and $t > \sqrt{(\mathbf{m}/\tau)}\,(a-x)$, by § 8·51. When $x = 0$, the value of $I_0\sqrt{(\alpha z^2 + \beta z)}$ in (1) § 14·43 is unity, and we obtain

$$\xi \simeq (\mathbf{f}\omega/\mathbf{m})\left\{\frac{t^3}{3!} - \{t - a\sqrt{(\mathbf{m}/\tau)}\}^{\frac52}\sqrt{a}\,(\mathbf{m}/\tau)^{\frac14}\frac{2^{\frac74}}{1.3.5}\right\}, \tag{3}$$

provided ω is not too large and $t > a\sqrt{(\mathbf{m}/\tau)}$. The first terms in (2) and (3) hold as $t \to 0$, so ξ varies as t^3 initially. In practice $\tau \gg \mathbf{m}$, so ξ varies nearly as t^3 when the second terms become operative. If the diaphragm responded faithfully to the air-pressure variations, ξ would vary as t.

14·51. Transient oscillations in a loud speaker horn

[120]. The equation for propagation of sound waves of infinitesimal amplitude in a conduit or cellular type of exponential horn* is (see [233])

$$\frac{\partial^2\phi}{\partial x^2} + \beta\frac{\partial\phi}{\partial x} - \frac{1}{c^2}\frac{\partial^2\phi}{\partial t^2} = 0, \tag{1}$$

---

* See Fig. 95, where the conduit dimensions are small enough to ensure approximately uniform pressure across any section at audio frequencies.

where ϕ represents the velocity potential, c the velocity of sound in free air, t the time, β the flaring index in the formula $A_x = A_0 e^{\beta x}$, A_0 being the throat area and A_x that at abscissa x. At time $t = 0$, the air particles at the horn throat are incited to vibrate with axial velocity

$$\frac{\partial \xi}{\partial t} = -\left(\frac{\partial \phi}{\partial x}\right)_{x=0} = -\omega \xi_0 \sin \omega t. \tag{2}$$

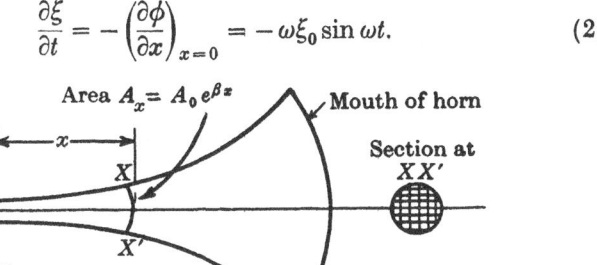

Area $A_x = A_0 e^{\beta x}$

Mouth of horn

Section at
XX'

X

x

O

X'

Throat
area $= A_0$

Fig. 95. Exponential loud speaker horn with conduits.

The problem is to find the sound pressure at abscissa x at any subsequent time, given that $p = \rho \partial \phi / \partial t$, p being the excess pressure due to sound, and ρ the density of the undisturbed air, reflection at the horn mouth being assumed negligible.

14·52. Solution of (1) §14·51. Owing to the general use of p for sound pressure and ϕ for velocity potential, in this problem we shall write σ for $\partial / \partial t$, and $\phi(x,t) \Rightarrow \psi(x, \sigma)$. By (1) §14·51, the transform equation for initial quiescence, is

$$\frac{d^2 \psi}{dx^2} + \beta \frac{d\psi}{dx} - \frac{\sigma^2}{c^2} \psi = 0, \tag{1}$$

its formal solution being

$$\psi = A\, e^{\lambda_1 x} + B\, e^{\lambda_2 x}, \tag{2}$$

where $\lambda_1, \lambda_2 = \tfrac{1}{2}\{-\beta \mp \sqrt{(\beta^2 + 4\sigma^2/c^2)}\}$. By hypothesis, reflection at the mouth of the horn is negligible, so we may consider that sound is propagated outwards only. Under this condition, the term $B e^{\lambda_2 x}$ in (2) is to be omitted for the reason given in [234], p. 75. Thus the appropriate solution of (1) is

$$\psi = A\, e^{-\frac{1}{2}\beta x}\, e^{-x\sqrt{(\beta^2 + 4\sigma^2/c^2)}/2}. \tag{3}$$

The boundary condition (2) § 14·51 gives

$$-(\partial \psi/\partial x)_{x=0} = -\omega^2 \xi_0 \sigma/(\sigma^2 + \omega^2). \tag{4}$$

By (3), (4), we get

$$A \tfrac{1}{2}\{\beta + \sqrt{(\beta^2 + 4\sigma^2/c^2)}\} = -\omega^2 \xi_0 \sigma/(\sigma^2 + \omega^2). \tag{5}$$

Multiplying both sides of (5) by $\{\beta - \sqrt{(\beta^2 + 4\sigma^2/c^2)}\}$, we obtain

$$A = \omega^2 c^2 \xi_0 \{\beta - \sqrt{(\beta^2 + 4\sigma^2/c^2)}\}/2\sigma(\sigma^2 + \omega^2). \tag{6}$$

Substituting this value of A in (3) yields

$$\phi = \omega^2 c^2 \xi_0 e^{-\frac{1}{2}\beta x} \{\beta - \sqrt{(\beta^2 + 4\sigma^2/c^2)}\} e^{-x\sqrt{(\beta^2 + 4\sigma^2/c^2)}/2}/2\sigma(\sigma^2 + \omega^2). \tag{7}$$

The sound pressure is $p = \dfrac{\rho \partial \phi(x, t)}{\partial t}$ (see [233], p. 17). Since

$\phi(x, t) \rightleftharpoons \psi(x, \sigma)$ and $\phi(x, 0) = 0$, $t \to 0$, $x \geqslant 0$, by (5) § 8·43

$$p \rightleftharpoons \rho \sigma \psi \tag{8}$$

$$= \rho \omega^2 c^2 \xi_0 e^{-\frac{1}{2}\beta x} \{\beta - \sqrt{(\beta^2 + 4\sigma^2/c^2)}\} e^{-x\sqrt{(\beta^2 + 4\sigma^2/c^2)}/2}/2(\sigma^2 + \omega^2), \tag{9}$$

provided ϕ is continuous in $(t - x/c) > 0$, $x \geqslant 0$, a point which the reader may verify using the formula for p found later on. Applying the Mellin theorem to (9), we obtain

$$p = \frac{K}{2\pi i} \int_{Br_1} \frac{e^{zt - x\sqrt{(\beta^2 + 4z^2/c^2)}/2} \{\beta - \sqrt{(\beta^2 + 4z^2/c^2)}\} dz}{z(z^2 + \omega^2)}, \tag{10}$$

where $K = \rho \omega^2 c^2 \xi_0 e^{-\frac{1}{2}\beta x}/2$. The singularities of the integrand are:

(i) two simple poles on the imaginary axis at $z = \pm i\omega$,

(ii) two branch points on the imaginary axis at $z = \pm \frac{1}{2}i\beta c$.

In practice $\omega > \frac{1}{2}\beta c$, since the cut-off point of the horn is given by $\omega_c = \frac{1}{2}\beta c$. In virtue of the factor $\{\beta - \sqrt{(\beta^2 + 4z^2/c^2)}\}$, the origin is not a pole of (10). For when z is small, we have

$$\beta\{1 - \sqrt{(1 + 4z^2/c^2\beta^2)}\}/z(z^2 + \omega^2)$$
$$\simeq \beta\{-2z^2/c^2\beta^2\}/z(z^2 + \omega^2) \simeq -2z/c^2\beta\omega^2, \tag{11}$$

so the integrand vanishes with z.

The contour Br_2 of Fig. 96 (a) is equivalent to Br_1, since the contributions to the integral from the arcs in the second and third quadrants $\to 0$ as the radius $\to \infty$ (see § 4·24 et seq.). Also the integrand is the same at P and Q.* Thus the contributions from the lines above and below the negative real axis cancel out. After passing round either of the poles, the integrand has its initial value, so the integrals along R and S cancel out, as also do those along T, U. Consequently the contour of Fig. 96 (b) is equivalent to that of Fig. 96 (a).

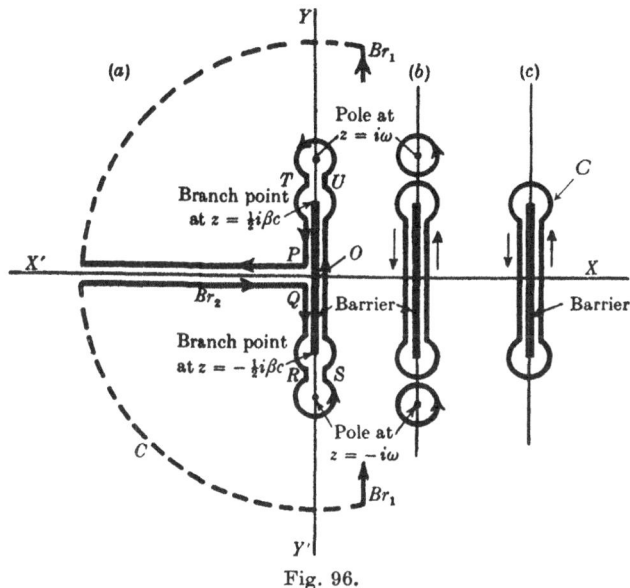

Fig. 96.

14·53. Steady state solution.† This is given by the contribution to (10) § 14·52 from the poles at $z = \pm i\omega$. We write $(z^2 + \omega^2) = (z - \omega e^{\frac{1}{2}i\pi})(z - \omega e^{-\frac{1}{2}i\pi})$, and proceed as in § 3·21.‡ With $k = \omega/c$, the pole at $z = \omega e^{\frac{1}{2}i\pi}$ yields

$$- \exp[i\omega t - x\sqrt{(\beta^2 + 4e^{i\pi}k^2)}/2]\,K\{\beta - \sqrt{(\beta^2 + 4e^{i\pi}k^2)}\}/2\omega^2 \quad (1)$$

$$= - \exp[i\omega t - x e^{\frac{1}{2}i\pi}\sqrt{(k^2 - \beta^2/4)}]\,K\{\beta - 2e^{\frac{1}{2}i\pi}\sqrt{(k^2 - \beta^2/4)}\}/2\omega^2. \quad (2)$$

* Since the integrand is single-valued on a vertical line PQ, the two parallel lines through P and Q may be replaced by one line along the negative real axis. Similarly at T, U, and R, S.

† When this *alone* is required, put $\phi = \phi_1 e^{i\omega t}$ in (1) § 14·51, and solve.

‡ When branch points occur, as in (10) § 14·52, to avoid ambiguity i is written in its exponential form. Although $i^2 = -1$, and $(-i)^2 = -1$, the respective square roots are $e^{\frac{1}{2}i\pi}$ and $e^{-\frac{1}{2}i\pi}$, since $i^2 = e^{i\pi}$ and $(-i)^2 = e^{-i\pi}$ (see Fig. 21 (c)).

Changing i to $-i$, the pole at $z = \omega\, e^{-\frac{1}{2}i\pi}$ yields

$$-\exp\left[-i\omega t - x\, e^{-\frac{1}{2}i\pi}\, \sqrt{(k^2 - \beta^2/4)}\right] K\{\beta - 2\, e^{-\frac{1}{2}i\pi}\, \sqrt{(k^2 - \beta^2/4)}\}/2\omega^2. \tag{3}$$

Adding (2), (3) and writing $m = \sqrt{(k^2 - \beta^2/4)}$, the steady sound pressure is

$$p_s = -\tfrac{1}{2}\rho c^2 \xi_0\, e^{-\frac{1}{2}\beta x} \left[\beta \cos(\omega t - mx) + 2m \sin(\omega t - mx)\right] \tag{4}$$

$$= -\rho c^2 k \xi_0\, e^{-\frac{1}{2}\beta x} \sin\left[(\omega t - mx) + \tan^{-1}(\beta/2m)\right]. \tag{5}$$

14·54. Transient solution. We have now to integrate (10) § 14·52 round the contour enclosing the two branch points at $z = \pm \tfrac{1}{2}i\beta c$, in order to find the transient oscillation due to the horn. This oscillation is superimposed upon the steady oscillation given by (5) § 14·53. The requisite contour is depicted in Fig. 96 (c). The greatest value of $|z|$ on the contour is $\tfrac{1}{2}\beta c$. Now, if attention is confined to a frequency band where $\omega^2 \gg \tfrac{1}{4}\beta^2 c^2$, then ω^2 can be written for $(\omega^2 + z^2)$ in the integral. Omitting K/ω^2, (10) § 14·52 becomes

$$I = \frac{1}{2\pi i} \int_C e^{zt - \sqrt{(x/c)(z^2 + a^2)}} \{\beta - (2/c)\sqrt{(z^2 + a^2)}\} \frac{dz}{z}, \tag{1}$$

with $a = \tfrac{1}{2}\beta c$. From (6) § 5·32

$$J_0[a\,\sqrt{(t^2 - x^2/c^2)}] = \frac{1}{2\pi i} \int_{Br_2 \text{ or } C} e^{u - a^2(t^2 - x^2/c^2)/4u}\, \frac{du}{u}, \tag{2}$$

u being the complex variable on the contour Br_2, which is equivalent to C in (1). Substituting*

$$u = (t - x/c)\, \tfrac{1}{2}\{z + \sqrt{(z^2 - a^2)}\}$$

in (2), and proceeding as in § 8·7, leads to

$$J_0[a\,\sqrt{(t^2 - x^2/c^2)}] = \frac{1}{2\pi i} \int_{Br_2 \text{ or } C} e^{zt - (x/c)\sqrt{(z^2 + a^2)}} \frac{dz}{\sqrt{(z^2 + a^2)}}. \tag{3}$$

---

* This transformation may be justified by considering Br_2 in its more general form, as in Fig. 24 (a), and adopting the argument of § 9 Appendix 5. Then C is equivalent to Br_1, so long as the singularities are within the closed contour having an internal barrier as in Fig. 96 (c).

Since $J_0(u) = I_0(iu)$, this result could have been obtained immediately from (7) § 8·7 on writing ia for a. Differentiating both sides of (3) with respect to x/c, we get

$$(ax/c)\frac{J_1[a\sqrt{(t^2 - x^2/c^2)}]}{\sqrt{(t^2 - x^2/c^2)}} = -\frac{1}{2\pi i}\int_C e^{zt-(x/c)\sqrt{(z^2+a^2)}}\,dz. \tag{4}$$

This integral converges uniformly with respect to x, $0 \leqslant x \leqslant x_1$, so that differentiation under the sign in (3) is valid. Differentiating both sides of (4) with respect to x/c gives

$$a\frac{\partial}{\partial(x/c)}\left\{\frac{x}{c}\frac{J_1[a\sqrt{(t^2 - x^2/c^2)}]}{\sqrt{(t^2 - x^2/c^2)}}\right\} = \frac{1}{2\pi i}\int_C e^{zt-(x/c)\sqrt{(z^2+a^2)}}\sqrt{(z^2+a^2)}\,dz, \tag{5}$$

the integral being uniformly convergent with respect to x, $0 \leqslant x \leqslant x_1$.

Writing $x/c = b$, $v = a\sqrt{(t^2 - x^2/c^2)}$, we have $\partial v/\partial b = -a^2 b/v$, so from the left-hand side of (5)

$$a^2\frac{\partial}{\partial b}\left[b\frac{J_1(v)}{v}\right] = a^2\frac{J_1(v)}{v} + a^2 b\frac{\partial}{\partial b}\left[\frac{J_1(v)}{v}\right] \tag{6}$$

$$= a^2\frac{J_1(v)}{v} + a^2 b\left[-\frac{J_1(v)}{v^2} + \frac{J_1'(v)}{v}\right]\frac{\partial v}{\partial b} \tag{7}$$

$$= a^2\frac{J_1(v)}{v} + \frac{a^4 b^2}{v^2}\left[\frac{J_1(v)}{v} - J_1'(v)\right]. \tag{8}$$

Using the recurrence relation (see [234])

$$J_1'(v) = J_1(v)/v - J_2(v), \tag{9}$$

(8) may be written in the form

$$a^2\frac{\partial}{\partial b}\left[b\frac{J_1(v)}{v}\right] = a^2\frac{J_1(v)}{v} + a^4 b^2\frac{J_2(v)}{v^2}. \tag{10}$$

Thus by (4), (5) and (10)

$$a^2(\beta b + 2/c)\frac{J_1(v)}{v} + (2a^4 b^2/c)\frac{J_2(v)}{v^2}$$

$$= \frac{1}{2\pi i}\int_C e^{zt-b\sqrt{(z^2+a^2)}}\{\beta - (2/c)\sqrt{(z^2+a^2)}\}\,dz. \tag{11}$$

It is permissible to integrate the right-hand side of (11) with respect to t, so introducing the factor K/ω^2 from above, the transient sound pressure is given by the expression

$$\frac{K}{\omega^2}\left[\frac{\beta^2 c(2+\beta x)}{4}\int_t^\infty \frac{J_1(v)}{v}\,dt+\frac{\beta^4 x^2 c}{8}\int_t^\infty \frac{J_2(v)}{v^2}\,dt\right]$$

$$=\frac{(K/\omega^2)}{2\pi i}\int_C e^{zt-(x/c)\sqrt{(z^2+a^2)}}\{\beta-(2/c)\sqrt{(z^2+a^2)}\}\frac{dz}{z}. \quad (12)$$

The limits t and ∞ were chosen for the integrals, since the *transient* sound pressure $\to 0$ as $t\to\infty$. It is left as an exercise for the reader to show that the contour integral in (12) is evanescent when $t\to\infty$. Inserting the value of K in the left-hand side of (12), yields

$$p_t = \tfrac{1}{8}\rho\beta^2 c^3\xi_0\, e^{-\frac{1}{2}\beta x}\left[(2+\beta x)\int_t^\infty \frac{J_1(v)}{v}\,dt+\frac{\beta^2 x^2}{2}\int_t^\infty \frac{J_2(v)}{v^2}\,dt\right], \quad (13)$$

which is valid if $\omega^2 \gg \tfrac{1}{4}\beta^2 c^2$.

The preceding analysis, although complicated, is given to illustrate certain aspects of contour integration. (13) and the alternative form at (3) § 14·55 may be derived more easily by inverting

$$\phi(\sigma) = e^{-(x/c)(\sigma^2+a^2)}\{\beta-(2/c)\sqrt{(\sigma^2+a^2)}\}, \quad (14)$$

from (1), using Laplace transform procedure (see [236], p. 193). This is left as an exercise for the reader.

14·55. Alternative form of (13) §14·54. Formula (13) § 14·54 may be expressed in a form more convenient for computation. Thus $\displaystyle\int_t^\infty = \int_0^\infty - \int_0^{x/c} - \int_{x/c}^t$, the values of the first two integrals being given at (1) and (2) below. From [120]

$$(2+\beta x)\left[\int_0^\infty \frac{J_1(v)}{v}\,dt - \int_0^{x/c}\frac{J_1(v)}{v}\,dt\right] = \frac{8+4\beta x}{\beta^2 cx}(1-e^{-\frac{1}{2}\beta x}),$$

$$\tag{1}$$

and $\quad\tfrac{1}{2}\beta^2 x^2\left[\int_0^\infty \frac{J_2(v)}{v^2}\,dt - \int_0^{x/c}\frac{J_2(v)}{v^2}\,dt\right]$

$$=\frac{2x}{c}\left[\frac{e^{-\frac{1}{2}\beta x}-1}{(\tfrac{1}{2}\beta x)^2}+\frac{e^{-\frac{1}{2}\beta x}}{\tfrac{1}{2}\beta x}+\frac{1}{2}\right]. \quad (2)$$

The third integral, whose limits are x/c and t, is found by taking numerical values, plotting a curve and finding its area up to various values of t, or by numerical integration. The sum of (1) and (2) is $(4/\beta + x)/c$. Whence (13) § 14·54 may be written in the form

$$p_t = \tfrac{1}{8}\rho_0\beta^2c^3\xi_0\, e^{-\frac{1}{2}\beta x}\Bigg[\,(4/\beta c) + (x/c)$$

$$-\,(2+\beta x)\int_{x/c}^{t}\frac{J_1(v)}{v}\,dt - \tfrac{1}{2}\beta^2x^2\int_{x/c}^{t}\frac{J_2(v)}{v^2}\,dt\,\Bigg]. \qquad (3)$$

14·56. Total sound pressure, transient + steady oscillation. At any time $t \geqslant x/c$, the sound pressure at abscissa x is due to the transient and steady state oscillations, so by (5) § 14·53 and (3) § 14·55,

$$p = p_t + p_s = \tfrac{1}{2}\rho_0 c^2\xi_0\, e^{-\frac{1}{2}\beta x}\Bigg\{\tfrac{1}{4}\beta^2c\Bigg[\,(4/\beta c) + (x/c)$$

$$-\,(2+\beta x)\int_{x/c}^{t}\frac{J_1(v)}{v}\,dt - \tfrac{1}{2}\beta^2x^2\int_{x/c}^{t}\frac{J_2(v)}{v^2}\,dt\,\Bigg]$$

$$-\,2k\sin\left[(\omega t - mx) + \tan^{-1}\beta/2m\right]\Bigg\}. \qquad (1)$$

The factor $e^{-\frac{1}{2}\beta x}$, which occurs in the formulae for sound pressure, is due to attenuation caused by expansion of the wave front as it travels down the horn towards the mouth.

14·57. Wave form of horn transient. The form of damped oscillation represented by (3) § 14·55 has been computed using the following practical data:

Cut-off frequency of horn $n_c = \beta c/4\pi = 31\cdot8$ cycles per sec.

Flaring index $\qquad\qquad\qquad\beta = 0\cdot0117$ cm.$^{-1}$

Abscissa at which pressure is calculated $\qquad\qquad\qquad\qquad x = 427$ cm.

Velocity of sound at 18° C. $\qquad c = 3\cdot43 \times 10^4$ cm. sec.$^{-1}$

$$\tfrac{1}{2}\beta x = 2\cdot5.$$

$$a = \tfrac{1}{2}\beta c = 200.$$

The calculated transient is shown in Fig. 97. Prior to time $t = x/c$, there is no sound pressure at x, but at $t = x/c$ (this being the time taken by the sound to reach x) the pressure *due to the transient* rises precipitately to the value

$$\tfrac{1}{8}\rho_0 \beta^2 c^2 \xi_0 \, e^{-\frac{1}{2}\beta x} \, (4/\beta + x).$$

This is obtained from (3) § 14·55 by substituting $t = x/c$ in the upper limits of the integrals. The pressure commences to decay immediately. As shown in § 14·6, when $t > 0·2\,\mathrm{sec}$. the oscillation takes the form $t^{-\frac{3}{2}} \sin (\tfrac{1}{2}\beta ct + \tfrac{1}{4}\pi)$. The occurrence of a damped oscillation may be regarded as peculiar to an exponential horn, since with a conical horn the decay is purely exponential, which implies an infinite period of oscillation.

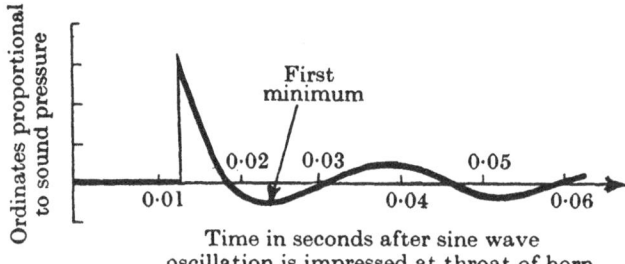

Fig. 97. Transient oscillation 427 cm. from throat of
exponential loud speaker horn.

In practice, whatever the value of x, the *total* sound pressure starts from zero at $t = x/c$ and tends ultimately to the value given by (5) § 14·53. Initially the phase of the steady state oscillation is such that the transient pressure is neutralized when $t = x/c$. Decay of the transient is accompanied by variation in sound pressure amplitude. Transient sounds in speech or music entail a band frequency spectrum. Modification therein caused by the horn may be determined by analysis of the type given above, provided the boundary condition at $x = 0$ is expressible as a transform, as exemplified at (4) § 14·52.

14·6. Asymptotic formula for (1) § 14·54. The integrand of (1) § 14·54 has two branch points at $z = \pm ia$, and the appropriate contour (a Br_2 type) for obtaining an asymptotic formula is illustrated in Fig. 98 (see § 5·351).

The integral may be written in the form

$$I = \left\{ \frac{(\beta + 2\partial/\partial x)}{2\pi i} \right\} \int_C e^{zt-(x/c)\,\sqrt{(z^2+a^2)}} \frac{dz}{z} \quad (1)$$

$$= (\beta + 2\partial/\partial x)\,I_0, \quad (2)$$

differentiation under the sign being valid, since the resulting integral converges uniformly with respect to x in the closed interval $0 \leqslant x \leqslant x_1$.

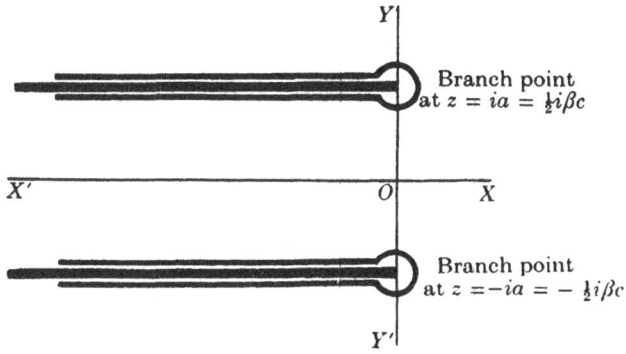

Fig. 98.

We move the origin to the point $z = ia$ by writing $\zeta = z - ia$ or $z = \zeta + ia$ in I_0, and for the upper contour in Fig. 98

$$I_1 = \frac{(e^{iat}/\omega^2)}{2\pi i} \int_{Br_2} e^{\zeta t - \sqrt{(2ia)}\,(x/c)\,\sqrt{\zeta}\,\sqrt{(1-i\zeta/2a)}} \frac{d\zeta}{(\zeta + ia)}. \quad (3)$$

When t is relatively large, and a is large enough, the main contribution to (3) occurs in the range where $|\,i\zeta/2a\,| \ll 1$. Dividing above and below by ia, and expanding the denominator of the integrand,

$$I_1 \sim \frac{e^{iat}}{ia\omega^2} \frac{1}{2\pi i} \int_{Br_2} e^{\zeta t - b\sqrt{\zeta}} \, d\zeta [1 + i\zeta/a - \zeta^2/a^2 - i\zeta^3/a^3 + \dots], \quad (4)$$

where $b = (x/c)\sqrt{(2ai)}$. Expanding the exponential term $e^{-b\sqrt{\zeta}}$, (4) becomes

$$I_1 \sim \frac{e^{iat}}{ia\omega^2}\frac{1}{2\pi i}\int_{Br_s} e^{\zeta t}d\zeta\left[1 - b\sqrt{\zeta} + \frac{b^2\zeta}{2!} - \frac{b^3\zeta^{\frac{3}{2}}}{3!} + \ldots\right]$$

$$\times[1 + i\zeta/a - \zeta^2/a^2 - i\zeta^3/a^3 + \ldots], \quad (5)$$

so, to the required degree of approximation,

$$I_1 \sim \frac{e^{iat}}{ia\omega^2}\frac{1}{2\pi i}\int_{Br_s} e^{\zeta t}d\zeta(1 - b\sqrt{\zeta}) \quad (6)$$

$$= \frac{b\,e^{iat}}{ia\omega^2\,2\sqrt{\pi}\,t^{\frac{3}{2}}} = x\,e^{i(\frac{1}{2}\beta ct+\frac{1}{4}\pi)}/i\omega^2\sqrt{(\pi\beta)}\,(ct)^{\frac{3}{2}}. \quad (7)$$

The integral along the lower contour in Fig. 98 is found from (7) by changing i to $-i$, which gives

$$I_2 \sim -x\,e^{-i(\frac{1}{2}\beta ct+\frac{1}{4}\pi)}/i\omega^2\sqrt{(\pi\beta)}\,(ct)^{\frac{3}{2}}. \quad (8)$$

Adding (7) and (8), we obtain

$$I_3 \sim \frac{2x}{\omega^2\sqrt{(\pi\beta)}\,(ct)^{\frac{3}{2}}}\sin(\tfrac{1}{2}\beta ct + \tfrac{1}{4}\pi). \quad (9)$$

Introducing the factor $K/\omega^2 = \rho c^2\xi_0\,e^{-\frac{1}{2}\beta x}/2$ from § 14·54, and $(\beta + 2\partial/\partial x)$ from (1), the asymptotic formula for the transient sound pressure is

$$p_t \sim \frac{\rho_0\sqrt{c}\,\xi_0}{\sqrt{(\pi\beta)}}e^{-\frac{1}{2}\beta x}(2 + \beta x)\frac{\sin(\frac{1}{2}\beta ct + \frac{1}{4}\pi)}{t^{\frac{3}{2}}}. \quad (10)$$

Thus at abscissa x the transient sound pressure decays according to the law $\dfrac{\sin(\frac{1}{2}\beta ct + \frac{1}{4}\pi)}{t^{\frac{3}{2}}}$, provided t and a are large enough.

Integral (3) may be evaluated also by making ζ small and applying (3) § 5·222. This is left as an exercise for the reader.

XV

DIFFUSION OF HEAT: ABSORPTION OF MOISTURE

15·11. Homogeneous slab. We consider the simple case of a rectangular slab of material of uniform thickness when the effect of heat loss at the edges can be neglected. Theoretically this entails a slab of infinite extent, but in practice adequate accuracy can be obtained when the edge area is small compared with that of the main surfaces. Let

θ = temperature difference between the surface at A and that of a parallel plane through CE (Fig. 99) distant x from the outer surface at B;

q = heat flux or quantity of heat passing through the plane at CE per unit area in unit time;

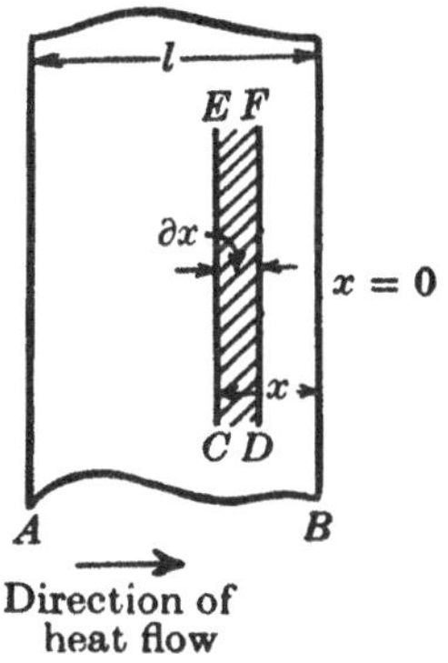

R = thermal or heat resistance per unit length of material in the direction of flow per unit area (thermal resistivity), i.e. the temperature drop in unit length when the heat flux is unity; $1/R$ is the thermal conductivity.

Fig. 99.

C = heat capacitance, i.e. the number of heat units to raise a block of unit area and unit length 1 degree in temperature = specific heat × density.

Then the fall in temperature in the length ∂x, when a quantity of heat q per unit area flows in the direction of the arrow in unit time, is

$$\partial\theta = Rq\,\partial x,$$

$$q = \frac{1}{R}\frac{\partial\theta}{\partial x},\tag{1}$$

the slope being positive since the temperature falls with decrease in x. Now the quantity of heat per unit area which enters C in unit time exceeds that which leaves the plane through D. This excess is the heat required to raise the temperature of the lamina of thickness ∂x, at a rate $\partial\theta/\partial t$. The amount of heat is

$$\partial q = C\partial x \frac{\partial\theta}{\partial t},$$

so

$$\frac{\partial q}{\partial x} = C\frac{\partial\theta}{\partial t}. \tag{2}$$

From (1),

$$\frac{\partial^2\theta}{\partial x^2} = R\frac{\partial q}{\partial x}. \tag{3}$$

Substituting in (3) for $\partial q/\partial x$ from (2), we obtain

$$\frac{\partial^2\theta}{\partial x^2} = CR\frac{\partial\theta}{\partial t}. \tag{4}$$

Equation (4) is identical in form with (9) § 13·21, provided $L = G = 0$, so the diffusion of heat in an infinite homogeneous slab or plate follows the same law as electrical transmission in a uniform unloaded cable. The quantity CR is usually written $1/k$, where

k = thermal conductivity/(density × specific heat)

  = diffusivity, diffusion coefficient, or thermometric conductivity.

15·12. Refrigerator. A household refrigerator consists essentially of a rectangular box $3 \times 2 \times 2$ ft., the walls being made of thermal insulating material, e.g. cork, 4 in. thick. Originally the inner and outer temperatures are 20° C. At a certain instant the temperature of the inner surface is lowered to $-10°$ C., and it is desired to maintain it at this value. Find the rate of extraction of heat by the refrigerating mechanism, assuming that the problem can be treated on the same lines as an infinite slab, one side being at 20° C., the other at $-10°$ C.

The appropriate differential equation for this problem is (4) § 15·11, which we shall solve by the method of § 13·13. Multiplying throughout by $p\,e^{-pt}$ and integrating from $t = 0$ to $+\infty$, we get

$$p \int_0^\infty e^{-pt}\left[\frac{\partial^2\theta(x,t)}{\partial x^2}\right]dt - \mathbf{CR}p \int_0^\infty e^{-pt}\left[\frac{\partial\theta(x,t)}{\partial t}\right]dt = 0. \quad (1)$$

If the order of differentiation and integration in the first member of (1) may be changed, then

$$\frac{\partial^2}{\partial x^2}\left[p \int_0^\infty e^{-pt}\theta\,dt\right] - \mathbf{CR}p \int_0^\infty e^{-pt}\,d(\theta) = 0. \quad (2)$$

Putting $p \int_0^\infty e^{-pt}\theta\,dt = \phi$, i.e. $\theta \rightleftharpoons \phi$, and integrating the second member of (2) by parts, gives

$$\frac{d^2\phi}{dx^2} - \mathbf{CR}\{[p\theta\,e^{-pt}]_{t=0}^\infty + p\phi\} = 0. \quad (3)$$

The initial condition is $\theta = 20°\,\mathrm{C}$. as $t \to -0$. Thus if $\theta\,e^{-pt} \to 0$ as $t \to +\infty$, with $\lambda = \sqrt{(p\mathbf{CR})}$, (3) becomes

$$\frac{d^2\phi}{dx^2} - \lambda^2\phi = -20\lambda^2, \quad (4)$$

of which the formal solution is

$$\phi = A\sinh\lambda x + B\cosh\lambda x + 20. \quad (5)$$

Now $\theta \rightleftharpoons \phi$, and if θ is constant, $\theta = \phi$. Accordingly the boundary conditions may be written (i) $\phi = 20°\,\mathrm{C}$. at $x = 0$ for $t > 0$; (ii) $\phi = -10°\,\mathrm{C}$. at $x = l$ for $t > 0$. Inserting (i), (ii) in (5), we obtain $B = 0$, and $A = -30/\sinh\lambda l$. With these values of A, B, (5) gives

$$\phi = 20 - 30\sinh\lambda x/\sinh\lambda l. \quad (6)$$

The heat flux through unit area of the insulation per unit time is $q \rightleftharpoons \psi(p)$. Since θ decreases with increase in x, by (1) §15·11

$$\psi = -(1/\mathbf{R})\,d\phi/dx. \quad (7)$$

Applying (7) to (6), at the inner surface where $x = l$, we have

$$\psi = 30\sqrt{(\mathbf{C}/\mathbf{R})}\sqrt{p}\cosh lb\sqrt{p}/\sinh lb\sqrt{p}, \quad (8)$$

with $b = \sqrt{(CR)}$. Using the Mellin theorem, with $\theta_0 = 30^\circ$ C. (8) gives

$$q = \frac{\theta_0 \sqrt{(C/R)}}{2\pi i} \int_{Br_1} \frac{e^{zt} \cosh lb \sqrt{z}\, dz}{\sqrt{z} \sinh lb \sqrt{z}}. \qquad (9)$$

Apart from e^{zt}, the integrand of (9) may be written

$$\frac{1 + l^2 b^2 z/2! + l^4 b^4 z^2/4! + \ldots}{lbz + l^3 b^3 z^2/3! + \ldots}. \qquad (10)$$

When z is small (10) is approximately $1/lbz$, so there is a simple pole at the origin, whose contribution to (9) is

$$q_1 = \theta_0 \sqrt{(C/R)}/lb = \theta_0/Rl. \qquad (11)$$

Poles occur also when $\sinh lb \sqrt{x} = 0$, i.e. $lb \sqrt{z} = in\pi$, or $z = -n^2\pi^2/l^2 CR$, where $n = 1, 2, \ldots$, there being a limit point at infinity. Proceeding as in §§ 3·237, 4·72, we obtain

$$q_2 = \frac{2\theta_0}{Rl} \sum_{n=1}^{\infty} e^{-n^2\pi^2 t/l^2 CR} \qquad (12)$$

for $t > 0$. Hence adding (11) and (12), for $t > 0$

$$q = \frac{\theta_0}{Rl} \left[1 + 2 \sum_{n=1}^{\infty} e^{-n^2\pi^2 t/l^2 CR} \right], \qquad (13)$$

this being the rate of extraction of heat from unit area at the inner surface of the refrigerator. The series in (13) is uniformly convergent in every closed interval $0 < t_1 \leqslant t \leqslant t_2$.

15·13. Solution in alternative form. When t is small, (13) § 15·12 is not convenient for computation, so we shall find an alternative expression. Writing a for lb, and K for $\theta_0 \sqrt{(C/R)}$ in (9) § 15·12, we get

$$q = \frac{K}{2\pi i} \int_{Br_1} \frac{(e^{a\sqrt{z}} + e^{-a\sqrt{z}})}{(e^{a\sqrt{z}} - e^{-a\sqrt{z}})} \frac{e^{zt}\, dz}{\sqrt{z}}. \qquad (1)$$

Multiplying the integrand above and below by $e^{-a\sqrt{z}}$, (1) becomes

$$q = \frac{K}{2\pi i} \int_{Br_1} (1 + e^{-2a\sqrt{z}})(1 - e^{-2a\sqrt{z}})^{-1} e^{zt} \frac{dz}{\sqrt{z}} \qquad (2)$$

$$= \frac{K}{2\pi i} \int_{Br_1} [(1 + e^{-2a\sqrt{z}})(1 + e^{-2a\sqrt{z}} + e^{-4a\sqrt{z}} + \ldots)] \frac{e^{zt}\, dz}{\sqrt{z}}, \qquad (3)$$

provided $|e^{-2a\sqrt{z}}| < 1$. Thus

$$q = \frac{K}{2\pi i} \int_{Br_1} \left[1 + 2 \sum_{n=1}^{\infty} e^{-2na\sqrt{z}} \right] \frac{e^{zt} dz}{\sqrt{z}} \tag{4}$$

$$= \theta_0 \sqrt{(C/\pi Rt)} \left[1 + 2 \sum_{n=1}^{\infty} e^{-n^2 l^2 CR/t} \right], \qquad (t > 0) \tag{5}$$

by (4) § 5·222. The *series* in (5) is uniformly convergent in $0 \leqslant t \leqslant t_1$, and may be used for computation if t is not too large. (5) above and (13) § 15·12 are identical for $t > 0$,* the latter being suitable for computation when t is large. For t small, (5) gives

$$q \simeq \theta_0 \sqrt{(C/\pi Rt)}, \tag{6}$$

so that as $t \to 0$, $q \to \infty$. Moreover, when the temperature of the inner surface is lowered instantaneously to $-10°$ C., the momentary rate of heat extraction must be infinite! (see end of § 15·23).

When t is large enough for a substantially steady state to have been attained, by (13) § 15·12 the thermal 'current' density is

$$q_0 = \theta_0/Rl. \tag{7}$$

Since Rl is the thermal resistance per unit area of the insulating material of thickness l, (7) is in effect the thermal analogue of Ohm's law. It is used extensively in engineering calculations pertaining to the design of heating systems, refrigerators, lagging and furnaces.

15·21. Boundary conditions for slab as functions of t.

Let

(i) $\theta(x, t) = \xi_1(t) \rightleftharpoons \phi_1(p)$ at $x = l, t > 0$,

(ii) $\theta(x, t) = \xi_2(t) \rightleftharpoons \phi_2(p)$ at $x = 0, t > 0$,

ξ_1, ξ_2 being arbitrary functions of t having transforms. For the initial condition take $\theta(x, t) = \theta_0$, a constant throughout the slab, as $t \to -0$. Then from (3) § 15·12, we obtain

$$d^2\phi/dx^2 - \lambda^2\phi = -\lambda^2\theta_0, \tag{1}$$

---

* Integrals like (9) § 15·12 involving hyperbolic functions, may be evaluated either by the theorem of residues or after expansion in exponentials. In the first case the exponential index is a positive and in the second a negative power of t, preceded by a minus sign in both cases. See also § 13·43 and [236], p. 114.

of which the complete solution is

$$\phi = A \sinh \lambda x + B \cosh \lambda x + \theta_0. \tag{2}$$

Substituting from (ii) into (2) gives

$$B = (\phi_2 - \theta_0). \tag{3}$$

Likewise (i) and (2) give

$$\phi_1 = A \sinh \lambda l + B \cosh \lambda l + \theta_0, \tag{4}$$

so $$A = [(\phi_1 - \theta_0) - (\phi_2 - \theta_0) \cosh \lambda l]/\sinh \lambda l. \tag{5}$$

Hence from (2), (3), (5)

$$\phi = \theta_0 + (\phi_1 - \theta_0) \frac{\sinh \lambda x}{\sinh \lambda l}$$

$$- (\phi_2 - \theta_0) \frac{\cosh \lambda l}{\sinh \lambda l} \sinh \lambda x + (\phi_2 - \theta_0) \cosh \lambda x \tag{6}$$

$$= \theta_0 + (\phi_1 - \theta_0) \frac{\sinh \lambda x}{\sinh \lambda l} + (\phi_2 - \theta_0) \frac{\sinh \lambda (l - x)}{\sinh \lambda l}. \tag{7}$$

$\theta(x, t)$ may be obtained by applying the Mellin theorem to (7). If l, the thickness of the slab $\to \infty$, and x is finite, (7) gives

$$\phi = \theta_0 + (\phi_2 - \theta_0) e^{\lambda(l-x)}/e^{\lambda l} \tag{8}$$

$$= \theta_0 + (\phi_2 - \theta_0) e^{-\lambda x}. \tag{9}$$

15·22. Example. Consider the case of a slab which is long enough for the effect of heat reflected from the far end, at an intermediate point x, to be negligible. Then as an analytical expedient we may let $l \to + \infty$, and use (9) § 15·21. For boundary conditions take

(i) $\theta = \theta_0$ as $l \to + \infty$, $t \geqslant 0$;

(ii) $\left. \begin{array}{l} \theta = \theta_0 + (\theta_1 - \theta_0) t/h \\ = \theta_1 \end{array} \right\}$ $\left. \begin{array}{l} 0 \leqslant t \leqslant h \\ h \leqslant t \end{array} \right\}$ at $x = 0.$

From (ii) by aid of Fig. 100

$$\phi_2 = \theta_0 + (\theta_1 - \theta_0)(1 - e^{-hp})/hp. \tag{1}$$

Inserting this into (9) § 15·21, we obtain

$$\theta(x,t) \bumpeq \phi(x,p) = \theta_0 + (\theta_1 - \theta_0)(1 - e^{-hp})e^{-\lambda x}/hp, \qquad (2)$$

so $\qquad \phi(x,p) = \theta_0 + (\theta_1 - \theta_0)(e^{-2a\sqrt{p}} - e^{-2a\sqrt{p-hp}})/hp, \qquad (3)$

where $2a = x\sqrt{(CR)}$. We shall invert (3) using transform procedure. Then by (33) Appendix 10 and (3) § 8·55

$$\operatorname{erfc}(a/\sqrt{t}) \bumpeq e^{-2a\sqrt{p}}, \quad \operatorname{erfc}\{a/\sqrt{(t-h)}\} \bumpeq e^{-2a\sqrt{p-hp}}. \qquad (4)$$

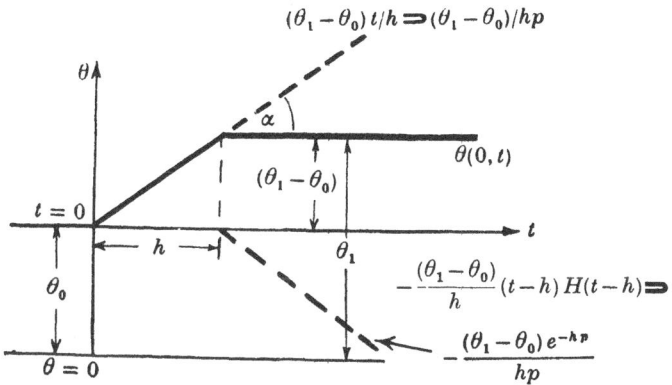

Fig. 100. The datum line for the transforms is $\theta = \theta_0$.

Applying § 8·45, we obtain

$$\theta(x,t) = \theta_0 + \frac{\theta_1 - \theta_0}{h}\left[\int_0^t \operatorname{erfc}(a/\sqrt{t})\,dt - \int_h^t \operatorname{erfc}\{a/\sqrt{(t-h)}\}\,dt\right], \qquad (5)$$

$t > h$, the lower limit in the second integral being h, since its integrand is zero for $t < h$ (see § 8·51 et seq.). If $\theta(0,t)$ increased with increase in t beyond $t = h$, this integral would be absent. Since $\theta(0,t) = \theta_1$ a constant, for $t \geqslant h$, the integral is needed for this condition to be satisfied. As an exercise the reader may obtain (5) using the Mellin theorem, (6) § 5·221 and § 8·51.

15·23. Evaluation of integrals in (5) § 15·22. Put $v = (a/\sqrt{t})$, then by (8) § 5·212

$$\operatorname{erfc} v = \frac{2}{\sqrt{\pi}}\int_{at^{-\frac12}}^\infty e^{-v^2}\,dy, \qquad (1)$$

so $\qquad d(\operatorname{erfc} v)/dt = (a/\sqrt{\pi})e^{-a^2/t}t^{-\frac32}. \qquad (2)$

Taking the first integral in (5) § 15·22 by parts, using (2), we get

$$\int_0^t \operatorname{erfc} v\, dt = t \operatorname{erfc} v - \frac{a}{\sqrt{\pi}} \int_0^t t(e^{-a^2/t} t^{-\frac{3}{2}})\, dt \qquad (3)$$

$$= t \operatorname{erfc} v - \frac{2a}{\sqrt{\pi}} \int_0^t e^{-a^2/t} d(\sqrt{t}), \qquad (4)$$

$$= t \operatorname{erfc} v - 2a \sqrt{(t/\pi)}\, e^{-v^2} + \frac{2a^3}{\sqrt{\pi}} \int_0^t e^{-a^2/t} t^{-\frac{3}{2}}\, dt. \qquad (5)$$

Writing $u^2 = a^2/t$ in the integral on the right-hand side of (5), it becomes

$$\frac{4a^2}{\sqrt{\pi}} \int_{at^{-\frac{1}{2}}}^{\infty} e^{-u^2}\, du = 2a^2 \operatorname{erfc} v. \qquad (6)$$

Hence
$$\int_0^t \operatorname{erfc} v\, dt = (t + 2a^2) \operatorname{erfc} v - 2a \sqrt{(t/\pi)}\, e^{-v^2}, \qquad (7)$$

with $v = x \sqrt{(\mathbf{CR}/t)}/2$.

For the second integral in (5) § 15·22 put $(t - h) = \tau$, and we get

$$\int_0^{t-h} \operatorname{erfc} v_1\, d\tau, \quad \text{where} \quad v_1 = x \sqrt{(\mathbf{CR}/\tau)}/2,$$

so its value is obtained by writing $(t - h)$ for t in the right-hand side of (7). Finally, therefore, the evaluation yields

$$\theta(x, t) = \theta_0 + \{(\theta_1 - \theta_0)/h\} [(t + 2a^2) \operatorname{erfc} v - 2a \sqrt{(t/\pi)}\, e^{-v^2}$$
$$- H(t - h) \{(t - h + 2a^2) \operatorname{erfc} v_1 - 2a \sqrt{[(t - h)/\pi]}\, e^{-v_1^2}\}], \quad (8)$$

which may be computed using tabular values of $\operatorname{erfc} v$, etc.

When x is finite > 0, as $t \to +\infty$, $\operatorname{erfc} v \to 1$, so

$$\theta(x, \infty) = \theta_0 + \{(\theta_1 - \theta_0)/h\} [(t + 2a^2) - (t - h + 2a^2)]_{t \to +\infty} \qquad (9)$$

$$= \theta_1, \qquad (10)$$

as would be expected during the steady state. When t is finite > 0, as $x (= l) \to +\infty$, all terms of [ ] in (8) $\to 0$, since by § 5·23 $\operatorname{erfc} v \sim e^{-v^2}/\sqrt{\pi} v$, so

$$\theta(\infty, t) = \theta_0, \qquad (11)$$

as in boundary condition (i). For $x = 0$, $0 < t < h$, (8) yields

$$\theta(0, t) = \theta_0 + (\theta_1 - \theta_0) t/h, \qquad (12)$$

as in boundary condition (ii), while for $x = 0$, $h < t$,

$$\theta(0, t) = \theta_1, \qquad (13)$$

as in boundary condition (ii). For any finite x, when $t = 0$, (8) gives

$$\theta(x, 0) = \theta_0, \qquad (14)$$

so the initial condition is satisfied. It may be remarked that (14) holds for $x = 0$, because the temperature there is *continuous* at $t = 0$. A discontinuous case is examined below.

In Fig. 100 as $h \to +0$, $\alpha \to \frac{1}{2}\pi$, and in the limit, boundary condition (ii) becomes $\theta = \theta_1$ at $x = 0$, $t > 0$. The right-hand side of (8) may now be expressed as

$$\lim_{h \to 0} \{\theta_0 + (\theta_1 - \theta_0)[f(x, t) - f(x, t-h)]/h\}, \tag{15}$$

the second member being the derivative of $f(x, t)$ with respect to t. Hence differentiating

$$f(x, t) = (t + 2a^2)\,\text{erfc}\,v - 2a\,\sqrt{(t/\pi)}\,e^{-v^2} \tag{16}$$

with respect to t, we obtain

$$\theta(x, t) = \theta_0 + (\theta_1 - \theta_0)\,\text{erfc}\,v, \tag{17}$$

a result which the reader should establish.

The foregoing boundary condition entails a discontinuity in θ at $x = 0$, since it changes precipitately from θ_0 to θ_1 when $t = 0$. If we let $t \to 0$ in $v = x\,\sqrt{(\mathbf{CR}/t)}/2$ for any $x > 0$, then $\text{erfc}\,v \to 0$, and $\theta = \theta_0$. But if we let $x \to 0$, the value of v in the limit depends upon the way $x \to 0$, so *it is not unique*. Hence, so far as (17) is concerned, the value of θ at $x = 0$ is indeterminate, in virtue of the discontinuity in the $(\theta\text{-}t)$ relationship. In practice this boundary condition would entail an infinite heat flux (!) momentarily at $t = 0$, but actually the temperature of the heat source falls. The reduction may be taken into account by assigning a suitable value to h. If, however, the $(\theta\text{-}t)$ curve at $x = 0$ is not linear, but has a transform, the problem of a slab with finite rate of rise of temperature of one face, may be solved by aid of § 15·21.

15·24. Heating of automobile brakes.
An automobile travelling at 60 m.p.h. is brought to rest at a uniform rate in a distance of 120 ft. Find the rise in temperature of the working faces of the four brake drums when the car stops, using the following data: effective diameter of tyre tread = 2 ft. 6 in.; diameter of each brake drum = 1 ft. 3 in.; width of working face of drum = $1\frac{1}{2}$ in.; total weight of car = 4000 lb.; thermal conductivity of material of brake drums = 0·5 watt per cm. per 1° C.; heat capacitance = 4·8 watt sec. per cm.^{3} per 1° C. A brake drum is to be regarded as a hollow cylinder, the whole of the braking being attributed to an equal frictional force on each drum. There is additional braking due to windage, etc., but this will be ignored as it amounts approximately to only 3 per cent of the weight of the vehicle.

The time to stop the car is (distance/average speed) $= \frac{120}{44}$ $= 2 \cdot 73$ sec. During this short interval, there will be no appreciable alteration in the temperature of the inner non-working face of a drum, so the thickness may be assumed very great. Since the ratio (radius/thickness) is probably of the order $20/1$, the effect of curvature and edge effect may be ignored and the result for plane faces used. Moreover, the drum will be regarded as a large slab of metal. The electrical analogue is a uniform unloaded transmission line of very great length. From (5) § 15·11 we take the transform solution for $\theta = 0$ at $t = 0$, and absence of reflection from the back of the slab,* namely,

$$\theta \Rightarrow \phi = A\, e^{-\lambda x}, \tag{1}$$

$$q = -\frac{1}{R}\frac{\partial \theta}{\partial x} \Rightarrow \phi_1 = \frac{A\lambda}{R}\, e^{-\lambda x}, \tag{2}\dagger$$

where $\lambda = \sqrt{(pCR)}$, θ is the temperature rise, q the heat flow, C the capacitance, and R the resistance, these relating to the *total* working area of one brake drum.

When the brakes are applied at $t = 0$, the rate of energy dissipation per drum is $q = q_0$ watts, while at $t = t_1$ the car is at rest, so $q = 0$. Since the retardation is constant, the heat dissipation at time t $(0 \leqslant t < t_1)$ is

$$q = q_0(t_1 - t)/t_1. \tag{3}$$

The transform of (3) is $q_0(t_1 - 1/p)/t_1$ and this is the value of ϕ_1 in (2) at the braking surface where $x = 0$. Substituting in (2) we get

$$A = q_0 R(t_1 - p^{-1})/t_1 \lambda = q_0 \sqrt{(R/C)}\,[p^{-\frac{1}{2}} - t_1^{-1}p^{-\frac{3}{2}}]. \tag{4}$$

Inserting this value of A in (1) and making $x = 0$, the transform for the temperature at the braking surface is

$$\phi = q_0 \sqrt{(R/C)}\,(p^{-\frac{1}{2}} - t_1^{-1}p^{-\frac{3}{2}}). \tag{5}$$

Inverting (5) we obtain

$$\theta = 2q_0 \sqrt{(tR/C\pi)}\,(1 - 2t/3t_1). \tag{6}$$

* The reader should work out the present example when reflection from the inner surface of the drum is taken into account; assume a thickness of $\frac{1}{4}$ inch.

† The minus sign is due to the braking surface being taken as $x = 0$, so q decreases with increase in x.

15·25. Numerical evaluation of (6) **§15·24.** We have now to calculate the values of C, R and q_0:

C = heat capacitance × braking surface area per drum

$$= 4\cdot8 \times \pi \times 15 \times 1\cdot5 \times 2\cdot54^2$$

$$= 2190. \tag{1}$$

R = 1/area of drum × conductivity

$$= 1/455 \times 0\cdot5 = 4\cdot4 \times 10^{-3}. \tag{2}$$

To calculate q_0 we must first find the kinetic energy of the car at 60 m.p.h. This is

$$\text{K.E.} = \frac{1}{2}\frac{Wv^2}{g} = \frac{1}{2}\cdot\frac{4000}{32} \times 88^2$$

$$= 4\cdot84 \times 10^5 \text{ ft. lb.} \tag{3}$$

The distance travelled by a point on a brake-drum surface is 60 ft., since the drum radius is one-half that of the tyre tread. The work done in stopping the car is $4\cdot84 \times 10^5$ ft. lb., so the constant retarding force is

$$f = 4\cdot84 \times 10^5/60$$

$$= 8000 \text{ lb. approx.} \tag{4}$$

At 60 m.p.h. the rate of heat supply to *one* drum is

$$q_0 = 2000 \times 88 = 1\cdot76 \times 10^5 \text{ ft. lb. per sec.}$$

$$= 1\cdot356 \times 1\cdot76 \times 10^5 = 2\cdot38 \times 10^5 \text{ watts.} \tag{5}$$

Substituting for the various quantities in (6) § 15·24 we find that when the car stops, the temperature rise of the brake drums is

$$\theta = 2 \times 2\cdot38 \times 10^5 \bigg/ \sqrt{\left(\frac{2\cdot73 \times 4\cdot4 \times 10^{-3}}{\pi \times 2190}\right)} (1 - 2/3)$$

$$= 210^\circ\,\text{C}.$$

15·26. Maximum temperature of brake drums. It might be thought that 210° is the highest temperature reached by the working faces of the brake drums, but this is not so, for by differentiating (6) § 15·24 and equating to zero we find that the temperature is a maximum when $t = \tfrac{1}{2}t_1$. The maximum temperature is then $\sqrt{2}$ times the final value, i.e.

$$\theta_{max.} = 210\sqrt{2} = 300°\,C.\ \text{approx.}$$

The occurrence of the maximum temperature is due to the variable heat supply to the drums. When the brakes have been applied for $1.35\,\text{sec.}$ the temperature is $300°\,C.$, but after this the speed is such that the rate of generation of heat is less than the rate of diffusion into the metal.

15·31. Annealing of steel rod. A cylindrical rod of tool steel, 4 cm. diameter, 32 cm. long, is heated to $920°\,C.$ and cooled slowly in an annealing oven until the surface temperature is $20°\,C.$ If the rate of fall in surface temperature is constant, find a formula for the temperature at any radius r during cooling.

Let θ = temperature, r = radius, t = time and k = diffusivity. When the ratio length/radius of a cylinder is large, end-effects can be neglected and the diffusion assumed to be radial throughout. The appropriate differential equation is then

$$\frac{\partial^2\theta}{\partial r^2} + \frac{1}{r}\frac{\partial\theta}{\partial r} - \frac{1}{k}\frac{\partial\theta}{\partial t} = 0, \tag{1}$$

θ being independent of the angular position about the axis. The solution of (1) for the initial condition $\theta = \theta_0 = 920°\,C.$ at $t = 0$ is given by (6) § 13·13 with $B = 0$, so we have

$$\phi = AJ_0[ir\,\sqrt{(p/k)}] + 920. \tag{2}$$

At the surface $r = a$ and $\theta = 920 - 900t/t_1$, t_1 being the time taken for the temperature to reach $20°\,C.$ By (4) § 13·13, in terms of ϕ this boundary condition is

$$\phi = p\int_0^\infty e^{-pt}\,(920 - 900t/t_1)\,dt \tag{3}$$

$$= 920 - 900/pt_1. \tag{4}$$

Inserting (4) into (2) and writing $r = a$, we obtain

$$A = -900/t_1\,pJ_0[ia\,\sqrt{(p/k)}].$$

Substituting from (5) into (2) leads to

$$\phi = 920 - 900 J_0[ir \sqrt{(p/k)}]/t_1 p J_0[ia \sqrt{(p/k)}]. \tag{6}$$

Applying the Mellin theorem to (6) gives

$$\theta = 920 - \frac{(900/t_1)}{2\pi i} \int_{Br_1} \frac{e^{zt} J_0[ir \sqrt{(z/k)}] \, dz}{z^2 J_0[ia \sqrt{(z/k)}]}. \tag{7}$$

Writing $a/\sqrt{k}$ for a and $r/\sqrt{k}$ for r in (6) § 4·74, we obtain

$$\theta = 920 - \frac{900}{t_1} \left[t + \frac{r^2 - a^2}{4k} + \frac{2a^2}{k} \sum_{n=1}^{\infty} \frac{e^{-\alpha_n^2 kt/a^2} J_0(\alpha_n r/a)}{\alpha_n^3 J_1(\alpha_n)} \right]. \tag{8}$$

Thus at time $t > 0$ the temperature at radius r is

$$\theta_r = 900 \left[1 - (1/t_1) \left\{ t + \frac{r^2 - a^2}{4k} \right. \right.$$

$$\left. \left. + \frac{2a^2}{k} \sum_{n=1}^{\infty} \frac{e^{-\alpha_n^2 kt/a^2} J_0(\alpha_n r/a)}{\alpha_n^3 J_1(\alpha_n)} \right\} \right] + 20° \text{C.} \tag{9}$$

15·32. The series in (9) § 15·31 is uniformly convergent in every closed interval $0 \leqslant t \leqslant t_1$ for $0 \leqslant r \leqslant a$, and therefore represents a continuous function therein. Now when $t = 0$, $\theta_r = 920°$ C. for all values of r, so we must have

$$\sum_{n=1}^{\infty} \frac{J_0(\alpha_n r/a)}{\alpha_n^3 J_1(\alpha_n)} = \tfrac{1}{8}(1 - r^2/a^2), \tag{1}$$

a relationship established in § 4·74. When $t = t_1$ and $r = a$, the surface temperature of the rod is $\theta_a = 20°$ C., since by hypothesis $J_0(\alpha_n) = 0$ in (9) § 15·31. The axial temperature can be found from (9) § 15·31 by putting $t = t_1$ and $r = 0$, which gives

$$\theta_{\text{axis}} = \frac{225a^2}{kt_1} - \frac{1800a^2}{kt_1} \sum_{n=1}^{\infty} \frac{e^{-\alpha_n^2 kt_1/a^2}}{\alpha_n^3 J_1(\alpha_n)} + 20° \text{C.} \tag{2}$$

From (2) it is seen that the greater t_1 and the smaller a, the radius of the rod, the more nearly does the temperature at the centre approach that at the surface.

15·41. Absorption of moisture by a slab. When a dry sheet of absorbent material is situated in an atmosphere under constant conditions of temperature and humidity, it absorbs moisture until a substantially equilibrium state is attained. The amount absorbed depends upon the material. If the sheet is now immersed in a liquid, say water, moisture is absorbed again, but at a much more rapid rate than before. After a time, absorption practically ceases, and a second equilibrium state is attained. Referring to Fig. 101, the problem is to determine the moisture content at any point distant x from the central plane of the slab at time t after immersion in the liquid.

To formulate the problem mathematically, it is necessary to make three assumptions: (1) that the main absorption surfaces are very large compared with the areas of the edges or the latter are sealed to prevent access of moisture; (2) that the phenomenon of absorption can be expressed by the same differential equation as the diffusion of heat; (3) that at the main surfaces, the rate of absorption of moisture is proportional to the difference between (a) the maximum concentration of moisture corresponding to the equilibrium state, and (b) the concentration at the surface at time t.

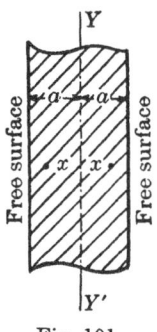

Fig. 101.

The latter assumption is the analogue of Newton's law of cooling, in which the rate of heat loss is proportional to the difference in temperature between a body and that of its surroundings.

15·42. Solution of problem [116]. By the first two hypotheses in § 15·41, (4) § 15·11 gives

$$\frac{\partial^2 \theta}{\partial x^2} = \frac{1}{k}\frac{\partial \theta}{\partial t}, \tag{1}$$

where $1/k$ is written for CR. In (1) θ is the moisture concentration at a plane distant x from the central plane, at time t, and k the diffusion coefficient or diffusivity. This latter is the mass of liquid in gm. sec.$^{-1}$ flowing across 1 cm.^{2} when the concentration gradient $\partial \theta/\partial x$ is 1 gm. cm.$^{-3}$ per cm. length along the direction of flow (assuming c.g.s. units).

As $t \to -0$, $\theta = 0$ throughout the sheet, so with $\theta(x, t) \Rightarrow \phi_1(x, p)$, the transform equation for (1) is

$$d^2\phi_1/dx^2 - \lambda^2\phi_1 = 0, \qquad (2)^*$$

with $\lambda^2 = p/k$. Since the sheet is symmetrical about the central plane (see Fig. 101), the solution must be an even function of x. Hence

$$\theta \Rightarrow \phi_1 = A \cosh \lambda x. \qquad (3)$$

From the third hypothesis in § 15·41, the condition at the surface is expressed by

$$\partial Q/\partial t = c(\theta_m - \theta_a), \qquad (4)$$

where θ_m is the maximum or equilibrium concentration in the sheet, θ_a is the surface concentration at time t, and Q the quantity of moisture in the whole sheet per unit area of exposed surface, in excess of that contained initially. Since $Q = 0$ at $t = 0$, with $Q \Rightarrow \phi_2$, by (5) § 8·43 and (3), (4) above,

$$\partial Q/\partial t \Rightarrow p\phi_2 = c(\theta_m - A \cosh \lambda a), \qquad (5)$$

where $\theta_a \Rightarrow A \cosh \lambda a$ from (3), so

$$\phi_2 = (c/p)(\theta_m - A \cosh \lambda a). \qquad (6)$$

Also
$$Q = 2\int_0^a \theta \, dx \Rightarrow \phi_2 = 2\int_0^a \phi_1 \, dx, \qquad (7)$$

so by (3), (7)
$$\phi_2 = \frac{A \sinh \lambda a}{\lambda}. \qquad (8)$$

Equating ϕ_2 in (6), (8) gives

$$A = c\theta_m/[2(p/\lambda)\sinh \lambda a + c \cosh \lambda a]. \qquad (9)$$

Inserting this value of A in (3) and applying the Mellin theorem,

$$\theta = \frac{c\theta_m}{2\pi i}\int_{Br_1} \frac{e^{zt}\cosh g\sqrt{z}\,dz}{z[2\sqrt{(kz)}\sinh h\sqrt{z} + c\cosh h\sqrt{z}]}, \qquad (10)$$

* If $\theta = \theta_0 > 0$ as $t \to -0$, $-p\theta_0/k$ must be added to the right-hand side of (2), and θ_0 to that of (3). Sheets are usually 'conditioned' prior to being tested, in an atmosphere at say 65° F. and 70 per cent. relative humidity, so they contain a little moisture initially. The reader should obtain the solution of the problem using the above initial condition.

where $g = x/\sqrt{k}$, $h = a/\sqrt{k}$. From (7) and (10), we obtain

$$Q = \frac{2c\theta_m}{2\pi i} \int_{Br_1} \frac{e^{zt} \sinh h \sqrt{z}\, dz}{z[2z \sinh h \sqrt{z} + c \sqrt{(z/k)} \cosh h \sqrt{z}]}. \tag{11}$$

Dividing the integrand above and below by $\sinh h \sqrt{z}$, writing $h \sqrt{z} = \sqrt{\zeta}$ and replacing ζ by z thereafter, putting $\beta = 2k/ca$ and $Q_m = 2a\theta_m$,* (11) becomes

$$Q = \frac{Q_m}{2\pi i} \int_{Br_1} \frac{e^{zkt/a^2}\, dz}{z[\beta z + \sqrt{z} \coth \sqrt{z}]}. \tag{12}$$

To evaluate (12) we consider first the singularities of the integrand. Expanding the hyperbolic function, the corresponding term gives

$$\sqrt{z} \coth \sqrt{z} = (1 + z/2! + z^2/4! + \ldots)/(1 + z/3! + z^2/5! + \ldots). \tag{13}$$

Since (13) $\to 1$ as $z \to 0$, there is a simple pole at the origin. The remaining singularities of the integrand are simple poles, and a limit point at infinity. The singularities occur when

$$\sqrt{z} \coth \sqrt{z} + \beta z = 0, \tag{14}$$

or

$$\coth \sqrt{z} = -\beta \sqrt{z}. \tag{15}$$

Writing $i \sqrt{z} = y$, (15) becomes

$$\cot y = \beta y. \tag{16}$$

A table giving the first four positive roots of (16) for various values of $\lambda^\circ = \tan^{-1} \beta$ is reproduced below, and will be useful for computation. If α_n is a positive root of (16), we have $i \sqrt{z} = \alpha_n$ and $z = -\alpha_n^2$, where $n = 1, 2, 3, \ldots$.

The contribution to (12) from the pole at the origin is

$$Q_m[e^{zkt/a^2}/(\beta z + \sqrt{z} \coth \sqrt{z})]_{z=0}, \tag{17}$$

and by aid of (13) this is Q_m. For the singularities at $z = -\alpha_n^2$, we first differentiate $\beta z + \sqrt{z} \coth \sqrt{z}$, in accordance with § 3·232,

* $\theta \to \theta_m$, $Q \to Q_m$ as $t \to \infty$, so (7) gives $Q_m = 2a\theta_m$.

thereby obtaining $\beta + \frac{1}{2}[z^{-\frac{1}{2}}\coth\sqrt{z} - \coth^2\sqrt{z} + 1]$. Thus the contribution due to these singularities is

$$Q_m \sum_{n=1}^{\infty} \left[\frac{e^{zkt/a^2}}{z\{\beta + \frac{1}{2}[z^{-\frac{1}{2}}\coth\sqrt{z} - \coth^2\sqrt{z} + 1]\}} \right]_{\substack{\coth\sqrt{z} = -\beta\sqrt{z} \\ z = -\alpha_n^2}} \quad (18)$$

$$= -Q_m \sum_{n=1}^{\infty} \frac{e^{-\alpha_n^2 kt/a^2}}{\alpha_n^2[\beta + \frac{1}{2}(-\beta + \beta^2\alpha_n^2 + 1)]} \quad (19)$$

$$= -2Q_m \sum_{n=1}^{\infty} \frac{e^{-\alpha_n^2 kt/a^2}}{\alpha_n^2(1 + \beta + \beta^2\alpha_n^2)}. \quad (20)$$

The first four roots of the equation $\cot y = y \tan \lambda$
for values of λ *from* $0°$ *to* $90°$ *at* $5°$ *intervals*

$\lambda°$	1st root	2nd root	3rd root	4th root
0	1·5708	4·7124	7·8540	10·9956
5	1·4451	4·3488	7·2865	10·2639
10	1·3390	4·0879	6·9665	9·9432
15	1·2481	3·9044	6·7860	9·7890
20	1·1686	3·7712	6·6737	9·7008
25	1·0977	3·6704	6·5975	9·6436
30	1·0330	3·5910	6·5420	9·6033
35	0·9728	3·5264	6·4995	9·5729
40	0·9157	3·4722	6·4655	9·5490
45	0·8603	3·4256	6·4373	9·5294
50	0·8057	3·3846	6·4133	9·5127
55	0·7506	3·3478	6·3923	9·4983
60	0·6939	3·3141	6·3735	9·4856
65	0·6341	3·2827	6·3564	9·4740
70	0·5690	3·2530	6·3405	9·4633
75	0·4956	3·2245	6·3255	9·4532
80	0·4079	3·1968	6·3111	9·4435
85	0·2915	3·1692	6·2970	9·4341
90	0·0000	3·1416	6·2832	9·4248

Adding the contribution from the pole at the origin to (20), we obtain

$$Q = Q_m \left[1 - 2 \sum_{n=1}^{\infty} \frac{e^{-\alpha_n^2 kt/a^2}}{\alpha_n^2(1 + \beta + \beta^2\alpha_n^2)} \right]. \quad (21)$$

In like manner it can be shown that

$$\theta = \theta_m \left[1 - 2 \sum_{n=1}^{\infty} \frac{e^{-\alpha_n^2 kt/a^2}[\cos(x\alpha_n/a)/\sin\alpha_n]}{\alpha_n(1 + \beta + \beta^2\alpha_n^2)} \right]. \quad (22)$$

15·43. Surface saturated on immersion [116]. If c is very great, the surface of the sheet becomes saturated almost immediately after immersion in the liquid. Here $\beta = 2k/ca$ is extremely small and will be taken as zero. When $\beta = 0$ in (16) § 15·42 we obtain the zeros of the new denominator of (12) § 15·42. Thus $\cot y = 0$, or $y = (2n-1)\pi/2$, and as $z = -y^2$, we have $-\alpha_n^2 = -(2n-1)^2 \pi^2/4$, $n = 1, 2, 3, \ldots$ Since $\beta = 0$, (21) § 15·42 gives

$$Q = Q_m \left\{ 1 - 2 \left[\sum_{n=1}^{\infty} \frac{e^{-\alpha_n^2 kt/a^2}}{\alpha_n^2} \right]_{\alpha_n^2 = (2n-1)^2 \pi^2/4} \right\} \tag{1}$$

$$= Q_m \{ 1 - (8/\pi^2) [e^{-\pi^2 kt/4a^2} + \tfrac{1}{9} e^{-9\pi^2 kt/4a^2} + \tfrac{1}{25} e^{-25\pi^2 kt/4a^2} + \ldots] \} \tag{2}$$

$$= Q_m \psi(\pi^2 kt/4a^2) \quad (\psi \text{ tabulated in } [116]). \tag{3}$$

It is of interest to find the form of Q when t is small. From (12) § 15·42 when $\beta = 0$, we take z very large in accordance with §§ 6·31, 6·32, so $\coth \sqrt{z} \to 1$. Then

$$Q \simeq \frac{Q_m}{2\pi i} \int_{Br_1} e^{zkt/a^2} \frac{dz}{\sqrt[3]{z}} \tag{4}$$

$$= Q_m \frac{2}{a} \sqrt{(kt/\pi)}, \tag{5}$$

so the Q-t curve commences as a parabola tangential to the Q axis.

Since $Q \propto t^{\frac{1}{2}}$, the *rate* of absorption is very large when t is small. The foregoing approximation is usually adequate for values of $\sqrt{(kt)}/a$ up to $1/\pi$ where the error is less than 6 parts in 10^4. When t is large enough (2) yields the approximation

$$Q \sim Q_m \left\{ 1 - \frac{8}{\pi^2} e^{-\pi^2 kt/4a^2} \right\}. \tag{6}$$

The preceding analysis applies equally, *mutatis mutandis*, to evaporation of liquid from a porous slab.

15·44. Data illustrating (5) § 15·43. The graphs of Fig. 102 were plotted using data obtained from experiments on two different fibre boards of the type for insoles in footwear. The test samples, 7·62 cm. square with sealed edges were 'conditioned' and then immersed in water. Each sample was removed from the water and weighed after the superfluous moisture had been blotted off gently. The Q-t relationship holds up to a point where the absorption is approximately 10 and 13·5 per cent, respectively, of the original weight of the sample. Thereafter (5) § 15·43 is inapplicable.

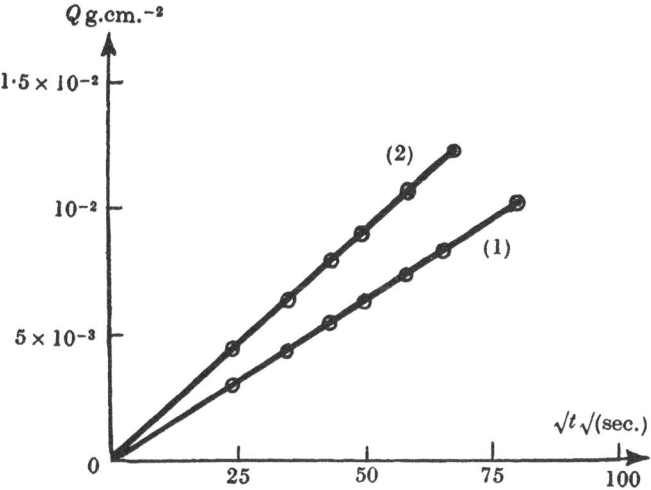

Fig. 102. Initial form of absorption-time curve for fibre boards, plotted on $\sqrt{t}$ base. (1) Mass of board 11·21 gm., thickness 0·173 cm. (2) Mass 11·30 gm., thickness 0·185 cm.; exposed area in both cases 116 cm.²

PROBLEMS

1. What are the singularities of the following functions in the finite part of the z-plane:

(a) $1/(z-a)$; (b) $1/z^2$; (c) $\sin(1/z)$; (d) $\sin z/z^2$;

(e) $(z^2+1)/(z-a)(z+b)(z-c)$; (f) $1/[1+(1/z^3)]$?

Obtain the Laurent expansion near each singularity in order to check your answers.

[(a) Simple pole at $z = a$; (b) double pole at $z = 0$; (c) essential s. at $z = 0$; (d) s.p. at $z = 0$; (e) s.p. at $z = a$, $-b$, c provided a, b, $c \neq \pm i$; (f) s.p. at $z = -1$, $(1 \pm i\sqrt{3})/2$.]

2. What are the singularities of the following functions in the finite part of the z-plane (consider first branch only):

(a) $z^{\frac{1}{3}}$, (b) $1+\sqrt{z}$; (c) $\sqrt{(1+z)}$; (d) $\sqrt{(1+z^2)}$; (e) $1/(z+\sqrt{z})$;

(f) $1/\{z\sqrt{(1+z)}\}$; (g) $e^{zt}z/\sqrt{(z^2+1)}\{z+\sqrt{(z^2+1)}\}$; (h) $z+\sqrt{(z^2+1)}$;

(j) $\sin\sqrt{(1+z)}$; (k) $e^{-\sqrt{(z^2+a^2)}}$; (l) $1/\sqrt{z}\sinh\sqrt{z}$; (m) $z^{\frac{1}{3}}\cosh z$?

In each case use an appropriate test to check your answer.

[(a) Branch point at $z=0$; (b) b.p. at $z=0$; (c) b.p. at $z=-1$; (d) b.p. at $z=\pm i$; (e) b.p. at $z=0$; (f) s.p. at $z=0$, b.p. at $z=-1$; (g) b.p. at $z=\pm i$; (h) b.p. at $z=\pm i$; (j) b.p. at $z=-1$; (k) b.p. at $z=\pm ia$; (l) s.p. at $z=-n^2\pi^2$, n a positive integer; (m) b.p. at $z=0$.]

3. (i) Show (a) that $\sinh\sqrt{z}$ has a branch point at $z=0$; (b) that $\cosh\sqrt{z}$ has no branch point there, but has an essential s. at $z=\infty$. [Put $z=1/\zeta$, expand and let $\zeta\to0$.]

(ii) If n is a positive integer $\geqslant 1$, show that z^n has no singularity in the finite part of the z-plane, but has a pole of the nth order at $z=\infty$. [Put $z=1/\zeta$ and let $\zeta\to0$.]

(iii) Verify that $w=z\,e^z$ satisfies the Cauchy-Riemann conditions at all points in the finite part of the z-plane. Also that u, v, its real and imaginary parts, satisfy Laplace's equation. Obtain the first derivative of w using the method of § 1·7. [$e^z(1+z)$.]

(iv) Demonstrate that if $z\neq0$ or ∞, $w=\sqrt{z}$ satisfies the Cauchy-Riemann conditions in $-\pi<\theta<\pi$. Using the method of § 1·7 show that the first derivative is $1/2\sqrt{z}$.

4. (a) What is the phase change in the function $w=z^{\frac{1}{3}}$ in passing from the lower to the upper side of the barrier in Fig. 31? (b) What is the phase change of $w=\sqrt{[z(z-a)(z-b)]}$ if z traverses contours enclosing one, two, and three branch points (see Fig. 8)? (c) Answer (b) if the index were $\frac{1}{3}$. (d) What is the phase change in $w=1/\sqrt{(1-z^4)}$ in passing counterclockwise round each branch point in Fig. 36? State the value of z on each pair of parallel lines in terms of x, y as in Fig. 25(b).

5. What are the singularities of the following functions in the finite part of the z-plane:

(a) $1/z\log z$; (b) $1/(1+\log z)$; (c) $\log(1+z)^\nu$; (d) $\log(1+z^2)$;

(e) $\log\{z+\sqrt{(z^2+a^2)}\}$; (f) $\sqrt{(z+a)}/z\log z$; (g) $(a+\sqrt{z})/\sqrt{(z-b)}$;

(h) $z+\sqrt{(z^2+a^2)}$; (j) $z^\nu e^{1/z}$; (k) $\exp(z^{-\nu})$; (l) $1/[(az)^{\frac{1}{3}}+1]$;

(m) $\log\{1/(z+1)\}$.

In each case use an appropriate test to check your answer.

[(a) b.p. at $z = 0$, s.p. at $z = 1$; (b) b.p. at $z = 0$, s.p. at $z = e^{-1}$; (c) $\log(1+z)^\nu = \nu \log(1+z)$ has b.p. at $z = -1$; (d) b.p. at $z = \pm i$; (e) b.p. at $z = 0$ and $\pm ia$; (f) b.p. at $z = 0$, $-a$, s.p. at $z = 1$; (g) b.p. at $z = 0$, b; (h) b.p. at $z = \pm ia$; (j) b.p. at $z = 0$ if ν non-integral, e.s. at $z = 0$, ν integral; (k) b.p. at $z = 0$ if ν non-integral, e.s. at $z = 0$ if ν a positive integer > 0; (l) b.p. at $z = 0$, and s.p. on fourth branch, angle ranges for z are $(2n-3)\pi < \theta < (2n-1)\pi$, $n = 1, 2, \ldots, 6$, so the fourth branch is $5\pi < \theta < 7\pi$, and since $(az)^{\frac{1}{2}} = -1$, $e^{\frac{1}{2}i\theta} = -1$, so $\theta = 6\pi$; (m) b.p. at $z = -1$.]

6. Solve $e^{1/(z-a)} = \alpha + i\beta$, α, β real. $[z = a + 1/\{\frac{1}{2}\log(\alpha^2+\beta^2) + i(\theta + 2n\pi)\}$, n integral. Put $(\alpha + i\beta) = r e^{i(\theta + 2n\pi)}$.]

7. (i) Evaluate $\int z^2 dz$ along the following contours: (a) the straight line from $z = 0$ to $8 + 6i$; (b) the real axis from $z = 0$ to 8 and a straight line from $(x, y) = (8, 0)$ to $(8, 6)$; (c) a semicircle above the line (a), whose centre is at its mid-point. $[(a), (b), (c), (-352 + 936i)/3.]$

(ii) Evaluate $\int \cos z\, dz$ along the straight line from $z = 0$ to $3 + 4i$. [$\sin 3 \cosh 4 + i \cos 3 \sinh 4$.]

(iii) Show that $\int_{z_0}^{z_1} \dfrac{f'(z)\, dz}{f(z)} = \log \dfrac{f(z_1)}{f(z_0)}$, if $f(z)$ is continuous, differentiable, and does not vanish on the path of integration between the points z_0, z_1, i.e. there is no singularity on or enclosed by the path.

8. (i) Show by the principle of analytical continuation that

$$\cosh 2z = 2 \cosh^2 z - 1$$

at all points in the finite part of the z-plane.

(ii) Show by the principle of analytical continuation that if $|z| < 1$,
$\int_0^z \dfrac{dz}{1+z} = z - \dfrac{z^2}{2} + \dfrac{z^3}{3} - \dfrac{z^4}{4} + \ldots$; also that except $z = 0$, $e^{1/z} = \sum\limits_{r=0}^{\infty} \dfrac{1}{r!\, z^r}$.

9. Obtain the Laurent expansion of $z^3 e^{1/z}$ near the origin, where it has an e.s.; what is (a) the residue there, (b) the principal part of the expansion?

$$\left[z^3 + z^2 + \frac{z}{2!} + \frac{1}{3!} + \frac{1}{4!\, z} + \frac{1}{5!\, z^2} + \ldots; \text{ residue } \frac{1}{4!}; \frac{1}{4!\, z} + \frac{1}{5!\, z^2} + \ldots \right]$$

10. Obtain the Laurent expansion of $1/(e^z - 1)$ near $z = 0$.

$$\left[\frac{1}{z} - \frac{1}{2} + \frac{z}{12} - \frac{z^3}{720} + \ldots \text{. The singularity is a s.p with residue unity.} \right]$$

11. Evaluate the following round a contour enclosing all singularities in the finite part of the z-plane:

$$(a)\ \frac{1}{2\pi i}\int \frac{dz}{(z-a)(z-b)};\quad (b)\ \frac{1}{2\pi i}\int \frac{z\,dz}{(z-a)^2};$$

$$(c)\ \frac{1}{2\pi i}\int \frac{z^3\,dz}{(z+a)(z+b)^2};\quad (d)\ \frac{1}{2\pi i}\int \frac{dz}{z^2(z-a)};$$

$$(e)\ \frac{1}{2\pi i}\int \frac{e^{zt}\,dz}{(z-a)};\quad (f)\ \frac{1}{2\pi i}\int \frac{e^{zt}\,dz}{(z+a)(z-b)};$$

$$(g)\ \frac{1}{2\pi i}\int \frac{e^{zt}\,dz}{(z-a)^2};\quad (h)\ \frac{1}{2\pi i}\int \frac{e^{zt}\,dz}{(z^2+1)};$$

$$(j)\ \frac{1}{2\pi i}\int \frac{e^{zt}z\,dz}{(z+a)(z-b)^2}.$$

$[(a)\ 0;\ (b)\ 1,$ write $(z-a)=\zeta;\ (c)\ (-a^3+3ab^2-2b^3)/(a-b)^2;\ (d)\ 0;$ $(e)\ e^{at};\ (f)\ (e^{bt}-e^{-at})/(a+b);\ (g)\ t\,e^{at};\ (h)\ \sin t;\ (j)\ e^{bt}(bt+1)/(a+b)-(a\,e^{-at}+b\,e^{bt})/(a+b)^2.]$

12. Evaluate $\dfrac{1}{2\pi i}\displaystyle\int_C \dfrac{e^{zt}\,dz}{(1-e^{-hz})}$, C being a finite circle, centre $z=0$, $\left(r=\dfrac{2m\pi}{h}+\eta\right)$, passing *between* two pairs of poles, one pair on each side of the imaginary axis. $\left[\dfrac{1}{h}\left\{1+2\displaystyle\sum_{n=1}^{m}\cos\dfrac{2\pi nt}{h}\right\}.\right]$

13. Find Σ residues at the singularities of the following functions:

$(a)\ e^{zt}/\sinh mz;\quad (b)\ e^{zt}/\cosh mz;\quad (c)\ e^{z-t^2/4z}/z.$

$\left[(a)\ \dfrac{1}{m}+\dfrac{2}{m}\displaystyle\sum_{n=1}^{\infty}(-1)^n\cos\dfrac{\pi nt}{m};\quad (b)\ \dfrac{2}{m}\displaystyle\sum_{n=0}^{\infty}(-1)^n\sin\dfrac{(2n+1)\pi t}{2m};\right.$

$\left.(c)\ \displaystyle\sum_{r=0}^{\infty}(-1)^r\dfrac{(\frac{1}{2}t)^{2r}}{(r!)^2},\text{ this being the series for the Bessel function } J_0(t).\right]$

14. Show that $\displaystyle\int_{-\infty}^{\infty}\frac{dx}{(1+x^2)^3}=\frac{3\pi}{8};$ see § 3·3.

15. Find (a) the residue of $e^{zt}/z^3(1-e^{-\pi z})$ at $z=0$, $(b)\ \Sigma$ residues at all singularities on the imaginary axis except $z=0$.

$\left[\dfrac{1}{12\pi}(2t^3+3\pi t^2+\pi^2 t);\ -\dfrac{1}{4\pi}\displaystyle\sum_{n=1}^{\infty}\dfrac{\sin 2nt}{n^3}.\right]$

16. Integrate $\dfrac{1}{2\pi i}\displaystyle\int \dfrac{e^{zt}dz}{\sqrt{(z^2+1)}}$ round the dumb-bell contour of Fig. 25 (a).

$$\left[\sum_{r=0}^{\infty}(-1)^r\frac{(\frac{1}{2}t)^{2r}}{(r!)^2}=J_0(t).\right]$$

17. (i) Evaluate $\dfrac{1}{2\pi i}\displaystyle\int_{Br_1}\dfrac{\theta^{z-1}dz}{(z-a)(z-b)}$. (ii) Show that

$$e^{-t}=\frac{1}{2\pi i}\int_{Br_1}\Gamma(z)\,t^{-z}dz.$$

$[(\theta^{a-1}-\theta^{b-1})/(a-b).]$

18. Evaluate $\displaystyle\int_{-i\infty}^{i\infty}e^{z^2t}dz$, and show that the contour $\pm i\infty$ is equivalent to a semicircle whose radius $\to\infty$ on either side of it. $[i\sqrt{(\pi/t)}.]$

19. Evaluate $\qquad$ (a) $\dfrac{1}{2\pi i}\displaystyle\int_{Br_1}\dfrac{e^{zt}dz}{z^2(1-e^{-\pi z})}$;

$\qquad\qquad\qquad\qquad$ (b) $\dfrac{1}{2\pi i}\displaystyle\int_{Br_1}\dfrac{e^{zt}\tanh z\,dz}{z}$.

$$\left[\frac{t^2}{2\pi}+\frac{t}{2}+\frac{\pi}{12}-\sum_{n=1}^{\infty}\frac{\cos 2nt}{2\pi n^2};\ \frac{4}{\pi}\sum_{n=0}^{\infty}\frac{1}{(2n+1)}\sin\left(\frac{2n+1}{2}\right)\pi t.\right]$$

20. Evaluate

$$\frac{b}{2\pi i}\int_{Br_1}\frac{e^{zt}\cosh a\sqrt{z}\,dz}{\sqrt{z}\sinh b\sqrt{z}}\quad(b>a>0),$$

and show that $\qquad 2\displaystyle\sum_{n=1}^{\infty}(-1)^{n+1}\cos\left(\frac{n\pi a}{b}\right)=1.$

$$\left[1+2\sum_{n=1}^{\infty}(-1)^n e^{-n^2\pi^2 t/b^2}\cos\left(\frac{n\pi a}{b}\right).\right]$$

21. Show, when $t\to h$ from the positive side, the form of the integral

$$\frac{1}{2\pi i}\int_{Br_1}\frac{e^{z(t-h)}dz}{(z^3+a_0z^2+b_0)(z^4+a_1z^2+b_1)}\quad\text{is}\quad(t-h)^6/6!.$$

22. What is the form of $\dfrac{1}{2\pi i}\displaystyle\int_{Br_1}\dfrac{e^{z(t-h)}\tanh\sqrt{z}\,dz}{z^{\frac{3}{2}}}$ when $t\to h$ from the side where $(t-h)>0$? $[2\sqrt{\{(t-h)/\pi\}}.]$

23. Show that $\displaystyle\sum_{n=0}^{\infty}\dfrac{(-1)^n(2n)!}{n!\,p^{2n}}$ is the asymptotic expansion of

$p\displaystyle\int_{\frac{1}{2}p}^{\infty}e^{\frac{1}{4}p^2-x^2}dx.$ [Write $v=x^2-p^2/4.$]

24. Show that

$$p\, e^{\frac14 p^2} \int_{\frac12 p}^{\infty} e^{-x^2}\, dx = p \int_0^{\infty} e^{-pt-t^2}\, dt = \frac{\sqrt{\pi}}{2}\, p\, e^{\frac14 p^2}\operatorname{erfc}(\tfrac12 p).$$

[Write $x = (t + \tfrac12 p)$.]

25. Verify that $\displaystyle\sum_{n=0}^{\infty} \frac{(-1)^n (2n)!}{t^{2n+1}}$ is the asymptotic expansion of

$$\int_0^{\infty} \frac{e^{-xt}\, dx}{(1+x^2)}.$$

26. Obtain the asymptotic expansion of $\displaystyle K_0(t) = \int_0^{\infty} e^{-t\cosh\theta}\, d\theta.$

$$\left[\text{Write } \cosh\theta = v+1;\; K_0(t) \sim \sqrt{\frac{\pi}{2t}}\, e^{-t} \left\{ 1 - \frac{1}{8t} + \frac{1^2.3^2}{2!\,(8t)^2} - \frac{1^2.3^2.5^2}{3!\,(8t)^3} + \dots \right\}. \right]$$

27. Show that

$$(a)\quad \frac{1}{2\pi i} \int_{Br_1} e^{zt} \tan^{-1}\left(\frac{1}{z}\right) dz = \frac{\sin t}{t},$$

$$(b)\quad \frac{1}{2\pi i} \int_{Br_1} e^{zt} \log\left(\frac{z+1}{z}\right) dz = \frac{1 - e^{-t}}{t}.$$

28. Show that $\displaystyle \frac{1}{2\pi i} \int_{Br_1} \frac{e^{zt-\sqrt{(1+z)}}\, dz}{\sqrt{(1+z)}} = \frac{e^{-t-1/4t}}{\sqrt{(\pi t)}}.$

[Write $\zeta = 1+z$; see §§ 5·221, 5·222.]

29. Evaluate $\displaystyle \frac{1}{2\pi i} \int \frac{e^{zt}\, dz}{\sqrt{z}\,(z-a)}$ round a Br_2 type contour enclosing both singularities. Show that the expansion obtained is asymptotic, $|R_n|$ being less than the modulus of the $(n+1)$th term. a real > 0.

$$\left[e^{at} a^{-\frac12} - \{1/a\sqrt{(\pi t)}\}\{1 - w^{-1} + 1.3 w^{-2} - 1.3.5 w^{-3} + \dots\} \right.$$
$$\left. + (-1)^{n+1}/\pi a^{n+1} \int_0^{\infty} \frac{e^{-xt} x^{n-\frac12}\, dx}{(1+x/a)};\; w = 2at. \right]$$

30. Evaluate the integral in problem **29** on the contour Br_1 and obtain an expansion in ascending powers of t. Show that it is absolutely convergent.

$$\left[2\sqrt{(t/\pi)} \left\{ 1 + \frac{w}{1.3} + \frac{w^2}{1.3.5} + \dots \right\};\; \left|\frac{u_{n+1}}{u_n}\right| = at/(n-\tfrac14) \to 0 \quad\text{as}\quad n \to \infty. \right]$$

31. When t is large show that:

(a) $\dfrac{1}{2\pi i}\displaystyle\int_{Br_1}\dfrac{e^{zt}dz}{\sqrt{z}\,(z^2+1)}\simeq\dfrac{1}{\sqrt{(\pi t)}}(1+\sqrt{2}\sin t);$

(b) $\dfrac{1}{2\pi i}\displaystyle\int_{Br_1}\dfrac{e^{zt}dz}{z\sqrt{\{(z+a)(z+b)\}}}\simeq\dfrac{1}{\sqrt{(ab)}},\quad R(a),\,R(b)>0:$

(c) $\dfrac{1}{2\pi i}\displaystyle\int_{Br_1}\dfrac{e^{zt}\sqrt{z}\,dz}{(z^2+1)}\simeq\sin(t+\tfrac14\pi)-1/2\sqrt{\pi}\,t^{\frac32};$

(d) $\dfrac{1}{2\pi i}\displaystyle\int_{Br_1}\dfrac{e^{zt-a\sqrt{\{(z+b)(z+c)\}}}\,dz}{z\sqrt{\{(z+b)(z+c)\}}}\simeq e^{-a\sqrt{(bc)}}/\sqrt{(bc)};$

$a,\,b,\,c$ real and positive. [Use §§ 6·41 et seq.]

32. Show that $\displaystyle\int_0^\infty J_\nu(t)\,dt=1,\ R(\nu)>-1.$

33. Verify that

(a) $\displaystyle\int_0^\infty e^{-ax}J_0\{\sqrt{(bx)}\}\,dx=\dfrac{e^{-b/4a}}{a},\quad a,\,b\text{ real}>0;$

(b) $\displaystyle\int_0^\infty e^{-b^2r^2}J_0(ar)\,r\,dr=\dfrac{e^{-a^2/4b^2}}{2b^2},\quad a\text{ real}>0,\,b\text{ real};$

(c) $\displaystyle\int_0^\infty e^{-ax}x^{\frac12\nu}J_\nu\{\sqrt{(bx)}\}\,dx=(\tfrac14 b)^{\frac12\nu}e^{-b/4a}/a^{\nu+1};$

(d) $\displaystyle\int_0^\infty J_0\{\sqrt{(x^2+t^2)}\}\,dt=\sqrt{(\tfrac12\pi x)}\,J_{-\frac12}(x)=\cos x.$

34. Show that the following are divergent on Br_1: (a) integral in § 7·34, (b) first integral in (2) § 7·35, (c) integral in § 7·45. If Br_1 is used, can the second integral in (3) § 7·35 be differentiated respecting t? [For (a), (b), (c) see Appendix 5; (a) subs. $(z+1)=\zeta$; (b) expand in inverse powers of ζ; (c) integrate along y-axis indented on right at $z=0$ by a semicircle of unit radius. No! the resulting integral diverges.]

35. Using the substitution $u=\tfrac12(t-a)\{z+\sqrt{(z^2+b^2)}\}$, verify that

(a) $\dfrac{abJ_1\{b\sqrt{(t^2-a^2)}\}}{\sqrt{(t^2-a^2)}}=\dfrac{1}{2\pi i}\displaystyle\int_{Br_2}e^{zt-a\sqrt{(z^2+b^2)}}\,dz;$

(b) $J_0\{b\sqrt{(t^2-a^2)}\}=\dfrac{1}{2\pi i}\displaystyle\int_{Br_1}\dfrac{e^{zt-a\sqrt{(z^2+b^2)}}\,dz}{\sqrt{(z^2+b^2)}}.$

[Put z in terms of u.]

36. Evaluate $\dfrac{1}{2\pi i}\displaystyle\int_{Br_1}\dfrac{e^{zt-b\sqrt{(z+a)}}dz}{z+a}$. Write $z=\zeta-a$. [See Appendix 7; $e^{-at}\,\mathrm{erfc}\,(b/2\sqrt{t})$.]

37. Evaluate $\dfrac{1}{2\pi i}\displaystyle\int_{Br_1}\dfrac{e^{zt}dz}{(z+z^{\frac12})}$. [Resolve into partial fractions and apply § 7·37 to one of them; $1-e^{t}\,\mathrm{erfc}\,\sqrt{t}$.]

38. Show that
$$\frac{1}{2\pi i}\int_{Br_2}\frac{e^{zt}dz}{(z\log z)}=e^{t}-\int_0^{\infty}\frac{e^{-xt}dx}{x\{\pi^2+(\log x)^2\}}.$$

[The singularities of the integrand are a simple pole at $z=1$, and a branch point at $z=0$. The equivalent contour is, therefore, a small circle round $z=1$, and that of Fig. 24 (b).]

39. Obtain the result
$$\frac{1}{2\pi i}\int_{Br_2}\frac{e^{zt}dz}{z(\log z)^2}=e^{t}(t-1)-2\int_0^{\infty}\frac{e^{-xt}\log x\,dx}{x\{\pi^2+(\log x)^2\}^2}.$$

[See Problem **38**.]

40. Using (1) § 8·11, find the transforms of the following:

(a) $\sin\omega t$; (b) $e^{-\alpha(t-b)}$; (c) $\sin(\omega t+\alpha)$; (d) $\sin^2\omega t$; (e) $\cos^2\omega t$;

(f) $\cos(\omega t-\beta)$; (g) $\sinh at$; (h) $\cosh^2 t$.

[(a) $\omega p/(p^2+\omega^2)$; (b) $p\,e^{-pb}/(p+a)$, not $e^{-ab}p/(p+a)$, because function starts when $t=b$ unless intended to start at $t=0$;

(c) $\{p/(p^2+\omega^2)\}(p\sin\alpha+\omega\cos\alpha)$; (d) $2\omega^2/(p^2+4\omega^2)$;

(e) $(p^2+2\omega^2)/(p^2+4\omega^2)$; (f) $p(p\cos\beta+\omega\sin\beta)/(p^2+\omega^2)$;

(g) $ap/(p^2-a^2)$; (h) $(p^2-2)/(p^2-4)$.]

41. Find the transforms of:

(a) $J_0(t-h)$; (b) $e^{-at}J_0(bt)$; (c) $e^{-at}J_\nu(bt)$; (d) $e^{-at}I_0(bt)$;

(e) $e^{-at}I_\nu(bt)$.

[(a) $e^{-ph}p/\sqrt{(p^2+1)}$; (b), (d) $p/\sqrt{\{(p+a)^2\pm b^2\}}$;

(c), (e) $p/\sqrt{\{(p+a)^2\pm b^2\}\{(p+a)+\sqrt{[(p+a)^2\pm b^2]}\}^\nu}$, $R(\nu)>-1$.]

42. Explain why the transform of $de^{t}/dt \neq p[p/(p-1)]$, and that of
$$J_1(t)\neq -p[p/\sqrt{(p^2+1)}].$$

[See § 8·43.]

43. Given $I_0\{\sqrt{(t^2 - b^2)}\} \Rightarrow p\, e^{-b\sqrt{(p^2-1)}}/\sqrt{(p^2 - 1)}$, find the transform of

$$I_0\{\sqrt{(t^2 - 2bt)}\} \quad (t > b).$$

$$[p\, e^{-b\{p+\sqrt{(p^2-1)}\}}/\sqrt{(p^2 - 1)}.]$$

44. Differentiate both sides of $p \displaystyle\int_0^\infty e^{-pt}\sin \omega t\, dt = \omega p/(p^2 + \omega^2)$, (a) with respect to p, (b) with respect to ω, and show that

$$t \sin \omega t \Rightarrow 2p^2\omega/(p^2 + \omega^2)^2, \quad t \cos \omega t \Rightarrow p(p^2 - \omega^2)/(p^2 + \omega^2)^2.$$

Using the same procedure, obtain the result $tJ_0(t) \Rightarrow p^2/(p^2 + 1)^{\frac{3}{2}}$. For justification of differentiation under the sign see [236], p. 173.

45. Expand in inverse powers of p, use (6) §8·12, and invert the transforms:

    (a) $p/\sqrt{(p + a)}$;   (b) $\sqrt{p}/\sqrt{(p + 2a)}$;   (c) $\sqrt{a}/\sqrt{(p + a)}$;

    (d) $p/\sqrt{(p^2 - a^2)}$;   (e) $1/\sqrt{\{p(p + 2a)\}}$;   (f) $p/(1 + \sqrt{p})$.

$[(a)$ $e^{-at}/\sqrt{(\pi t)}$;   (b) $e^{-at}I_0(at)$;   (c) erf $\sqrt{(at)}$;   (d) $I_0(at)$;

    (e) $\displaystyle\int_0^t e^{-at}I_0(at)\, dt$;   (f) $\dfrac{1}{\sqrt{(\pi t)}} - e^t \mathrm{erfc}\,\sqrt{t}.]$

46. Find the transform of n cycles of

    (a) $\sin \omega t$;   (b) $\sin(\omega t + \alpha)$, commencing $t = 0$.

$[(a)$ $\omega p(1 - e^{-2n\pi p/\omega})/(p^2 + \omega^2)$;

    (b) $p(1 - e^{-2n\pi p/\omega})(\omega \cos \alpha + p \sin \alpha)/(p^2 + \omega^2).]$

47. Find the transform of a periodic function having the value $A \sin \omega t$ from $t = 0$ to π/ω, and zero from $t = \pi/\omega$ to $2\pi/\omega$.

$$[A\omega p/(p^2 + \omega^2)(1 - e^{-p\pi/\omega}).]$$

48. In radio telephony the carrier wave is 'modulated' by a band of audio frequencies. If the former is represented by $\sin \omega_0 t$ and one of the latter by $\sin \omega t$, an expression of the type

$$E_0 \sin \omega_0 t (1 + k \sin \omega t)$$

is encountered, k being the modulation coefficient. Show that the transform is

$$E_0 \omega_0 p\{[1/(p^2 + \omega_0^2)] + 2k\omega p/[p^4 + 2p^2(\omega_0^2 + \omega^2) + (\omega_0^2 - \omega^2)^2]\}.$$

49. A p.d. $E = E_0 |\sin t|$ is applied to the circuit of Fig. 42. Show that, if the circuit is quiescent initially, the p.d. across R is

$$\frac{E_0}{CR_0}\left[\frac{2}{a\pi} - \frac{e^{-at}\coth(\tfrac{1}{2}a\pi)}{a^2+1} - \frac{4}{\pi}\sum_{n=1}^{\infty}\frac{\cos[2nt-\tan^{-1}(2n/a)]}{(4n^2-1)\sqrt{(4n^2+a^2)}}\right],$$

where $a = (R_0 + R)/CRR_0$.

50. Show that the transform of $f(t) = [a^2 - (t-a)^2]^{\nu-\frac{1}{2}}$ from $t = 0$ to $2a$ is

$$\sqrt{\pi}\,\Gamma(\nu+\tfrac{1}{2})\,(2a)^\nu p^{1-\nu}e^{-ap}I_\nu(ap) \quad (R(\nu) > -\tfrac{1}{2}).$$

51. Show that the transform of $f(t)$ over the range $t = 0$ to h is [133]

$$\phi(p) = \sum_{r=0}^{n-1} p^{-r}f^{(r}(0) + p^{-n+1}\int_0^h e^{-pt}f^{(n)}(t)\,dt - e^{-ph}\sum_{r=0}^{n-1}p^{-r}f^{(r)}(h).$$

52. Show that if

(i) $\phi(p) = p\displaystyle\int_b^\infty e^{-pt}f(t-b)\,dt,$ then $f(t-b) = \dfrac{1}{2\pi i}\displaystyle\int_{Br_1}\dfrac{e^{zt}\phi(z)\,dz}{z}$;

(ii) $\phi(p) = p\displaystyle\int_0^\infty e^{-pt}f(t)\,dt,$ then $f(t-b) = \dfrac{1}{2\pi i}\displaystyle\int_{Br_1}\dfrac{e^{z(t-b)}\phi(z)\,dz}{z}.$

53. Show that if $f(s) = \dfrac{1}{2\pi i}\displaystyle\int_{Br_1}\dfrac{e^{zs}\phi(z)\,dz}{z},$

then $\displaystyle\int_0^\infty f(s)\,J_0\{2\sqrt{(ts)}\}\,ds = \dfrac{1}{2\pi i}\displaystyle\int_{Br_1}e^{ut}\phi(1/u)\,du.$

[Substitute contour integrals for $f(s)$ and $J_0\{2\sqrt{(ts)}\}$.]

54. Using problem 53, show that [162]

$$p\phi(1/p) \Leftarrow \int_0^\infty f(s)\,J_0\{2\sqrt{(ts)}\}\,ds,$$

where $f(s) \Rightarrow \phi(p).$

55. If $f(s) \Rightarrow \phi(p)$ and $\phi(\sqrt{p}) = \sqrt{p}\displaystyle\int_0^\infty e^{-\sqrt{p}s}f(s)\,ds,$

show that [162] $\phi(\sqrt{p}) \Leftarrow \dfrac{1}{\sqrt{(\pi t)}}\displaystyle\int_0^\infty e^{-s^2/4t}f(s)\,ds.$

[Use the Laplace integral for $f(t)$ in 34 Appendix 10.]
State the conditions for validity of the results in Ex. 52–55.

56. By Appendix 8, $\phi_1(p)\,\phi_2(p)/p \Subset \int_0^t f_1(t-\lambda)\,f_2(\lambda)\,d\lambda$:

(a) taking $f_1 = f_2 = J_0(t)$, show that $\int_0^t J_0(t-\lambda)\,J_0(\lambda)\,d\lambda = \sin t$ [162];

show that   (b) $\int_0^t J_1(t-\lambda)\,J_0(\lambda)\,d\lambda = J_0(t) - \cos t$;

(c) $\int_0^t J_0(t-\lambda)\sin\lambda\,d\lambda = tJ_1(t)$;

(d) $\int_0^t J_0(t-\lambda)\,J_\nu(\lambda)\,d\lambda/\lambda = J_\nu(t)/\nu \quad (R(\nu) > 0)$.

57. Show that $\int_0^t f(t-\lambda)f(\lambda)\,d\lambda \Rightarrow \dfrac{\phi^2(p)}{p}$.

58. Show that $e^{-at}\int_0^t f(\tau)\,d\tau \Rightarrow \dfrac{p}{(p+a)^2}\phi(p+a)$.

59. Using § 8·11 and the transform of $J_\nu(t)$, show that

$$\int_0^\infty e^{-t\sinh\theta}J_\nu(t)\,dt = e^{-\nu\theta}\operatorname{sech}\theta \quad (R(\nu) > -1,\ \theta > 0).$$

60. Solve $\dfrac{d^2y}{dt^2} + s^2 y = f\sin\omega t$:

(a) for $y' = y = 0$ at $t = 0$;   (b) for $y' = y_1$, $y = 0$ at $t = 0$.

[(a) $f\{(\omega/s)\sin st - \sin\omega t\}/(\omega^2 - s^2)$;   (b) add $(y_1/s)\sin st$ to (a).]

61. Solve $\dfrac{d^2y}{dt^2} + a\dfrac{dy}{dt} + by = f\cos\omega t$, if $b > \tfrac{1}{4}a^2$ and $y' = y = 0$ at $t = 0$.

$[f\{\sin(\omega t + \theta_1) - (\sqrt{b}/\beta)\,e^{-\alpha t}\sin(\beta t + \theta_2)\}/\sqrt{[(\omega^2 - b)^2 + a^2\omega^2]},$

$\theta_1 = \tan^{-1}(b - \omega^2)/a\omega, \quad \theta_2 = \tan^{-1}\beta(b - \omega^2)/\alpha(b + \omega^2),$

$\alpha = a/2, \quad \beta = \sqrt{(b - \tfrac{1}{4}a^2)}.]$

62. Solve problem **61** subject to $y' = y_1$, $y = y_0$ at $t = 0$.

$\left[\text{Add } e^{-\alpha t}\{[(a - \alpha)y_0 + y_1]\dfrac{\sin\beta t}{\beta} + y_0\cos\beta t\} \text{ to the result in } \mathbf{61}.\right]$

63. Solve $\dfrac{d\xi}{dt} + a_1\dfrac{d\chi}{dt} + b_1\xi = f_1 e^{-at}$, $\quad \dfrac{d\chi}{dt} + a_2\dfrac{d\xi}{dt} + b_2\chi = 0$,

subject to quiescence initially, the constants being real and positive.

$$\left[\xi = \frac{f_1}{1-a_1 a_2}\left\{\frac{e^{-at}(b_2-a)}{(\alpha-a)(\beta-a)} + \frac{e^{-at}(b_2-\alpha)}{(a-\alpha)(\beta-\alpha)} + \frac{e^{-\beta t}(b_2-\beta)}{(a-\beta)(\alpha-\beta)}\right\},\right.$$

$$\chi = \frac{f_1 a_2}{1-a_1 a_2}\left\{\frac{a\,e^{-at}}{(\alpha-a)(\beta-a)} + \frac{\alpha\,e^{-\alpha t}}{(a-\alpha)(\beta-\alpha)} + \frac{\beta\,e^{-\beta t}}{(a-\beta)(\alpha-\beta)}\right\},$$

$$\left.\frac{\alpha}{\beta} = \frac{b_1+b_2}{2(1-a_1 a_2)}\left[-1 \pm \sqrt{\left(1 - \frac{4b_1 b_2(1-a_1 a_2)}{(b_1+b_2)^2}\right)}\right]; \left(\frac{b_1+b_2}{1-a_1 a_2}\right)^2 > \frac{4b_1 b_2}{1-a_1 a_2}.\right]$$

64. What is $\phi_3(p)$, the transform of the initial conditions function, for the differential equation

$$\frac{d^4\xi}{dt^4} + a\frac{d^3\xi}{dt^3} + b\frac{d^2\xi}{dt^2} + c\frac{d\xi}{dt} + g\xi = \chi(t), \text{ if } \dddot{\xi} = \xi_3, \ddot{\xi} = \xi_2, \dot{\xi} = \xi_1, \xi = \xi_0 \text{ at } t = 0?$$

$$[\xi_0(p^4 + ap^3 + bp^2 + cp) + \xi_1(p^3 + ap^2 + bp) + \xi_2(p^2 + ap) + \xi_3 p.]$$

65. In problem **64**, if $a = 10$, $b = 35$, $c = 50$, $g = 24$, $\chi(t) = A$, find ξ subject to quiescence initially.

$$\left[\xi = A\left\{\frac{1}{4!} - \frac{e^{-t}}{3!} + \frac{e^{-2t}}{4} - \frac{e^{-3t}}{3!} + \frac{e^{-4t}}{4!}\right\}.\right]$$

66. Solve $\dfrac{d^2 y}{dt^2} + a\dfrac{dy}{dt} = t\,e^{-t}$ for $\dot{y}$, if $\dot{y} = y = 0$ at $t = 0$; $a \neq 1$.

$$\left[\dot{y} = \frac{e^{-t}}{(a-1)}\left[t - \frac{1}{a-1}\right] + \frac{e^{-at}}{(a-1)^2}.\right]$$

67. Solve problem **66**, using the product theorem in § 3, Appendix 8.

68. The wave form of an impulse is the first half-cycle of $\sin \omega t$, the maximum ordinate being $\frac{1}{2}\omega$. What is the transform when $\omega \to \infty$? [p.]

69. What contour integrals correspond to the functions in Figs. 48 (c), 49 (d), 49 (h)?

$$\left[\frac{1}{2\pi i}\int_{Br_1} e^{zt}\frac{dz}{z}\left(\frac{1}{z} + h\right); \frac{1}{2\pi i}\int_{Br_1} e^{zt}(1 - e^{-zh})^2\frac{dz}{hz^2}; \frac{1}{2\pi i}\int_{Br_1} e^{z(t-h)}\frac{dz}{\sqrt{(z^2+1)}}.\right]$$

70. A mass m resting on a horizontal plane is acted upon by an impulsive force of the type in Fig. 49 (e), peak value f_0. If the force which resists sliding is r per unit velocity, find the displacement of m at any time after the impulse has ceased.

$$[f_0 m(e^{-rt/m} - e^{-r(t-h)/m} + hr/m)/r^2.]$$

71. Solve problem **70** for the impulse of Fig. 49 (d).

$$[(f_0 h/r)\{1 - (m/hr)^2 e^{-rt/m}(1 - e^{rh/m})^2\}.]$$

72. A p.d. having the form of Fig. 49 (b) is applied to an inductance L in series with a resistance R, the circuit being quiescent initially. What is the current at any time

$$nh < t < (n+1)h,$$

n an even integer? $[(E_0/R)\{\tfrac{1}{2}(1 - e^{-Rt/L}) - e^{-R(t-h)/L}(e^{nhR/L} - 1)/(e^{hR/L} + 1)\}.]$

73. An impulse $SI(t)$ is applied to the circuit of Fig. 62 (a) quiescent initially. What is the current when $t > 0$, if $1/LC > R^2/4L^2$?

$$[(S/\alpha L)e^{-\kappa t}(\alpha \cos \alpha t - \kappa \sin \alpha t) = (S\omega/\alpha L)e^{-\kappa t} \cos(\alpha t + \theta),$$

$$\kappa = R/2L, \ \alpha = \sqrt{(1/LC - R^2/4L^2)}, \ \theta = \tan^{-1}(\kappa/\alpha).]$$

74. An impulse generator has a wave form which is 'the first' half-cycle of a sine wave $E_0 \sin \omega t$. It is applied to an uncharged capacitance C in series with a resistance R. Find the current during the impulse, and after it has ceased. Neglect generator impedance.

$$\left[\frac{E_0 \omega C}{\sqrt{(1 + \omega^2 C^2 R^2)}} \{\sin[\omega t + \tan^{-1}(1/\omega CR)] - e^{-at}/\sqrt{(1 + \omega^2 C^2 R^2)}\} \right.$$

$$(0 < t < \pi/\omega);$$

$$\left. -\frac{E_0 \omega C e^{-at}}{1 + \omega^2 C^2 R^2}(1 + e^{\pi a/\omega}), \quad t > \pi/\omega, \ a = 1/CR. \right]$$

75. A force $f_0(e^{-at} + b)$ is applied to a body weighing W lb. fixed to one end of a spring of stiffness s lb. per foot, the spring being held securely at its other end. If the resistance to motion on a horizontal plane is r lb. per unit velocity (ft. sec.$^{-1}$), find ξ the displacement of the body, initially at rest, when $t > 0$ and interpret the result physically $(s > r^2/4m)$.

$$\left[\xi = f_0 \left\{ \frac{b}{s} + \frac{e^{-at}}{(ma^2 - ra + s)} - \frac{e^{-at}}{\beta \sqrt{m}} \left[\frac{b}{\sqrt{s}} \sin[\beta t + \tan^{-1}(\beta/\alpha)] \right. \right. \right.$$

$$\left. \left. \left. - \frac{\sin[\beta t - \tan^{-1}\{\beta/(a-\alpha)\}]}{\sqrt{(ma^2 - ra + s)}} \right] \right\}, \right.$$

$$\alpha = \frac{r}{2m}, \quad \beta = \sqrt{\left(\frac{s}{m} - \frac{r^2}{4m^2}\right)}, \quad m = \frac{W}{g}. \right]$$

76. A p.d. $E_0 \sin \omega t$ is applied to the circuit of Fig. 42 at time $t = 0$. Find the p.d. across R, if the circuit is quiescent initially.

$$\left[\frac{\omega E_0}{CR_0} \left\{ \frac{e^{-at}}{(a^2 + \omega^2)} - \frac{\cos [\omega t + \tan^{-1}(a/\omega)]}{\omega \sqrt{(a^2 + \omega^2)}} \right\}, \text{ where } a = (R + R_0)/CRR_0. \right]$$

77. A battery of voltage E is connected to an inductance L of resistance R in series with another inductance L_1 of resistance R_1. What is the current when $t > 0$, if the circuit is quiescent initially? If $R_1 L_1$ is short-circuited when the current is substantially steady, find the current in each individual circuit thereafter.

$$\left[\frac{E}{R + R_1} \{1 - e^{-(R + R_1)t/(L + L_1)}\}. \quad \frac{E}{R} \left\{ 1 - \left(\frac{R_1}{R + R_1} \right) e^{-Rt/L} \right\}; \quad \frac{E}{R + R_1} e^{-R_1 t/L_1}. \right]$$

78. If we write $F = $ stress, $\eta = $ viscosity, $N = $ shear modulus, the transform equation for an elasto-viscous material, e.g. flour dough, gelatine solution, subjected to a deforming force, is $F(a\sqrt{p} + 1) = \eta \dot{S}_0$, where $a = \sqrt{(\eta/N)}$. If $\dot{S}_0$ is a suddenly applied *constant* deformation velocity, find the resulting stress. Show that $F \simeq \dfrac{2\eta \dot{S}_0 \sqrt{(t/\pi)}}{a}$ when t is small [71].

$$[F = \eta \dot{S}_0 \{1 - e^{t/a^2} \operatorname{erfc} \sqrt{(t/a^2)}\}.]$$

79. Using § 11·51, determine the deflection of a uniform horizontal beam length l, built in at both ends, and carrying a distributed load varying linearly from zero at $x = 0$ to w per unit length at $x = l$.

80. Solve the problem in **79** if the beam were hinged freely at both ends.

81. Determine $\xi(x)$ in § 11·53 if the beam were hinged freely at both ends.

82. Masses m_1, m_2 are on a smooth horizontal plane. Starting from the left, m_1 is connected to a rigid frame by a uniform coil spring stiffness s_1, thence to m_2 by a spring stiffness s, thence to the frame by a spring s_2, the whole system being constrained to move coaxially. When quiescent the springs are unstrained and during motion their masses may be neglected. If m_2 is in its equilibrium position, and m_1 is displaced by $x_0 > 0$ when released at $t = 0$, find the subsequent displacement of each mass and the frequencies of the normal modes of vibration.

$$[x_1 = x_0 \{(\omega_1^2 - \alpha_2^2) \cos \omega_1 t - (\omega_2^2 - \alpha_2^2) \cos \omega_2 t\}/(\omega_1^2 - \omega_2^2),$$

$$x_2 = -x_0 \beta_2^2 (\cos \omega_1 t - \cos \omega_2 t)/(\omega_1^2 - \omega_2^2),$$

$$\omega_1^2, \, \omega_2^2 = (\alpha_1^2 + \alpha_2^2) \tfrac{1}{2} \{1 \pm \sqrt{[1 - 4(\alpha_1^2 \alpha_2^2 - \beta_1^2 \beta_2^2)/(\alpha_1^2 + \alpha_2^2)^2]}\},$$

$$\alpha_1^2 = (s + s_1)/m_1, \quad \alpha_2^2 = (s + s_2)/m_2, \quad \beta_1^2 = s/m_1, \quad \beta_2^2 = s/m_2.]$$

83. If, in problem **82**, m_1 is driven by an axial force $f \cos \omega t$, determine the displacements of m_1 and m_2 corresponding to the *forced* vibrations.

$[f(\alpha_2^2 - \omega^2) \cos \omega t / m(\omega_1^2 - \omega^2)(\omega_2^2 - \omega^2), \text{ and } f\beta_2^2 \cos \omega t / m(\omega_1^2 - \omega^2)(\omega_2^2 - \omega^2).]$

84. A uniform horizontal shaft of negligible inertia is fixed at one end and suitably supported on bearings. The torque-stiffness of a length l is γ per radian twist. Two discs, each having moment of inertia I about the axis, are keyed to the shaft at distances l, $2l$ from the fixed end. At $t = 0$ a torque $T = T_0(\cos \omega t + 0.1 \cos 3\omega t)$ is applied to the outer disc. Neglecting loss, find the angular displacement and velocity of each disc when $t > 0$, and also the frequencies of the normal modes of vibration.

85. Solve problem **84** if the inner disc experiences a damping torque k per unit angular velocity.

86. An electric train has three coaches, each of mass m, coupled together, being driven by the leading coach. A coupling unit may be regarded as a coil spring of stiffness s alongside a viscous damper. The latter introduces a resistive force r per unit velocity, and the electrical analogue is a capacitance in series with a resistance. The train starts from rest due to a tractive force f between the wheels of the first coach and the rails. Determine the displacement of each coach during a short time interval when f is constant. What is the least value of r to prevent oscillation?

87. Find the secondary current in § 12·52 corresponding to an applied impulse $SI(t)$. $\left[\dfrac{S(n_2/n_1)}{L(a-b)}(a\,e^{-at} - b\,e^{-bt}). \; S \text{ has dimensions volts} \times \text{time.}\right]$

88. If the p.d. in § 13·23, applied at $x = 0$, $t = 0$, has the form $E_0 \cos \omega t$, the cable being quiescent initially, i.e. at earth potential throughout, show that

$$E = E_0\left\{e^{-\alpha x}\cos(\omega t - \alpha x) - \frac{1}{\pi}\int_0^\infty e^{-ut}\sin[x\sqrt{(u\mathbf{CR})}]\frac{u\,du}{(u^2 + \omega^2)}\right\},$$

where $\alpha = \sqrt{(\omega \mathbf{CR}/2)}$, $z = u\,e^{\pm i\pi}$ on the parallel lines of the contour Br_2 of Fig. 24 (b).

[The integral term represents the transient. In practice $\omega \ll 100$, and if t is large enough, the bulk of the integral will be obtained when $\omega \gg u$. Thus the asymptotic formula for the transient is found by evaluating

$$\frac{E_0}{\pi\omega^2}\int_0^\infty e^{-ut}u\sin(a\sqrt{u})\,du, \text{ where } a = x\sqrt{(\mathbf{CR})}.$$

Now $\qquad \displaystyle\int_0^\infty e^{-ut}\sin(a\sqrt{u})\,du = (a\sqrt{\pi}/2t^{\frac{3}{2}})\,e^{-a^2/4t},$

and differentiation of each side with respect to t leads to the asymptotic formula, namely, $(E_0 a\,e^{-a^2/4t}/8\sqrt{\pi}\,\omega^2 t^{\frac{5}{2}})\{(-a^2/t) + 6\}$, $t \gg 0.]$

89. If in § 13·23 a resistance R were connected between the battery and the cable at $x = 0$, find the p.d. at any $x > 0$.

$[E_x = E_0\{\text{erfc}\,(a/2\sqrt{t}) - e^{bt+a\sqrt{b}}\,\text{erfc}\,[(a/2\sqrt{t}) + \sqrt{(bt)}]\}$, where $a = x\sqrt{(CR)}$, $b = R/CR^2$. On comparison with (4) § 13·23, the effect of R is to reduce the p.d. applied to the cable at $x = 0$, the reduction being given by the second term. For t large, its value $\sim -e^{-a^2/4t}/\sqrt{(\pi bt)}$—see § 5·23—$x$ finite. Since erfc (0) $= 1$, $E_x \to E_0$ as $t \to \infty$.]

90. The far end of a short unloaded cable length l is insulated from earth, and the whole is at earth potential. At $x = 0$, $t = 0$, a battery of voltage E_0 is applied between the cable and earth. What is the potential at any point x when $t > 0$? At $x = l$, $I_1 = 0$, so by (2) § 13·21, $\partial E/\partial x = 0$.

$$\left[E_x = E_0\left\{1 + \frac{4}{\pi}\sum_{n=0}^{\infty}\frac{(-1)^{n+1}}{(2n+1)}e^{-(2n+1)^2\pi^2 t/4l^2 CR}\cos\left[(2n+1)\,\pi(l-x)/2l\right]\right\}.\right.$$

The integrand of the contour integral has the factor

$$\cosh\,(l-x)\sqrt{(z/k)}/\cosh l\sqrt{(z/k)}, \quad k = 1/CR.$$

When l is large enough, this reduces to $e^{-x\sqrt{(z/k)}}$, and the result at (5) § 13·23 follows.]

91. Determine the current in problem **90**.

$$\left[I_x = \frac{2E_0}{Rl}\sum_{n=0}^{\infty}(-1)^n e^{-(2n+1)^2\pi^2 t/4l^2 CR}\sin\,[(2n+1)\,\pi(l-x)/2l], t > 0.\right.$$

By (2) § 13·21 with $L = 0$, $I_x = -(1/R)\,\partial E/\partial x$, and in virtue of uniform convergence, if $t > 0$ the series in problem **90** may be differentiated term by term. At $t = 0$, $I_{x=0}$ is infinite momentarily, for the reason given in § 13·23.]

92. Show that if a p.d. $E_0 H(t)$ is applied to the sending end of a very long distortionless transmission line ($\beta = 0$) at $t = 0$, the current at any point x is $I_x = E_0\sqrt{(C/L)}\,e^{-\alpha x/v}H(t-x/v)$.

[Write $\beta = 0$ in (7) § 13·33.]

93. In § 13·33 if the p.d. at the sending end were $\xi_2(t) \Rightarrow \phi_2(p)$, show that

$$I_x = \sqrt{(C/L)}\left\{e^{-\alpha x/v}\xi_2(t-x/v) + \beta\int_{x/v}^{t}\xi_2(t-\mu)\,e^{-\alpha\mu}\left[\frac{\mu I_1(\beta w)}{w} - I_0(\beta w)\right]d\mu\right\},$$

$t \geqslant x/v$, where $w = \sqrt{(\mu^2 - x^2/v^2)}$.

$[I_x \Rightarrow \sqrt{(C/L)}\,(p+2b)\,\phi_2\,e^{-\lambda x}/v\lambda$. Write $\phi_1\phi_2/p = e^{-\lambda x}\phi_2/v\lambda$, so that $\phi_1 = p\,e^{-\lambda x}/v\lambda \Leftarrow e^{-\alpha t}I_0[\beta\sqrt{(t^2 - x^2/v^2)}]$ by (1) and (5) § 13·33. Then by

(3), (4) Appendix 8, $2b\phi_2 e^{-\lambda x/v\lambda} \Subset 2b \int_{x/v}^{t} \xi_2(t-\mu) e^{-\alpha\mu} I_0(\beta w) d\mu$, the lower limit being x/v instead of zero. If $\xi_2(t)$ is such that differentiation under the integral sign is permissible ([236], p. 172), then

$$p\phi_2 e^{-\lambda x/v\lambda} \Subset \frac{\partial}{\partial t} \int_{x/v}^{t} \xi_2(t-\mu) e^{-\alpha\mu} I_0(\beta w) d\mu$$

$$= \xi_2(t-x/v) e^{-\alpha t/v} I_0(0) + \int_{x/v}^{t} \xi_2(t-\mu) \frac{\partial}{\partial\mu} [e^{-\alpha\mu} I_0(\beta w)] d\mu.$$

Taking $(2b-\alpha) = -\beta$, the desired result is obtained.]

94. Verify that in problem **93** the p.d. at any point x, $t \geqslant x/v$, is

$$E_x = e^{-\alpha x/v} \xi_2(t-x/v) + (\beta x/v) \int_{x/v}^{t} e^{-\alpha\mu} \xi_2(t-\mu) I_1(\beta w) \frac{d\mu}{w}.$$

[Use the product theorem as in problem **93**.]

95. In problems **93**, **94** when $\beta = 0$, signals are transmitted down the cable with reduction in amplitude but without alteration of shape. Show that the impedance of the cable is $Z_x = E_x/I_x = \sqrt{(L/C)}$ at any x, irrespective of the form of the applied p.d. $\xi_2(t)$, provided that it satisfies the requirement in problem **93**.

96. Obtain an expression for the current in § 13·521 if the applied p.d. were $SI(t)$ at $t = 0$, the filter being quiescent initially. Plot the current-time curve for $m = 2$. $[I_m = (2S/L) J_{2m}(\omega_c t).]$

97. What is the current in § 13·553 for the conditions in problem **96**? [Differentiate (8) § 13·553 with regard to t and multiply by S/E_0. See § 10·33.]

98. Solve (1) § 14·51 if at $x = 0$, $t = 0$, the air particle velocity were

$$-\left(\frac{\partial\phi}{\partial x}\right)_{x=0} = \omega\xi_0 \cos\omega t.$$

99. Solve (1) § 14·51 if $-\left(\frac{\partial\phi}{\partial x}\right)_{x=0} = -\omega\xi_0 e^{-\alpha t} \sin\omega t$. [The right-hand side of (5) § 14·51 is now $-\omega^2\xi_0\sigma/\{(\sigma+a)^2+\omega^2\}.]$

100. Solve (1) § 14·54 using x/c and t instead of t and $+\infty$, as the limits for the real integrals in (12).

101. What is the form of (1) § 14·56 when t is small?

102. Solve the problem in § 14·51 if the horn were conical. The area at abscissa x is $A = A_0(x+x_0)^2$, x_0 being the distance of the throat, of area A_0, from the vertex (fictitious). The horn equation is

$$\frac{\partial^2\phi}{\partial x^2} + \frac{2}{(x+x_0)}\frac{\partial\phi}{\partial x} - \frac{1}{c^2}\frac{\partial^2\phi}{\partial t^2} = 0.$$

$$\left[p = \frac{\rho_0 c\omega^2\xi_0 x_0}{(x+x_0)\,b} \left\{ \frac{a\,e^{-a\tau}}{b} - \cos\{\omega\tau - \tan^{-1}(\omega/a)\} \right\}, \right.$$

$$\left. a = c/x_0, \quad b = \sqrt{\{\omega^2 + (c^2/x_0^2)\}}, \quad \tau = t - x/c. \right]$$

PART IV

APPENDICES AND LIST
OF REFERENCES

APPENDIX 1

The modulus of a definite integral cannot exceed the product of the
maximum modulus M and the length of the path l

Referring to Fig. 11, let $f(z_{0m}), f(z_{1m})$, ... be the values of the integrand at the mid-points of the arcs $(z_0 z_1), (z_1 z_2)$, ..., and let $\delta z_0, \delta z_1$, ... represent the lengths of these arcs. Then, since the sum of the moduli of n complex numbers is greater than or equal to the modulus of the sum, we have

$$|f(z_{0m})|\,|\,\delta z_0\,| + |f(z_{1m})|\,|\,\delta z_1\,| + ... \geqslant |f(z_{0m})\,\delta z_0 + f(z_{1m})\,\delta z_1 + ...|. \quad (1)$$

If M is the maximum value of $|f(z_{nm})|$, then

$$M\{|\,\delta z_0\,| + |\,\delta z_1\,| + ...\} \geqslant |f(z_{0m})\,\delta z_0 + f(z_{1m})\,\delta z_1 + ...| \quad (2)$$

or $\qquad\qquad Ml \geqslant \text{r.h.s. of (2)}, \qquad\qquad (3)$

l being the length of the contour C over which integration is effected. Hence in the limit when the number of divisions of arc tends to infinity

$$Ml \geqslant \left|\int_C f(z)\,dz\right|. \quad (4)$$

APPENDIX 2

Proof that $\sin\theta \geqslant 2\theta/\pi$ when $0 \leqslant \theta \leqslant \tfrac{1}{2}\pi$

Let $y = \sin\theta/\theta$, then $\dfrac{dy}{d\theta} = \dfrac{(\theta\cos\theta - \sin\theta)}{\theta^2}$. (1)

Now $\qquad \dfrac{d}{d\theta}(\theta\cos\theta - \sin\theta) = -\theta\sin\theta \leqslant 0 \qquad (2)$

if $0 \leqslant \theta \leqslant \tfrac{1}{2}\pi$. Thus $(\theta\cos\theta - \sin\theta)$ decreases from zero with increase in θ in this range, so $dy/d\theta < 0$ in $0 < \theta \leqslant \tfrac{1}{2}\pi$. Hence y decreases in this range, being unity when $\theta = 0$, so the above result follows.

APPENDIX 3

Asymptotic series

Consider the following series obtained by partial integration:

$$f(z) = \left[\frac{1}{z} - \frac{1}{z^2} + \frac{2!}{z^3} - \frac{3!}{z^4} + ... + \frac{(-1)^{n-1}(n-1)!}{z^n}\right] + (-1)^n n!\, e^z \int_z^\infty \frac{e^{-t}dt}{t^{n+1}} \quad (1)$$

$$= S_n + R_n$$

$$= e^z \int_z^\infty \frac{e^{-t}dt}{t}, \quad (z > 0), \quad (2)$$

where R_n is the integral on the right-hand side of (1), which represents the remainder after n terms of the series have been taken, this latter sum being S_n. The ratio of the modulus of the $(n+1)$th to that of the nth term is

$$\left| \frac{u_{n+1}}{u_n} \right| = \left| \frac{n!}{z^{n+1}} \middle/ \frac{(n-1)!}{z^n} \right| = \frac{n}{|z|} . \tag{3}$$

When $n > |z|$, $n/|z| > 1$, so that (1) considered as a power series is divergent for all values of z. At the beginning of the expansion, where $|z| > n$, the moduli of the terms decrease steadily, until when

$$n \leqslant |z| < n+1$$

the numerical value of the term is less than that of any other. Beyond this point $n > |z|$ and the numerical values of the terms increase steadily. To illustrate this effect, let z in (1) have the value 5. Then

$$f(5) = \frac{1}{5} - \frac{1}{5^2} + \frac{2!}{5^3} - \frac{3!}{5^4} + \frac{4!}{5^5} - \frac{5!}{5^6} + \frac{6!}{5^7} - \frac{7!}{5^8} + \cdots$$

$$= 0 \cdot 2 - 0 \cdot 04 + 0 \cdot 016 - 0 \cdot 0096 + 0 \cdot 00768 - 0 \cdot 00768$$

$$+ 0 \cdot 009216 - 0 \cdot 0129024 + R_8, \tag{4}$$

R_8 being the remainder after eight terms have been taken.

The terms in (4) decrease numerically until the fifth and sixth are reached when the value is less than that of any other term. These two terms are, therefore, the smallest numerically. After the sixth term the terms increase steadily in absolute value as illustrated in Fig. 103. The correct value of $f(5)$ is shown by the broken line. The effect of adding the series (4) term by term is to bring the result closer and closer to $f(5)$, up to and including the first smallest term. Thereafter the error increases steadily, being alternately positive and negative. The correct value of $f(5)$ can be considered to be approached asymptotically at the first smallest term, where the error is least.

We shall now show that $|R_5|$ is less than the fifth term. Let S_5 be the sum of the first five terms, then from (1)

$$f(5) - S_5 = R_5 = -5! \, e^5 \int_5^\infty \frac{e^{-t}dt}{t^6} = -5! \int_5^\infty \frac{e^{5-t}dt}{t^6} , \tag{5}$$

or $\qquad \qquad \displaystyle |R_5| < 5! \int_5^\infty \frac{dt}{t^6} = \frac{5!}{5^6} = 0 \cdot 00768, \tag{6}$

since $e^{5-t} \leqslant 1$ for $t \geqslant 5$.

The result in (6) is the numerical value of the fifth term, so $|R_5|$ is less than this. Thus by retaining only those terms as far as and including the first smallest, the value of the function can be calculated with an error less than that term. For the general case where n terms are taken

$$|R_n| = n! \int_z^\infty \frac{e^{z-t}dt}{t^{n+1}} < \frac{(n-1)!}{z^n} . \tag{7}$$

Hence as $|z| \to \infty$, $|z^{n-1}R_n| \to 0$. This is the definition of an asymptotic series, i.e. that as the modulus of the variable tends to infinity, then for a fixed number n, $|z^{n-1}R_n|$ is evanescent. In the present example, R_n is numerically less than the nth and $(n+1)$th terms, but with some series it is possible to prove the latter case only.

Series (1) is unsuitable for reasonably accurate computation of $f(z)$, unless $|z| > 10$. The asymptotic series for the Bessel functions are good examples of expansions where a few terms give reasonable accuracy for moderate values of $|z|$ (see § 5·351 and [234], p. 70).

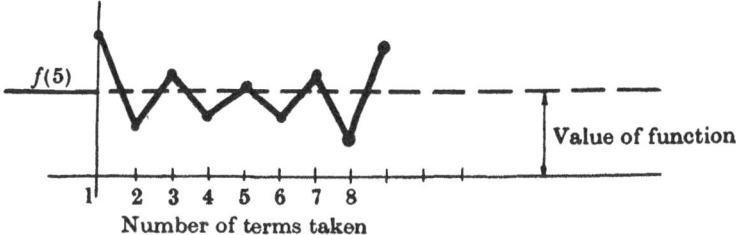

Fig. 103.

It is useful to obtain an approximate estimate of the position of the smallest term. This occurs roughly when the nth and $(n+1)$th terms are either equal or most nearly so. From (3) this happens when $n = |z|$, which is corroborated in the numerical case (4), where $n = 5$ and $|z| = 5$. The formulae for the $(n-1)$th and nth terms can also be equated in the above case, which gives $(n-1) = |z|$, or $n = 1 + |z|$, so $n = 6$. This is substantiated in (4), because the fifth and sixth terms are equal numerically.

Terms in an asymptotic expansion may take the form $\Gamma(\alpha)/x^\beta$, α, β being functions of n, e.g. $\alpha = \beta = (2n - \frac{1}{2})$; or $\alpha = (n - \frac{1}{2})$, $\beta = (2n - 2)$. In finding the minimum term, the formula for it may often be expressed approximately, e.g. $n \simeq a + bx^m$, a, b being constants, m a positive integer $\geqslant 1$. In this case the number of the minimum term increases with increase in x, so that numerical calculation up to that term becomes steadily more laborious. If, however, the error in stopping at the rth term, $r < n$, is numerically less than that term (x being large enough), adequate accuracy will probably be obtained before the minimum term is reached.

An asymptotic expansion can be integrated term by term, but in general it is not permissible to differentiate such an expansion (see [243]).

Example. A large cylindrical mass is fixed coaxially to a very long uniform helical spring of negligible inertia, whose upper end is held so that the spring hangs vertically. There is a resistance to axial motion of the spring per unit length, causing a damping force proportional to the velocity. A constant downward force is applied axially to the free end of the mass. The resultant force on the latter is obtained by evaluating an integral of the form

$$I = \int_0^\infty e^{-xt}\sqrt{x}\,(1+a^3x^3)^{-1}\,dx. \tag{8}$$

Find an asymptotic expansion so that the value of the integral can be computed for large values of t.

This is analogous to the case of an inductance connected between the battery and the sending end of a very long submarine telegraph cable, as shown in Fig. 34 (b). The p.d. applied to the cable is obtained by evaluating an integral of type (8).

Expanding $(1+a^3x^3)^{-1}$, and including a term for the remainder (which is expressed exactly in this case), we get

$$I = \int_0^\infty e^{-xt}\sqrt{x}\bigg[1 - a^3x^3 + a^6x^6 - a^9x^9 + \ldots + (-1)^{n-1}(ax)^{3(n-1)}$$

$$+ \frac{(-1)^n(ax)^{3n}}{1+a^3x^3}\bigg]\,dx \tag{9}$$

$$= \frac{\Gamma(\tfrac{3}{2})}{t^{\tfrac{3}{2}}} - \frac{a^3\Gamma(\tfrac{9}{2})}{t^{\tfrac{9}{2}}} + \frac{a^6\Gamma(\tfrac{15}{2})}{t^{\tfrac{15}{2}}} - \ldots$$

$$+ \frac{(-1)^{n-1}a^{3(n-1)}\Gamma(3n-\tfrac{3}{2})}{t^{3n-\tfrac{3}{2}}} + (-1)^n a^{3n}\int_0^\infty \frac{e^{-xt}x^{3n+\tfrac{1}{2}}}{1+a^3x^3}\,dx. \tag{10}$$

The remainder after n terms, namely,

$$R_n = (-1)^n a^{3n}\int_0^\infty \frac{e^{-xt}x^{3n+\tfrac{1}{2}}}{1+a^3x^3}\,dx, \tag{11}$$

is less numerically than

$$a^{3(n-1)}\int_0^\infty e^{-xt}x^{3n-\tfrac{5}{2}}dx = \frac{a^{3(n-1)}\Gamma(3n-\tfrac{3}{2})}{t^{3n-\tfrac{3}{2}}}. \tag{12}$$

Now (12) is equal numerically to the nth term, so the error which is incurred in stopping the series (10) at the nth term is less than that term. Also as $t \to \infty$, $|t^{3n-3}R_n| \to 0$ when n is fixed, so (10) is a true form of asymptotic expansion. When t is large, the major part of integral (8) is contributed for values of x near the origin, owing to the rapid decrease in e^{-xt} with increase in x. The same argument holds for a denominator $\sqrt{(1+a^3x^3)}$. Here a binomial expansion may be used when most of the integral is contributed before $|a^3x^3| = 1$. This procedure may often be applied, e.g. in deriving the asymptotic expansion of the Bessel function $J_0(z)$ in § 5·351. The remainder, after a fixed number of terms n, decreases rapidly with increase in t, since t occurs in the exponential index. The greater t, the smaller e^{-xt}, so the value of the integral contributed when $|ax| > 1$ decreases with increase in t. It is frequently impossible to express the remainder exactly as at (11).

To illustrate the point that the bulk of the integral is obtained before $|ax| = 1$, we consider a practical case where $a = 0\cdot014$. Then $ax \simeq 1$ when $x = 70$, and with $t = 0\cdot5$ second, $e^{-xt} = e^{-35}$, so the ratio of the integrand of (8) when $x = 70$ to that when $x = 1$ is approximately $4\cdot2\,e^{-34\cdot5}$, which is of the order 5×10^{-15}.

APPENDIX 4

The Mellin inversion theorem

1. *Theorem* 1. Let

(a) $\phi(z)$ be analytic in the strip $\alpha < x < \beta$, α and β being real (Fig. 104);

(b) $\displaystyle\int_{x-i\infty}^{x+i\infty} |\phi(z)|\, dz = \int_{-\infty}^{\infty} |\phi(x+iy)|\, dy$ converge;

(c) $\phi(z) \to 0$ uniformly as $|y| \to \infty$ in the strip $\alpha < x < \beta$;

(d) θ be real and positive:

If $\qquad f(\theta) = \dfrac{1}{2\pi i}\displaystyle\int_{c-i\infty}^{c+i\infty} \theta^{-z}\phi(z)\, dz,$ (1)

then will $\qquad \phi(z) = \displaystyle\int_{0}^{\infty} \theta^{z-1}f(\theta)\, d\theta.$ (2)

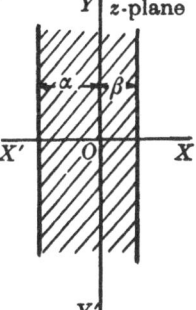

Fig. 104.

2. *Proof.* By hypotheses (a), (c) it follows that the contour $c \pm i\infty$ may be situated anywhere within the strip, i.e. all contours within the strip are equivalent. Thus if

$$\alpha < c_1 < c_2 < \beta,$$

the contours $c_1 \pm i\infty$ and $c_2 \pm i\infty$ are equivalent (Fig. 105). Substituting from (1) into (2), we get

$$\int_{0}^{\infty} \theta^{z-1}d\theta\, \frac{1}{2\pi i}\int_{c-i\infty}^{c+i\infty} \theta^{-z}\phi(z)\, dz \qquad (3)$$

$$= \int_{0}^{1} \theta^{z-1}d\theta\, \frac{1}{2\pi i}\int_{c_1-i\infty}^{c_1+i\infty} \theta^{-z_1}\phi(z_1)\, dz_1$$

$$+ \int_{1}^{\infty} \theta^{z-1}d\theta\, \frac{1}{2\pi i}\int_{c_2-i\infty}^{c_2+i\infty} \theta^{-z_2}\phi(z_2)\, dz_2, \qquad (4)$$

the range of θ being split, while two different but equivalent contours have been used for z.

Changing the order of integration, the right-hand side of (4) becomes

$$\frac{1}{2\pi i}\int_{c_1-i\infty}^{c_1+i\infty}\phi(z_1)\,dz_1\int_0^1\theta^{(z-z_1)-1}\,d\theta$$

$$+\frac{1}{2\pi i}\int_{c_2-i\infty}^{c_2+i\infty}\phi(z_2)\,dz_2\int_1^\infty\theta^{-(z_2-z)-1}\,d\theta. \qquad (5)$$

The changed order is permissible, since by hypothesis

$$\int_{c-i\infty}^{c+i\infty}|\phi(z)|\,dz$$

is convergent and

$$R(z-z_1)=(c-c_1)>0,\quad R(z_2-z)=(c_2-c)>0,$$

the latter conditions entailing convergence of the θ integrals in (5).* Evaluating these integrals, we have

$$\frac{1}{2\pi i}\int_{c_1+i\infty}^{c_1-i\infty}\frac{\phi(z_1)\,dz_1}{(z_1-z)}+\frac{1}{2\pi i}\int_{c_2-i\infty}^{c_2+i\infty}\frac{\phi(z_2)\,dz_2}{(z_2-z)}. \qquad (6)$$

Since $\phi\to0$ as $|y|\to\infty$, the integrals along the lines $c\pm i\infty$ vanish (Fig. 105). The only singularity within the contour is a simple pole at $z_1=z_2=z$. Hence (6) is equivalent to

$$\frac{1}{2\pi i}\int_C\frac{\phi(\zeta)\,d\zeta}{(\zeta-z)}=\phi(z), \qquad (7)$$

on evaluating at the pole $\zeta=z$.

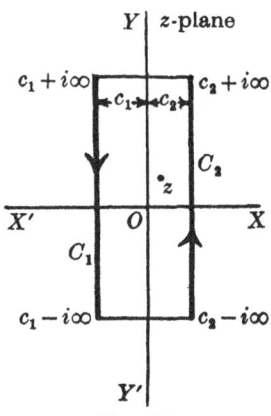

Fig. 105.

3. *Theorem 2.* For θ real and positive, $\alpha<R(z)<\beta$, let $f(\theta)$ be continuous or piecewise continuous,† and integral (2) be absolutely convergent. Then (1) follows from (2).

---

* If the two members of (4) are χ_1, χ_2, respectively, then

$$|\chi_1|\leqslant\frac{1}{2\pi}\int_{-\infty}^\infty|\phi(c_1+iy)|\,dy\int_0^1\theta^{z-z_1-1}\,d\theta,$$

$$|\chi_2|\leqslant\frac{1}{2\pi}\int_{-\infty}^\infty|\phi(c_2+iy)|\,dy\int_1^\infty\theta^{z-z_2-1}\,d\theta,$$

so the integrals in (4) are absolutely convergent. Accordingly the order of integration may be changed (see [221]).

† See Fig. 41 for an example. The function is single-valued and continuous between its discontinuities, which are finite, and limited in number in a finite interval, i.e. the function is continuous in stretches.

4. *Proof.* Substituting from (2) into (1) and writing θ_1 for θ in the former, we get

$$\frac{1}{2\pi i} \int_{c-i\infty}^{c+i\infty} \theta^{-z} dz \int_0^\infty \theta_1^{z-1} f(\theta_1) d\theta_1. \tag{8}$$

Taking $\theta = e^u$, $\theta_1 = e^v$, (8) becomes

$$\frac{1}{2\pi i} \int_{c-i\infty}^{c+i\infty} e^{-uz} dz \int_{-\infty}^\infty e^{vz} f(e^v) dv \tag{9}$$

$$= \frac{1}{2\pi} \int_{-\infty}^\infty e^{-u(c+iy)} dy \int_{-\infty}^\infty e^{v(c+iy)} f(e^v) dv \tag{10}$$

$$= \frac{e^{-uc}}{2\pi} \int_{-\infty}^\infty dy \int_{-\infty}^\infty e^{(v-u)iy} e^{vc} f(e^v) dv \tag{11}$$

$$= e^{-uc} \cdot e^{uc} f(e^u) = f(\theta), \tag{12}$$

by Fourier's integral theorem, which completes the proof. In the two proofs given above, we have shown that under the conditions stipulated, (a) the solution of (1) as an integral equation for $\phi(z)$ is given by (2); (b) the solution of (2) as an integral equation for $f(\theta)$ is given by (1). The arrangement is, therefore, a reciprocal one known as the Mellin inversion theorem. $\phi(z)$ and $f(\theta)$ are called Mellin transforms.

5. *Change of variable.* If in (1), (2) we write $\theta = e^{-t}$, t being real, and in (2) put p for z and $g(t)$ for $f(e^{-t})$, we get

$$g(t) = \frac{1}{2\pi i} \int_{c-i\infty}^{c+i\infty} e^{zt} \phi(z) dz, \tag{13}$$

$$\phi(p) = \int_{-\infty}^\infty e^{-pt} g(t) dt, \tag{14}$$

which is another form of the Mellin inversion theorem.

6. *Lower limit zero in* (14). In the majority of practical applications a lower limit $t = 0$ is required in (14), so we shall now determine the condition under which this alteration is permissible. Writing $\theta = e^{-t}$ in (5), we obtain

$$\frac{1}{2\pi i} \int_{c_1-i\infty}^{c_1+i\infty} \phi(z_1) dz_1 \int_0^\infty e^{-(z-z_1)t} dt$$
$$+ \frac{1}{2\pi i} \int_{c_2-i\infty}^{c_2+i\infty} \phi(z_2) dz_2 \int_{-\infty}^0 e^{(z_2-z)t} dt. \tag{15}$$

If the second double integral in (15) vanishes, the lower limit in (14) must be zero. The value of the double integral is equal to the second member in expression (6), so the appropriate condition for the altered lower limit is that this integral shall vanish, i.e.

$$\int_{c_1-i\infty}^{c_1+i\infty} \frac{\phi(z_2)\,dz_2}{(z_2-z)} = 0. \tag{16}$$

In technical applications the singularities of $\phi(z_2)$ usually lie upon or to the left of the imaginary axis, although there are cases where they lie on its right, but at a finite distance therefrom. For $\phi(z_2)$ write $\phi(\zeta)/\zeta$, and let all the singularities lie to the left of $c_1 \pm i\infty$ $(c_1 > 0)$. Since the contours $c_2 \pm i\infty$ and C enclose no singularity (Fig. 106), by Cauchy's theorem the integrals along the two paths taken in the same direction are equal, i.e.

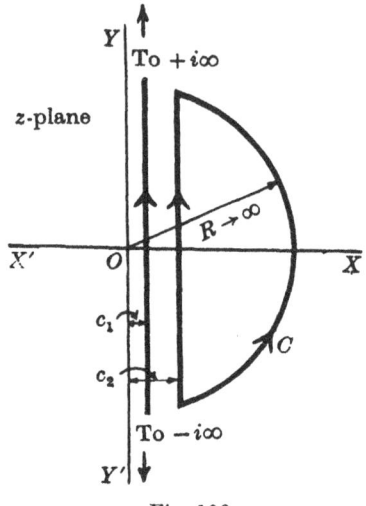

Fig. 106.

$$\int_{c_1-i\infty}^{c_1+i\infty} \frac{\phi(\zeta)\,d\zeta}{\zeta(\zeta-z)} = \int_C \frac{\phi(\zeta)\,d\zeta}{\zeta(\zeta-z)}. \tag{17}$$

Hence if the latter integral is zero, the condition for using a lower limit $t = 0$ in (14) is fulfilled. The integral round C vanishes provided $|\phi(\zeta)/\zeta| \to 0$ uniformly, with respect to phase ζ, as $|\zeta| \to \infty$, $-\tfrac{1}{2}\pi \leqslant \text{phase } \zeta \leqslant \tfrac{1}{2}\pi$. Accordingly we have the following particular case of the Mellin inversion theorem, stated in two parts:

$1°$.   If
$$\phi(p) = p \int_0^\infty e^{-pt} f(t)\,dt, \tag{18}*$$

then will
$$f(t) = \frac{1}{2\pi i} \int_{c-i\infty}^{c+i\infty} e^{zt} \phi(z) \frac{dz}{z} = \frac{1}{2\pi i} \int_{Br_1} e^{zt} \phi(z) \frac{dz}{z}, \tag{19}$$
provided

(a) All singularities of $\phi(z)/z$ lie to the left of $c \pm i\infty$, $c > 0$;

(b) $\displaystyle\int_{c-i\infty}^{c+i\infty} \left|\frac{\phi(z)}{z}\right| dz = \int_{-\infty}^{\infty} \left|\frac{\phi(c+iy)}{(c+iy)}\right| dy$ converges; but this condition may be waived when z^ν, $0 < R(\nu) < 1$ is the highest power of z in the expansion of $\phi(z)$, e.g.

$$\frac{1}{2\pi i} \int_{c-i\infty}^{c+i\infty} e^{zt} z^{\nu-1} dz = 1/t^\nu\,\Gamma(1-\nu);$$

* In *Acta Mathematica*, **27**, 339, 1903, Lerch showed that there is only one continuous function $f(t)$ which satisfies (18) regarded as an integral equation. (See [236], § 1·16.)

(c) $|\phi(z)/z| \to 0$ uniformly with respect to phase z, as $|z| \to \infty$, in the range $-\frac{1}{2}\pi \leqslant \text{phase } z \leqslant \frac{1}{2}\pi$ (see footnote on p. 346 with $b = 0$). Integral (19) is then zero for $t < 0$, as shown in § 4·41.

$2°$. If
$$f(t) = \frac{1}{2\pi i} \int_{c-i\infty}^{c+i\infty} e^{zt} \phi(z) \frac{dz}{z}, \tag{19a}$$

then
$$\phi(p) = p \int_0^\infty e^{-pt} f(t)\, dt, \tag{18a}$$

provided

(a) $f(t)$ is a continuous or a piecewise continuous function of t real > 0;

(b) $R(p)$ finite > 0;

(c) Integral $(18a)$ is absolutely convergent.

7. *Change of origin.* Instead of (18), (19) we have the following:

If
$$\phi(p) = p \int_b^\infty e^{-pt} f[a\sqrt{(t^2 - b^2)}]\, dt, \tag{20}$$

then will
$$f[a\sqrt{(t^2 - b^2)}] = \frac{1}{2\pi i} \int_{Br_1} e^{zt} \phi(z) \frac{dz}{z}, \tag{21}$$

and vice-versa, provided that the above conditions are fulfilled. It is necessary, however, to modify $1°(c)$ and $2°(a)$ § 6. In the latter case $t > b$ real > 0, so we proceed to ascertain the alteration to $1°(c)$. Substituting from (21) into (20), we get (b real $\geqslant 0$)

$$p \int_b^\infty e^{-pt} dt \frac{1}{2\pi i} \int_{Br_1} e^{zt} \phi(z) \frac{dz}{z}. \tag{22}$$

By virtue of $1°(b)$, $2°(b)$ and if $R(p) > R(z)$ on Br_1, the order of integration may be changed, giving

$$\frac{p}{2\pi i} \int_{Br_1} \phi(z) \frac{dz}{z} \int_b^\infty e^{(z-p)t}\, dt \tag{23}$$

$$= -\frac{p}{2\pi i} \int_{Br_1} e^{(z-p)b} \frac{\phi(z)\, dz}{z(z-p)}. \tag{24}$$

Consider the closed contour formed by $Br_1 + C$, C being a large semicircle of radius R on the right of Br_1. Since $R(p) > R(z)$ on Br_1, the only singularity of the integrand of (24) within the contour is a simple pole at $z = p$. The value of (24), taken round the contour *clockwise*, is $\phi(p)$. Hence (24) represents $\phi(p)$, provided that the integral round $C \to 0$ as $R \to \infty$. By hypothesis p is finite, so the integral round $C \to 0$, provided that

$$\left|\frac{e^{zb}\,\phi(z)}{z}\right| \to 0 \text{ uniformly* as } |z| \to \infty \text{ in the angle range } -\tfrac{1}{2}\pi \leqslant \text{phase } z \leqslant \tfrac{1}{2}\pi.$$

Integral (21) is then zero for $t < b$. If $b = 0$, condition $1°(c)$ §6 is reproduced.

In this latter exposition, we have discriminated succinctly between p and z. So far as the convergence of (18) is concerned, we must have $R(p) > 0$ (see [236], ch. 1), but for (19) to represent $f(t)$, condition $1°(c)$ §6 must be satisfied, z being on the semicircle C. Moreover, although p and z may be regarded as complex numbers having entirely different values, $\phi(p)$ is identical in form with $\phi(z)$. Consequently if p be written for z in the above condition ($b \geqslant 0$), it is merely a conventional expedient when using the functional form.

APPENDIX 5

Transformation of contour

1. $z = \zeta t$, t being finite. In the integral calculus of real variables, it happens frequently that a change of variable is expedient. To take a simple example, suppose we evaluate formally the integral

$$I = \int_0^a \cos xt\,dx. \tag{1}$$

Writing $v = xt$, gives $dx = dv/t$, so

$$I = \frac{1}{t}\int_0^{at}\cos v\,dv = \frac{\sin at}{t}. \tag{2}$$

This can be regarded as a contour integral along the x or real axis from $x = 0$ to a. When the variable is changed, both x and t are real, so the v-axis is also the x-axis, which means that the contour is unaltered except in length.

Passing to the complex variable, suppose the contour to be the imaginary axis upon which $z = yi$ from $y = -\infty$ to ∞. Then if t is a positive real quantity and we write $z = \zeta t$ or $\zeta = z/t$, the contour in the ζ-plane is identical in type with that in the z-plane. If it were a circle of radius r in the z-plane, the contour in the ζ-plane would be a circle of radius r/t.

When t is complex, $z = r_1 e^{i\theta_1}$, $t = r_2 e^{i\theta_2}$, so $\zeta = \dfrac{r_1}{r_2} e^{i(\theta_1 - \theta_2)}$, r_1, θ_1 being variable while r_2, θ_2 are fixed. If the contour in the z-plane were the

---

* Tending uniformly to the limit means that if we select a real number δ, however small, the radius R of the semicircle on the right of $c \pm i\infty$ may be chosen so that $\left|\dfrac{e^{zb}\,\phi(z)}{z}\right| < \delta$ in the given range of phase z; i.e. in the first and fourth quadrants.

imaginary axis, then $\theta_1 = \frac{1}{2}\pi$, and if the phase angle of t were $\frac{1}{4}\pi$, i.e. $\theta_2 = \frac{1}{4}\pi$, the contour in the ζ-plane would be a straight line through O at an angle of $45°$ with the positive real axis. In fact the contour is the imaginary axis rotated clockwise through $45°$. Generally any figure in the z-plane transforms into a similar figure in the ζ-plane, whose orientation and dimensions are governed by the factor $\dfrac{1}{t} = \dfrac{e^{-i\theta_2}}{r_2}$. Thus a square in the z-plane, two of whose sides of length a were coincident with the axes, would transform to a square whose sides were of length a/r_2. The vertical sides would now be inclined at $-\theta_2$ to the imaginary axis. Since $\zeta = z/t$, the singularities of the function to be integrated are situated in the same relative positions in the ζ-plane as in the z-plane. The transformation does not call, therefore, for any alteration in the technique of integration or other operation to be performed. By way of a simple example consider the integral

$$ I = \int_C e^{z/t}\frac{dz}{z}, \tag{3} $$

where C is a circle of radius δ enclosing the pole at the origin. Substituting $z = \zeta t$, we have $dz/z = d\zeta/\zeta$, so

$$ I = \int_{C'} e^{\zeta}\frac{d\zeta}{\zeta} = 2\pi i, \tag{4} $$

by § 3·21.

The contour in the ζ-plane is a circle C' round O, its radius being δ/r, where $r = |t|$. If we write t for $1/t$ in (3), the result in (4) is obtained, for the transformation $z = \zeta/t$ is a legitimate modification of $z = \zeta t$, with $t \neq 0$.

2. $z = (\zeta - a)$, a being complex and finite. This is equivalent to a shift of origin to the point $z = -a$. Neither the contour nor the positions of the singularities are affected in relation to each other, so the transformation can be made without any alteration in technique. As an example take

$$ I = \int_C \frac{e^z\,dz}{(z+a)}, \tag{5} $$

where C is a closed contour surrounding the pole at $z = -a$.

Writing $\zeta = z+a$, we have $z = \zeta - a$ and $dz = d\zeta$, so

$$ I = \int_C e^{\zeta - a}\frac{d\zeta}{\zeta} = e^{-a}\int_C e^{\zeta}\frac{d\zeta}{\zeta} \tag{6} $$

$$ = 2\pi i\,e^{-a}, \tag{7} $$

by § 3·21. In this case the origin was moved to $z = -a$, the contour remaining *in situ*.

3. $z = \zeta - (a/t)$, where a is complex and finite, t being real and > 0. This case is identical with § 2 provided $t \neq 0$. When $t \geqslant \epsilon > 0$, the transformation holds since a/t is finite. In this type of transformation the contour is kept in the finite part of the plane by considering $t > 0$. When

dealing with technical applications, formulae obtained for $t > 0$ are often valid when $t = 0$.

4. $z = a\zeta + b$. This is a general linear transformation, and it is covered by § 1 and § 2. Clearly from § 3 we can also take $z = a\zeta + b/t$, in the sense that $t > 0$ as in § 3.

5. $z = \zeta^2$, this being a quadratic transformation. If the contour were the imaginary axis, then $z = iy$. By choosing the positive square root, we have $\zeta = \sqrt(iy)$ above and $\zeta = \sqrt(-iy)$ below the origin. Now

$$\sqrt{i} = \frac{1}{\sqrt{2}}(1 + i) = e^{\frac{1}{4}\pi i},$$

and

$$\sqrt{-i} = \frac{1}{\sqrt{2}}(1 - i) = e^{-\frac{1}{4}\pi i}$$

as shown in Fig. 21 (c).

Hence $\zeta = \sqrt{y}\, e^{\pm \frac{1}{4}\pi i}$, so the contour in the ζ-plane is that shown in Fig. 107 (a). If the contour in the z-plane is a circle of radius r about O, we write $z = r e^{i\theta}$. Now $\zeta = \sqrt{z} = \sqrt{r}\, e^{\frac{1}{2}i\theta}$, choosing the positive root. When θ varies from 0 to 2π, $\frac{1}{2}\theta$ varies from 0 to π, so the contour in the ζ-plane is a semicircle, of centre O and radius $\sqrt{r}$, situated above the real axis. To illustrate this transformation we shall evaluate

$$I = \int_C \frac{dz}{z}, \tag{8}$$

where C is a circle of radius 4 units about O in the z-plane. Then if $z = \zeta^2$, $dz = 2\zeta d\zeta$, so

$$I = 2 \int_{C'} \frac{d\zeta}{\zeta}, \tag{9}$$

C' being a semicircle 2 units radius, situated in the upper half of the ζ-plane. Writing $\zeta = \rho\, e^{i\phi}$, (9) becomes

$$I = 2i \int_0^\pi d\phi = 2\pi i, \tag{10}$$

which is equal to (8), as shown at (5) § 2·123.

We can now consider the Bromwich contour Br_1 which extends from $c - i\infty$ to $c + i\infty$ as illustrated in Fig. 107 (b). On Br_1, $z = c + iy$, so writing $z = \zeta^2$ gives $\zeta = \sqrt(c + iy)$, the positive root being chosen. This may also be written $\zeta = (c^2 + y^2)^{\frac{1}{4}} e^{\frac{1}{2}i\phi}$, where $\phi = \tan^{-1} y/c$. When $y = 0$, Br_1 crosses the real axis, so $\phi = 0$ and $\zeta = \sqrt{c}$. The representative point in the ζ-plane falls either to the right or to the left of Br_1 according as $c <$ or > 1. When $y = \pm \infty$, $\phi = \pm \frac{1}{2}\pi$, so $\zeta \sim \sqrt{y}\, e^{\pm \frac{1}{4}\pi i}$, and the contour Br_2 in the ζ-plane is asymptotic to the straight lines $\sqrt{y}\, e^{\pm \frac{1}{4}\pi i}$, as shown in Fig. 107 (c). It is curved near the origin and passes through the point $\zeta = \sqrt{c}$. If we put $\zeta = u + iv = \sqrt(c + iy)$, then $u^2 - v^2 = c$, so the contour is a hyperbola.

When this contour occurs in technical applications it is sometimes convenient to find an equivalent path to facilitate integration. To do so, we complete the contour by the arcs AB, CD and, say, the imaginary axis indented on the right if required to avoid a singularity. Then if the complete contour contains no singularity of the integrand (expressed in terms of ζ), and if the integrals along AB, CD vanish as the radius $\to \infty$, the integral along Br_3 is equal to that along the imaginary axis, both being described positively.

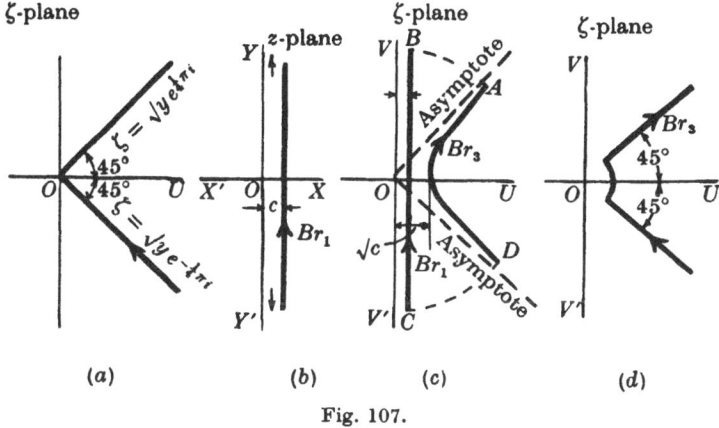

Fig. 107.

To illustrate the use of Br_3, we shall evaluate

$$I = \int_{Br_1} e^{-\sqrt{z}} \frac{dz}{z}, \qquad (11)$$

using the substitution $z = \zeta^2$. Then $dz/z = 2d\zeta/\zeta$, and (11) becomes

$$I = 2 \int_{Br_3} e^{-\zeta} \frac{d\zeta}{\zeta}. \qquad (12)$$

In this case when Br_1 is on the right of the imaginary axis, the integrals along the arcs vanish because

$$|e^{-\zeta}| \to 0, \quad (|\zeta| \to \infty),$$

so Br_1 is equivalent to Br_3. If the imaginary axis indented on the right at O is used as contour, its equivalence to Br_3 can be established by an argument akin to that in § 4·31. On the small semicircle $\zeta = \epsilon e^{i\theta}$, so (12) gives

$$I = 2i \int_{-\frac{1}{2}\pi}^{\frac{1}{2}\pi} e^{-\epsilon(\cos\theta + i\sin\theta)} d\theta = 2\pi i \quad (\epsilon \to 0). \qquad (13)$$

On the imaginary axis $\zeta = iy$, so (12) yields

$$2 \int_{-\infty}^{-\epsilon} e^{-iy} \frac{dy}{y} + 2 \int_{\epsilon}^{\infty} e^{-iy} \frac{dy}{y} = 2 \int_{\epsilon}^{\infty} (e^{-iy} - e^{iy}) \frac{dy}{y}. \tag{14}$$

Then making $\epsilon \to 0$, we get

$$-4i \int_{0}^{\infty} \sin y \frac{dy}{y} = -2\pi i, \tag{15}$$

by (13) § 3·3.

The value of the integral is the sum of (13) and (15), namely, zero. Integral (12) can also be evaluated directly along Br_3 taken to be a quadrant of a circle of small radius about O, and the two straight lines $\zeta = r e^{\pm \frac{1}{4} i \pi}$. On these lines making an angle of $45°$ at each side of the positive part of the x-axis, r is variable and θ constant (see Fig. 107 (d)). The original integral may, of course, be evaluated directly on the imaginary axis indented at O or along Br_2.

6. $z = a\zeta^2$ or $\zeta = \sqrt{(z/a)}$. Writing $z = r_1 e^{i\theta_1}$ and $a = r_2 e^{i\theta_2}$, we obtain $\zeta = \sqrt{(r_1/r_2)} e^{\frac{1}{2}(\theta_1 - \theta_2)}$. Taking the case of Br_1 in the z-plane, this will transform to a Br_3 type of contour, but its axis of symmetry will be veered round clockwise by $\frac{1}{2}\theta_2$. The value of $\sqrt{r_1}$ will be divided by $\sqrt{r_2}$. If the contour in the z-plane were a circle of radius r_1 about O (Fig. 108 (a)), it would transform to a semicircle of radius $\sqrt{(r_1/r_2)}$ about O in the ζ-plane, but its axis of symmetry would make an angle $-\frac{1}{2}\theta_2$ with the imaginary axis. For when θ_1 varies from 0 to 2π, $\frac{1}{2}(\theta_1 - \theta_2)$ varies from $-\frac{1}{2}\theta_2$ to $\pi - (\frac{1}{2}\theta_2)$, as shown in Fig. 108 (b).

7. $z = (\zeta - a/t)^2$, where a is complex, t being real and positive. This is merely a case of changing the origin in § 5 and is covered by §§ 2 and 3.

8. $z = (a\zeta - b/t)^2$, where a and b are complex, t being real and positive. This is the same as § 6 with a shift of origin, so it falls under §§ 6 and 3.

9. $z = \frac{1}{2}t\{\zeta + \sqrt{(\zeta^2 + a^2)}\}$ or $\zeta t = z - a^2 t^2/4z$, where a is complex, t being real and positive. When $|z|$, and therefore $|\zeta|$, is large enough, the second formula is $z \sim \zeta t$, so the transformation degenerates to type § 1. Considered in connection with Br_1, the smallest value of $|z|$ is c on the real axis. Then $\zeta t = c - a^2 t^2/4c$, or $\zeta = c/t - a^2 t/4c$ which has a positive real part if c is large enough. The Br_1 contour in the z-plane, therefore, transforms to an equivalent contour in the ζ-plane, so the process of integration is unaffected.

10. $z = \frac{1}{2}t\{\zeta + \sqrt{(\zeta^2 - a^2)}\}$ or $\zeta t = z + a^2 t^2/4z$. This is obviously of the same nature as § 9 above, so that further comment is unnecessary.

11. $z = \frac{1}{2}(\zeta + a^2/\zeta)$ or $\zeta = z + \sqrt{(z^2 - a^2)}$, where a is a finite complex number and the positive root is chosen. When $|z| \gg |a|$, we have $\zeta \sim 2z$, so the transformation is that of type § 1. Referring to the contour Br_1, the smallest value of $|z|$ is c, so $\zeta = c + \sqrt{(c^2 - a^2)}$, which has a positive real

part and lies on the right of the origin provided c is large enough. Consequently the path in the ζ-plane starts at $\zeta = 2c - i\infty$, passes through $\zeta = c + \sqrt{(c^2 - a^2)}$ and terminates at $\zeta = 2c + i\infty$, and this is reconcilable with Br_1.

12. $z^2 = \zeta$. On Br_1, $z = c + iy$, so $\zeta = (c^2 - y^2) + 2icy = u + iv$, and the contour in the ζ-plane is the parabola $v^2 = 4c^2(c^2 - u)$.

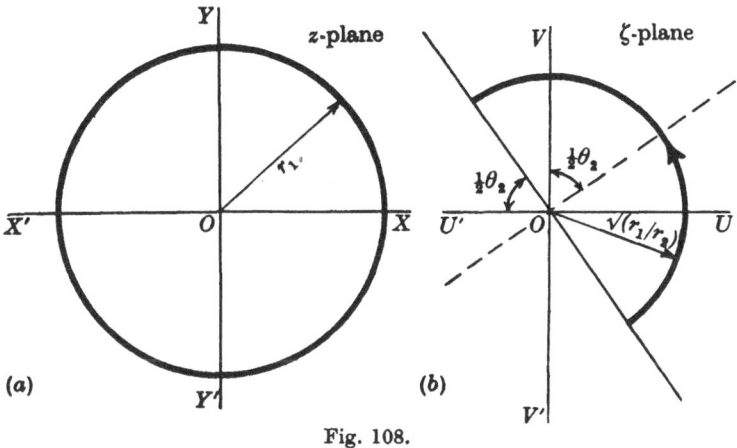

Fig. 108.

13. $z = a/\zeta$ or $\zeta = a/z$, a being real. Applying this to a circular contour about the origin, since $z = r e^{i\theta}$, $\zeta = (a/r) e^{-i\theta}$. Hence the contour in the ζ-plane is a circle of radius $|a|/r$ about O described in the opposite sense to that in the z-plane.

APPENDIX 6

Inversion of p^n on Br_2, $n \geqslant 1$

The Mellin inversion theorem is based on the path Br_1, but it holds for Br_2 also, provided the two paths are equivalent. For the integral

$$\frac{1}{2\pi i} \int \frac{e^{zt} dz}{z^{\nu+1}},$$

this occurs when $t > 0$ and $R(\nu) > -1$. The integral on Br_2 corresponding to p^n, $n \geqslant 1$, is

$$\frac{1}{2\pi i} \int_{Br_2} e^{zt} z^{n-1} dz,$$

and the contour may be replaced by a circle C round O, since the integrand has no singularity in the finite part of the z-plane. By Cauchy's theorem the value of the integral round C is zero for $t \geqslant 0$. Hence the inversion of p^n referred to Br_2 is zero, as also is the sum of any number of terms of this type. It may happen that such terms occur in the expansion of a function which itself is convergent on Br_1, but whose individual terms yield divergent integrals. When Br_2 is equivalent to Br_1, the integral of each term may be determinate on the former, as illustrated in §2 Appendix 7.

In the integral

$$f(t) = \frac{1}{2\pi i} \int_{Br_2} e^{zt} \phi(z) \frac{dz}{z}, \tag{1}$$

we may consider $\phi(p)$ to be the transform referred to Br_2. But since

$$\frac{1}{2\pi i} \int_{Br_2} e^{zt} z^{n-1} dz = 0, \tag{2}$$

when $n \geqslant 1$, it follows that the transform can be written as $\phi(p) + \sum_{r=0}^{n} a^r p^r$.

For instance $\qquad\qquad t^n/n! \rightleftharpoons p^{-n}, \tag{3}$

when referred to Br_1 or Br_2 provided $n \geqslant 0$. On the latter path we may write

$$t^n/n! \rightleftharpoons p^{-n} + a_0 + a_1 p + a_2 p^2 + \ldots + a_n p^n. \tag{4}$$

Moreover, $f(t)$ may have any number of transforms referred to Br_2. This is due to (1), regarded as an integral equation for $\phi(z)$, not having a unique solution. As an example, we find that the transform of $J_0(t)/t$ referred to Br_2, and obtained from the contour integral, is [125, 126, 128] for $t > 0$

$$J_0(t)/t \rightleftharpoons -p \log\{p + \sqrt{(p^2 + 1)}\}$$

$$= -p \log p - p \log 2 - \frac{1}{2^2 p} + \frac{1.3}{2.4^2 p^3} - \frac{1.3.5}{2.4.6^2 p^5} + \ldots. \tag{5}$$

The term $-p \log 2$ is, in effect, unnecessary, since its inverse is zero when referred to Br_2.

APPENDIX 7

Inversion of p^ν on Br_1 when $R(\nu) \geqslant 1$

1. The corresponding integral using Br_1 is

$$I = \frac{1}{2\pi i} \int_{Br_1} e^{zt} z^{\nu-1} dz. \tag{1}$$

This has no singularity in the finite part of the z-plane, so the imaginary axis may be used as the contour. Writing $z = iy$ from $-\infty$ to 0 and $z = iy$ from 0 to ∞ in (1) yields

$$I = -\frac{(-i)^\nu}{2\pi i} \int_0^\infty e^{-iyt} y^{\nu-1} dy + \frac{(i)^\nu}{2\pi i} \int_0^\infty e^{iyt} y^{\nu-1} dy \tag{2}$$

$$= \frac{1}{2\pi i} \int_0^\infty [e^{i(yt+\frac{1}{2}\nu\pi)} - e^{-i(yt+\frac{1}{2}\nu\pi)}] y^{\nu-1} dy \tag{3}$$

$$= \frac{1}{\pi} \int_0^\infty y^{\nu-1} \sin(yt + \tfrac{1}{2}\nu\pi) dy. \tag{4}$$

As y increases from 0, the integrand oscillates with steadily increasing amplitude; so the integral is divergent, if $R(\nu) > 1$. When $R(\nu) = 1$, (4) is oscillatory and therefore divergent, since it does not tend to a definite limit. Consequently when $R(\nu) \geqslant 1$, p^ν cannot be inverted with reference to Br_1. But if terms of the type p^ν occur in the expansion of an integral which is itself convergent on Br_1, the inverse may be obtained provided Br_1 is equivalent to Br_2, as shown in § 2 below.

So far as inversion referred to Br_2 is concerned, we have from (4) § 5·15

$$\frac{1}{2\pi i} \int_{Br_2} e^{zt} z^{\nu-1} dz = \frac{1}{t^\nu \Gamma(1-\nu)}, \tag{5}$$

for all values of ν. (5) cannot be inverted, using the Laplace integral, unless $R(\nu) < 1$, since this condition must be satisfied for validity of the Mellin theorem § 8·11.

2. *Example.* Evaluate

$$I = \frac{1}{2\pi i} \int_{Br_1} e^{zt - \sqrt{(az)}} \frac{dz}{z}, \tag{6}$$

a real > 0. Expanding $e^{-\sqrt{(az)}}$ gives

$$I = \frac{1}{2\pi i} \int_{Br_1} e^{zt} \left[1 - \sqrt{(az)} + \frac{az}{2!} - \frac{(az)^{\frac{3}{2}}}{3!} + \ldots \right] \frac{dz}{z}. \tag{7}$$

From § 1 all integrals after the second are divergent on Br_1, although (6) is convergent on this contour. Integral (6) is convergent on Br_2, which is equivalent to Br_1, since the integrals along the arcs in the second and third quadrants vanish as the radius $\to \infty$. Evaluating (7) on Br_2, all the positive integral powers of z give zero, while the remainder yield, with $a/t = w$,

$$I = 1 - \frac{2}{\sqrt{\pi}} \left[\frac{\sqrt{w}}{2} - \frac{w^{\frac{3}{2}}}{1! \, 2^3 . 3} + \frac{w^{\frac{5}{2}}}{2! \, 2^5 . 5} - \ldots \right] \tag{8}$$

$$= \operatorname{erfc} \tfrac{1}{2} \sqrt{w} \tag{9}$$

by (1) § 5·221.

The integral can be evaluated also on Br_2 by employing the real variable, as in § 5·221. Another method is to write $z = \zeta^2$, $dz = 2\zeta d\zeta$ and b for $\sqrt{a}$, then on Br_3 the contour in the ζ-plane, Fig. 107 (c), Appendix 5, we get

$$I = \frac{2}{2\pi i} \int_{Br_3} e^{\zeta^2 t - b\zeta} \frac{d\zeta}{\zeta}$$

$$= \operatorname{erfc} \tfrac{1}{2} (b/\sqrt{t}). \tag{10}$$

APPENDIX 8

The product theorem

1. This theorem states that if

$$\left.\begin{array}{c}\phi_1(p)\\ \phi_2(p)\\ \phi(p)\end{array}\right\} = p \int_0^\infty e^{-pt} \left\{\begin{array}{c}f_1(t)\\ f_2(t)\\ f(t)\end{array}\right\} dt, \tag{1}$$

and

$$\left.\begin{array}{c}f_1(t)\\ f_2(t)\\ f(t)\end{array}\right\} = \frac{1}{2\pi i} \int_{Br_1} e^{zt} \left\{\begin{array}{c}\phi_1(z)\\ \phi_2(z)\\ \phi(z)\end{array}\right\} \frac{dz}{z}, \tag{2}$$

then with

$$\phi(p) = \phi_1(p)\,\phi_2(p)/p, \tag{3}$$

$$f(t) = \int_0^t f_1(t-\lambda)\,f_2(\lambda)\,d\lambda. \tag{4}$$

Writing $(t-\lambda)$ for t in (2), we get

$$f_1(t-\lambda) = \frac{1}{2\pi i} \int_{Br_1} e^{p(t-\lambda)} \phi_1(p) \frac{dp}{p}. \tag{5}$$

By §4·41 this integral is zero when $\lambda > t$. Substituting from (5) into (4), we obtain

$$f(t) = \frac{1}{2\pi i} \int_0^\infty \int_{Br_1} e^{p(t-\lambda)} \phi_1(p) \frac{dp}{p} f_2(\lambda)\,d\lambda, \tag{6}$$

the upper limit ∞ being written for t, since the contour integral in (6) is zero from $\lambda = t$ to $\lambda = \infty$. If it is permissible to change the order of integration, (6) becomes

$$f(t) = \frac{1}{2\pi i} \int_{Br_1} e^{pt} \phi_1(p) \frac{dp}{p} \int_0^\infty e^{-p\lambda} f_2(\lambda)\,d\lambda \tag{7}$$

$$= \frac{1}{2\pi i} \int_{Br_1} e^{pt} \frac{\phi_1(p)\,\phi_2(p)}{p} \frac{dp}{p}, \tag{8}$$

from (1),

$$= \frac{1}{2\pi i} \int_{Br_1} e^{pt} \phi(p) \frac{dp}{p}, \tag{9}$$

by (3). Also by symmetry

$$f(t) = \int_0^t f_1(\lambda)\,f_2(t-\lambda)\,d\lambda. \tag{10}$$

See also [236], p. 38.

2. *Example* [74]. The solution of the problem of determining the use of aeroplane controls to eliminate the effect of gusts of wind, for instance action of the elevator (on the tail) to curb the variation of vertical velocity due to a horizontal gust, can with certain assumptions be made dependent upon the solution of an equation of the form (4) § 1. Here we have

$$f(t) = \sum_{r=1}^{4} a_r e^{k_r t} + K, \tag{1}$$

$$f_1(t) = \sum_{r=1}^{4} b_r e^{k_r t}. \tag{2}$$

The problem is to find $f_2(t)$, under the condition that $f(0) = 0$.
Inserting this condition in (1) gives

$$a_1 + a_2 + a_3 + a_4 + K = 0, \tag{3}$$

so

$$f(t) = \sum_{r=1}^{4} a_r(e^{k_r t} - 1). \tag{4}$$

The transform of (4) is

$$\phi(p) = \sum_{r=1}^{4} a_r \left[\frac{p}{p - k_r} - 1 \right] = \sum_{r=1}^{4} \frac{a_r k_r}{p - k_r}, \tag{5}$$

and that of (2) is

$$\phi_1(p) = \sum_{r=1}^{4} \frac{b_r p}{p - k_r}. \tag{6}$$

Since the solution of the problem depends upon an integral equation of type (4) § 1, we have from (3) § 1, (5), (6),

$$\phi_2(p) = p\phi(p)/\phi_1(p) \tag{7}$$

$$= \sum_{r=1}^{4} \frac{a_r k_r}{p - k_r} \bigg/ \sum_{r=1}^{4} \frac{b_r}{p - k_r} \tag{8}$$

$$= \frac{a_1 k_1 (p - k_2)(p - k_3)(p - k_4) + \dots}{b_1 (p - k_2)(p - k_3)(p - k_4) + \dots}. \tag{9}$$

Whence

$$f_2(t) = \frac{1}{2\pi i} \int_{Br_1} e^{zt} \phi_2(z) \frac{dz}{z}. \tag{10}$$

The integrand of (10) has a simple pole at $z = 0$, which gives

$$(a_1 + a_2 + a_3 + a_4)/(b_1 k_1 + b_2 k_2 + b_3 k_3 + b_4 k_4). \tag{11}$$

The contributions from the remaining poles are found by extracting the roots of the cubic equation in the denominator of (9), and proceeding in the usual way (see § 3·21).

When t is small, the form of the function $f_2(t)$ can be found by making p in (8) very large, so that $k_1, \ldots k_4$ are negligible (see §§ 4·61, 4·62). Then (8) can be written

$$\phi_2(p) = \left[\sum_{r=1}^{4} a_r k_r \Big/ \sum_{r=1}^{4} b_r \right] \quad (|\, p \,| \to \infty), \tag{12}$$

so by (10)
$$f_2(t) = \left[\sum_{r=1}^{4} a_r k_r \Big/ \sum_{r=1}^{4} b_r \right] \frac{1}{2\pi i} \int_{Br_1} e^{zt} \frac{dz}{z} \tag{13}$$

$$= \left[\sum_{r=1}^{4} a_r k_r \Big/ \sum_{r=1}^{4} b_r \right], \tag{14}$$

when t is small.

3. *Application of the product theorem to problem where the solution for the Heaviside unit function is known.* Let the solution of the differential equation of a system, quiescent when a disturbance of the form $H(t)$ is applied at $t = 0$, be

$$f_1(t) \Longrightarrow \phi_1(p), \tag{1}$$

and suppose that it is desired to know the solution when the form of disturbance is $f_2(t)$, *any* function of t, where

$$f_2(t) \Longrightarrow \phi_2(p). \tag{2}$$

By § 9·21 it can be shown that the solution is

$$\phi_3(p) = \phi_1(p)\,\phi_2(p) \Longleftarrow f_3(t), \tag{3}$$

so by § 1 above
$$f_3(t) = \frac{d}{dt} \int_0^t f_1(t - \lambda)\, f_2(\lambda)\, d\lambda \tag{4}$$

$$= \frac{d}{dt} \int_0^t f_1(\lambda)\, f_2(t - \lambda)\, d\lambda, \tag{5}$$

provided $\phi_3(p)$ satisfies 1° (c) § 6 Appendix 4. From above we obtain

$$f_3(t) = f_1(0)\, f_2(t) + \int_0^t f_2(\lambda)\, f_1'(t - \lambda)\, d\lambda \tag{6}$$

$$= f_1(0)\, f_2(t) + \int_0^t f_2(t - \lambda)\, f_1'(\lambda)\, d\lambda \tag{7}$$

by (4),
$$= f_1(t)\, f_2(0) + \int_0^t f_1(\lambda)\, f_2'(t - \lambda)\, d\lambda \tag{8}$$

$$= f_1(t)\, f_2(0) + \int_0^t f_1(t - \lambda)\, f_2'(\lambda)\, d\lambda \tag{9}$$

by (5).

In electrical problems where $f_1(t)$ represents a current, it may be written $A(t)$, which stands for the *indicial admittance* of the circuit at the point where the current wave is considered.

APPENDIX 9

Convergence of infinite series

1. In a broad sense, series in technical applications usually converge, but they may diverge for certain values of the variables and/or parameters concerned. Convergence should always be checked. When series are to be differentiated or integrated term by term, it is advisable to establish the validity of such operations to avoid the possibility of erroneous results. For information on these matters, references [215, 220, 240, 243] may be consulted. Convergence of some of the more important series in the text is considered below.

2. $\sum_{n=1}^{\infty} \dfrac{\cos n\theta}{n}$ is known to converge in every interval which excludes $\theta = 2s\pi$, s integral. When $\theta = 2s\pi$, the series becomes $\sum_{n=1}^{\infty}(1/n)$ which diverges to $+\infty$. Writing $(\pi+\theta)$ for θ, it follows that $\sum_{n=1}^{\infty}(-1)^n \dfrac{\cos n\theta}{n}$ converges in every interval which excludes $(2s+1)\pi$.

3. (6) §4·54: the series $\sum_{n=1}^{\infty} \dfrac{\sin(2\pi nt/h)}{n}$. Application of Dirichlet's test [215] shows that the series converges uniformly in $0 < h_1 \leqslant t \leqslant h_2 < h$,* $h < h_3 \leqslant t \leqslant h_4 < 2h$,* etc. In the neighbourhood of $t = 0$, nh, it is ordinarily convergent. Let $\theta = 2\pi t/h$, then the series is $\sum_{n=1}^{\infty} \dfrac{\sin n\theta}{n}$, which converges for all θ and, therefore, in the range $t \geqslant 0$. If for θ we write $(\pi+\theta)$, it follows that $\sum_{n=1}^{\infty}(-1)^n \dfrac{\sin n\theta}{n}$ converges in $t \geqslant 0$.

4. (10) §4·72: the series $\sum_{n=1}^{\infty}(-1)^n e^{-n^2 kt} \dfrac{\sin(n\pi r/a)}{n\pi r/a}$. Let $\pi r/a = \theta$, and the series becomes $(1/\theta) \sum_{n=1}^{\infty}(-1)^n e^{-n^2 kt} \dfrac{(\sin n\theta)}{n}$. If $r > 0$, then $\theta > 0$ and by §3 the series converges, since $e^{-n^2 kt} \leqslant 1$, if $k > 0$, $t \geqslant 0$. Applying Abel's test (see [215]), the original series converges uniformly in $0 \leqslant t \leqslant t_1$, for any θ except $2s\pi$ and their neighbourhoods, and, therefore, in $0 < r \leqslant a$.

5. (4) §4·73: the series $\sum_{n=1}^{\infty} e^{-\alpha_n^2 t/a^2} \dfrac{J_0(\alpha_n r/a)}{\alpha_n J_1(\alpha_n)}$. Since $J_0(\alpha_n) = 0$, if $r = a$

* This notation, applied to uniformly convergent series, signifies that the end points of the intervals are to be excluded. The convergence is uniform within and including the end points of the closed intervals, namely, $h_1 h_2$ and $h_3 h_4$.

the series converges to the value zero in the range $t \geqslant 0$. When u, v, n are large enough (see [234]),

$$J_0(u) \sim \sqrt{(2/\pi u)}\{\cos(u - \tfrac{1}{4}\pi) + O(u^{-1})\},$$

$$J_1(v) \sim \sqrt{(2/\pi v)}\{\cos(v - \tfrac{3}{4}\pi) + O(v^{-1})\},$$

$$\alpha_n \sim (n - \tfrac{1}{4})\,\pi.$$

Thus in $0 < r < a$, with $u = (\alpha_n r/a)$, $v = \alpha_n$, $\theta = \pi r/a$, $\beta = \tfrac{1}{4}\{(r/a) + 1\}\pi$,

$$\frac{J_0(\alpha_n r/a)}{\alpha_n J_1(\alpha_n)} \sim \frac{(-1)^{n-1}\sqrt{(a/r)}}{(n - \tfrac{1}{4})\pi}\cos(n\theta - \beta).$$

Accordingly we have to consider the convergence of

$$\sum_{n=m}^{\infty}(-1)^n\frac{\cos\beta\,\cos n\theta + \sin\beta\,\sin n\theta}{(n - \tfrac{1}{4})\pi},$$

when m is very large.

By § 2 the series in $\cos n\theta$ converges for all θ except $(2s + 1)\pi$. Since $0 < r < a$, θ cannot have this value, so the series converges. By § 3 the sine series converges for all θ and, therefore, in the interval $0 < r < a$. When $r = 0$, $J_0(\alpha_n r/a) = 1$, so

$$\frac{J_0(\alpha_n r/a)}{\alpha_n J_1(\alpha_n)} \sim (-1)^{n-1}/\sqrt{(2n - \tfrac{1}{2})}.$$

Now $\sum\limits_{n=m}^{\infty}\dfrac{(-1)^{n-1}}{\sqrt{(2n - \tfrac{1}{2})}}$ is an alternating series of decreasing terms which tend to zero as $n \to +\infty$, so by a known theorem (see [215]) it converges. Thus we have shown that $\sum\limits_{n=m}^{\infty}\dfrac{J_0(\alpha_n r/a)}{\alpha_n J(\alpha_n)}$ converges in the closed interval $0 \leqslant r \leqslant a$. Since $\sum\limits_{n=1}^{m-1}$ converges, so also does $\sum\limits_{n=1}^{\infty}$. Now $e^{-\alpha n^2 t/a^2}$ decreases steadily with increase in n for $t > 0$, and tends to unity for finite n as $t \to 0$. Hence by Abel's test (see [215]), the original series converges uniformly in $0 \leqslant t \leqslant t_1$, $0 < r_1 \leqslant r \leqslant a$.

6. (5) § 4·74: the series $\sum\limits_{n=1}^{\infty} e^{-\alpha n^2 t/a^2}\dfrac{J_0(\alpha_n r/a)}{\alpha_n^3 J_1(\alpha_n)}$. This is the series in § 5 with each term multiplied by $1/\alpha_n^2 \sim 1/\pi^2(n - \tfrac{1}{4})^2$, n large. Hence it has the same convergence properties as the series in § 5.

7. (9)§ 5·221: term by term integration of $e^{-xt}\sum\limits_{r=0}^{\infty}\dfrac{(-1)^r(ax)^{r+\frac{1}{2}}}{x(2r+1)!}$. By applying the '$M$' test (see [215, 236]), it may be shown that, excluding $r = 0$, the series is absolutely and uniformly convergent in the interval $0 \leqslant x \leqslant x_1$, a real > 0. It may also be shown that (see [215])

$$\int_0^{\infty} e^{-xt}\sum_{r=0}^{\infty}\frac{(ax)^{r+\frac{1}{2}}}{x(2r+1)!}\,dx = \int_0^{\infty} e^{-xt}\sinh\sqrt{(ax)}\frac{dx}{x} \tag{1}$$

converges if $t > 0$. Hence

$$\int_0^\infty e^{-xt} \sum_{r=1}^\infty (-1)^r \frac{(ax)^{r+\frac{1}{2}}}{x(2r+1)!}\, dx = \sum_{r=1}^\infty (-1)^r \int_0^\infty e^{-xt} \frac{(ax)^{r+\frac{1}{2}}}{x(2r+1)!}\, dx. \quad (2)$$

For the term corresponding to $r = 0$, we have

$$\sqrt{a} \int_0^\infty e^{-xt} x^{-\frac{1}{2}}\, dx = \sqrt{(\pi a/t)}, \quad (3)$$

so finally

$$\int_0^\infty e^{-xt} \sin \sqrt{(ax)}\, dx/x = \text{the r.h.s. of } (2) + (3), \quad (4)$$

term by term integration being valid.

8. (5) § 6·14: the series $\sum_{n=0}^\infty \dfrac{(-1)^n y^n}{\Gamma[\frac{1}{2}(4n+3)]}$, with $y = \omega^2 t^2$. Then

$$|u_{n+1}/u_n| = 4y/(4n+5)(4n+3) \to 0 \quad \text{as} \quad n \to +\infty$$

for all finite ωt, so the series is absolutely convergent. Being a power series, it is uniformly convergent in $0 \leqslant t \leqslant t_1$ (see [215]).

9. (6) § 8·62: the series $\sum_{n=1}^\infty \dfrac{\cos 2nt}{4n^2 - 1}$. Now $\sum_{n=1}^\infty \dfrac{|\cos 2nt|}{4n^2 - 1} \leqslant \sum_{n=1}^\infty \dfrac{1}{(4n^2-1)}$, and $\sum_{n=1}^\infty \dfrac{1}{n^2}$ is known to converge, so the original series converges absolutely. The convergence being independent of t is uniform (see 'M' test [215, 236]).

10. (12) § 8·63: the series

$$\sum_{n=1}^\infty \frac{\sin(n\theta - \beta_n)}{n\sqrt{(n^2\pi^2 + a^2 h^2)}}, \quad \beta_n = \tan^{-1}(n\pi/ah), \quad \theta = \pi t/h.$$

Then $\quad |\sin(n\theta - \beta_n)/n\sqrt{(n^2\pi^2 + a^2h^2)}| < 1/\pi n^2$, and $\sum_{n=1}^\infty (1/n^2)$

converges. Hence by the 'M' test (see [215, 236]) the original series is absolutely and uniformly convergent in every closed interval of θ.

11. (10) § 13·24: see § 4.

12. (1) § 13·25: the series $\sum_{n=1}^\infty (-1)^n e^{-n^2 kt} \cos n\theta$, $\theta = \pi x/l$. Since $|\cos n\theta| \leqslant 1$, we consider the series $\sum_{n=1}^\infty e^{-n^2 kt}$, $k > 0$, $t > 0$. Using the ratio test this converges. The 'M' test (see [215, 236]) shows that it is absolutely and uniformly convergent in $0 < t_1 \leqslant t \leqslant t_2$. When $t = 0$, it is oscillatory.

13. (2) § 13·25: see § 12.

14. (13) §14·12: the series $\sum\limits_{n=1}^{\infty} \dfrac{e^{-\alpha n^2 kt}}{\alpha_n^2}$. The sum may be written $S = S_m + R_m$, the remainder being $R_m = \sum\limits_{n=m+1}^{\infty} \dfrac{e^{-\alpha n^2 kt}}{\alpha_n^2}$. If m is large enough $\alpha_n \sim \pi(n - \tfrac{1}{4})$, so in the range $t \geqslant 0$, $R_m < \sum\limits_{n=m+1}^{\infty} \dfrac{1}{n^q}$, $1 < q < 2$, which converges, as also does the series S_m. Hence by the 'M' test (see [215, 236]) the original series is absolutely and uniformly convergent in $0 \leqslant t \leqslant t_1$.

15. Convergence of other series in Chapters xiv, xv may be investigated by aid of the preceding examples.

APPENDIX 10

Short list of p-multiplied Laplace transforms

The following abbreviations are used: $y = \sqrt{(t^2 - b^2)}$, $r = \sqrt{(p^2 + a^2)}$, $R = p + \sqrt{(p^2 + a^2)}$, $s = \sqrt{(p^2 - a^2)}$, $S = p + \sqrt{(p^2 - a^2)}$; $t > 0$, $b \geq 0$, a real and may have either sign in certain cases; $R(p) > 0$ usually. The list covers the requirements of the text. Additional transforms and definitions of various mathematical functions will be found in [235 a, b]. In these references P, Q correspond to R, S above. For various theorems see Chapter VIII, Appendices 4, 8.

	$\phi(p)$	$f(t)$	Condition
1	1	$H(t)$. Unit function	
2	p	$I(t)$. Impulsive function	
3	$\Gamma(\nu + 1)/p^\nu$	t^ν	$R(\nu) > -1$
4	$pa/(p^2 + a^2)$	$\sin at$	
5	$p^2/(p^2 + a^2)$	$\cos at$	
6	$a^2/(p^2 + a^2)$	$1 - \cos at$	
7	$p^2/(p^2 + a^2)^2$	$(t/2a)\sin at$	
8	$pa/(p^2 - a^2)$	$\sinh at$	
9	$p^2/(p^2 - a^2)$	$\cosh at$	
10	$p/(p + a)$	e^{-at}	
11	$a/(p + a)$	$1 - e^{-at}$	

	$\phi(p)$	$f(t)$	Condition
12	$p/(p+a)(p+b)$	$(e^{-bt}-e^{-at})/(a-b)$	$b \neq a$
13	$p/(p+a)^2$	$t e^{-at}$	
14	$p/(p^2+2ap+b^2)$	$e^{-at}\sin\sqrt{(b^2-a^2)}\,t/\sqrt{(b^2-a^2)}$	$b>a>0$
15	$p/(p^2+2ap+b^2)$	$e^{-at}\sinh\sqrt{(a^2-b^2)}\,t/\sqrt{(a^2-b^2)}$	$a>b>0$
16	$pb/[(p+a)^2+b^2]$	$e^{-at}\sin bt$	
17	$p(p+a)/[(p+a)^2+b^2]$	$e^{-at}\cos bt$	
18	$pb/[(p+a)^2-b^2]$	$e^{-at}\sinh bt$	
19	$p(p+a)/[(p+a)^2-b^2]$	$e^{-at}\cosh bt$	
20	$p/(p^3+a^3)$	$\dfrac{1}{3a^2}[e^{-at}+e^{\frac{1}{2}at}(\sqrt3\sin\tfrac{1}{2}\sqrt3\,at-\cos\tfrac{1}{2}\sqrt3\,at)]$	
21	$p^2/(p^3+a^3)$	$\dfrac{1}{3a}[-e^{-at}+e^{\frac{1}{2}at}(\sqrt3\sin\tfrac{1}{2}\sqrt3\,at+\cos\tfrac{1}{2}\sqrt3\,at)]$	
22	$2a^3p/(p^2+a^2)^2$	$\sin at - at\cos at$	
23	$(a^2-b^2)p^2/(p^2+a^2)(p^2+b^2)$	$\cos bt - \cos at$	$a \neq b$
24	$2a^3p/(p^4-a^4)$	$\sinh at - \sin at$	
25	$2a^2p^2/(p^4-a^4)$	$\cosh at - \cos at$	
26	e^{-ph}	$H(t-h)$	$t>h>0$
27	pe^{-ph}	$I(t-h)$	$t>h>0$

No.			
28	$1 - e^{-ph}$	$H(t) - H(t-h)$. Morse dot, or rectangular impulse, Fig. 50 (b)	
29	$1/(1 + e^{-ph})$	Semi-infinite sequence of Morse dots, Fig. 54 (b)	
30	$p/(1 - e^{-ph})$	Semi-infinite sequence of impulses $I(t)$ at interval h, Fig. 52 (a)	
31	$\dfrac{1}{2ph} - \dfrac{e^{-2ph}}{1 - e^{-2ph}}$	Saw-tooth wave form, Fig. 41	
32	$pa(1 + e^{-\pi p/a})/(p^2 + a^2)$	$\left.\begin{array}{l} f(t) = \sin at \\ \qquad = 0 \end{array}\right\}$ $\begin{array}{l} 0 \leqslant t \leqslant \pi/a, \\ \text{all other } t \end{array}$	
33	$e^{-a\sqrt{p}}$	$\mathrm{erfc}\,(a/2\sqrt{t})$	
34	$\sqrt{p}\,e^{-a\sqrt{p}}$	$e^{-a^2/4t}/\sqrt{(\pi t)}$	
35	$p\,e^{-a\sqrt{p}}$	$a\,e^{-a^2/4t}/2\sqrt{\pi}\,t^{\frac{3}{2}}$	
36	$e^{p^2/4a^2}\,\mathrm{erfc}\,(p/2a)$	$\mathrm{erf}\,at$	
37	e^{-br}/r	$\displaystyle\int_b^t J_0(ay)\,dt$	$t > b > 0$
38	$p\,e^{-br}/r$	$J_0(ay)$	$t > b > 0$
39	$e^{-bp} - e^{-br}$	$\displaystyle ab\int_b^t \frac{J_1(ay)}{y}\,dt$	$t > b > 0$
40	$p(e^{-bp} - e^{-br})$	$\dfrac{ab\,J_1(ay)}{y}$	$t > b > 0$

	$\phi(p)$	$f(t)$	Condition
41	$e^{-br} - e^{-b(p+a)}$	$ab \int_t^\infty \frac{J_1(ay)}{y}\, dt$	$t > b > 0$
42	e^{-bs}/s	$\int_b^t I_0(ay)\, dt$	$t > b > 0$
43	$p\,e^{-bs}/s$	$I_0(ay)$	$t > b > 0$
44	$e^{-bs} - e^{-bp}$	$ab \int_b^t \frac{I_1(ay)}{y}\, dt$	$t > b > 0$
45	$p(e^{-bs} - e^{-bp})$	$ab\,\frac{I_1(ay)}{y}$	$t > b > 0$
46	$e^{-\gamma\sqrt{[(p+2a)(p+2b)]}}$	$e^{-\alpha\gamma} H(t-\gamma) + \beta\gamma \int_\gamma^t e^{-\alpha t} \frac{I_1[\beta\sqrt{(t^2-\gamma^2)}]}{\sqrt{(t^2-\gamma^2)}}\, dt$	$\begin{cases} t > \gamma > 0 \\ \alpha = a+b,\ \beta = a-b \end{cases}$
47	$\dfrac{p\,e^{-\gamma\sqrt{[(p+2a)(p+2b)]}}}{\sqrt{[(p+2a)(p+2b)]}}$	$e^{-\alpha t} I_0[\beta\sqrt{(t^2-\gamma^2)}]$	$\begin{cases} t > \gamma > 0 \\ \alpha = a+b,\ \beta = a-b \end{cases}$
48	$\sqrt{\left(\dfrac{p+2a}{p+2b}\right)}\, e^{-\gamma\sqrt{[(p+2a)(p+2b)]}}$	$e^{-\alpha t} I_0[\beta\sqrt{(t^2-\gamma^2)}] + 2b \int_\gamma^t e^{-\alpha t} I_0[\beta\sqrt{(t^2-\gamma^2)}]\, dt$	$\begin{cases} t > \gamma > 0 \\ \alpha = a+b,\ \beta = a-b \end{cases}$
49	p/r	$J_0(at)$	
50	$a^\nu p_r/rR^\nu$	$J_\nu(at)$	$R(\nu) > -1$

51	$p/8$	$I_0(at)$	
52	$a^v p/8 S^v$	$I_v(at)$	$R(v) > -1$
53	$\sqrt{[a/(p+a)]}$	$\operatorname{erf}\sqrt{(at)}$	
54	$\sqrt{[(p+2b)/(p+2a)]}$	$e^{-\alpha t}I_0(\beta t) + (\alpha - \beta)\int_0^t e^{-\alpha t}I_0(\beta t)\,dt$	$\alpha = a+b,\ \beta = a-b$
55	$[pa + \sqrt{(p^2 a^2 + 1)}]^{-v}$	$v\int_0^t J_v(t/a)\,dt/t$	$R(v) > 0$

REFERENCES

Those marked † have lists of transforms, or may be consulted for methods of deriving transforms of various functions. In general the references are not applicable to the operational methods of Boole, Graves and Murphy. For these and additional references see [225, 239].

A. SCIENTIFIC PAPERS

1. ADAMS, E. P. Some applications of Heaviside's operational methods. *Proc. Amer. Phil. Soc.* **62**, 26, 1923.
2. BAERWALD, H. G. Über die Fortpflanzung von Signalen in dispergierenden Systemen. *Ann. Phys., Lpz.*, **6**, 295, 1930; **7**, 731, 1930; **8**, 565, 1931.
3. —— Some relations between transient phenomena in systems with similar frequency characteristics. *Phil. Mag.* **21**, 833, 1936.
4. BAKER, B. B. An extension of Heaviside's operational method of solving differential equations. *Proc. Edinb. Math. Soc.* **42**, 95, 1924.
5. BARANOV, V. Oscillations d'un disque circulaire plongé dans un liquid visqueux. *Cahiers de Physique*, cahier **15**, 43, 1943.
6. BATEMAN, H. Solution of system of differential equations in theory of radio-active transformation. *Proc. Camb. Phil. Soc.* **15**, 423, 1910.
7. —— *Partial differential equations of mathematical physics.* (1932.)
8. —— Operational equations. *Nat. Math. Mag. Amer.* **9**, 197, 1935.
9. BERG, E. J. Heaviside's operators in Engineering and Physics. *J. Franklin Inst.* **198**, 647, 1924.
10. —— Heaviside's operational calculus. *Gen. Elect. Rev.* **30**, 586, 1927; **31**, 93, 143, 212, 267, 395, 444, 504, 1928.
11. BROMWICH, T. J. I'A. Normal co-ordinates in dynamical systems. *Proc. Lond. Math. Soc.* (2), **15**, 401, 1916.
12. —— Examples of operational methods in mathematical physics. *Phil. Mag.* **37**, 407, 1919.
13. —— The problem of random flights. *Phil. Mag.* **42**, 432, 1921.
14. —— Symbolical methods in the theory of heat conduction. *Proc. Camb. Phil. Soc.* **20**, 411, 1921.
15. —— A certain series of Bessel functions. *Proc. Lond. Math. Soc.* (2), **25**, 103, 1926.
16. —— Some solutions of the electromagnetic equations, and of the elastic equations, with applications to the problem of secondary waves. *Proc. Lond. Math. Soc.* **28**, 438, 1928.
17. —— Heaviside's formula for alternating currents in cylindrical wires. *Phil. Mag.* **6**, 842, 1928.

18. BROMWICH, T. J. I'A. A new method for solving two dimensional problems of physical types. *Proc. Lond. Math. Soc.* **30**, 165, 1929.

19. —— Motion of a sphere in a viscous fluid. *Proc. Camb. Phil. Soc.* **25**, 369, 1929.

20. —— The application of operational methods to some electrical problems in diffusion. *Proc. Lond. Math. Soc.* **31**, 209, 1930.

21. —— An application of Heaviside's methods to viscous fluid motion. *J. Lond. Math. Soc.* **5**, 10, 1930.

22. BRYANT, L. W. and WILLIAMS, D. A. Application of operators to calculation of disturbed motion of an aeroplane. *Rep. Memor. Aero. Res. Comm.*, no. 1346, July 1930.

23. BUILDER, G. Amplification of transients. *Wireless Engr*, **12**, 246, 1935.

24. BUSH, V. Summary of Wagner's proof of Heaviside's formula. *Proc. Inst. Radio Engrs, N.Y.*, **5**, 377, 1917; **6**, 111, 1918.

25. —— Note on operational calculus. *J. Math. Phys.* **3**, 95, 1924.

26.† CAMPBELL, G. A. and FOSTER, R. M. *Bell Syst. Tech. Monogr.* B, no. 584, 1931.

27. CARSON, J. R. On a general expansion theorem for the transient oscillations of a connected system. *Phys. Rev.* **10**, 217, 1917.

28. —— Theory of transient oscillations in electrical networks and transmission lines. *Trans. Amer. Inst. Elect. Engrs*, **38**, 345, 1919.

29. —— Theory and calculation of variable systems. *Phys. Rev.* **17**, 116, 1921.

30. CARSON, J. R. and ZOBEL, O. J. Transient oscillations in electric wave filters. *Bell Syst. Tech. J.* **2**, 1, 1923.

31. CARSON, J. R. Building up of sinusoidal currents in long lump loaded lines. *Bell Syst. Tech. J.* **3**, 558, 1924.

32. —— Die Behandlung der Telegraphengleichung (auch unter Berücksichtigung der Stromverdrängung) nach der Operatorenmethode. *Elekt. Nachr.-Tech.* **2**, 359, 1925.

33. —— Selective circuits and static interference. *Bell Syst. Tech. J.* **4**, 265, 1925.

34. —— Heaviside operational calculus. *Bell Syst. Tech. J.* **1**, 43, 1922; **4**, 685, 1925; **5**, 50, 1926; **5**, 336, 1926.

35. —— Heaviside operational calculus. *Bull. Amer. Math. Soc.* **32**, 43, 1926.

36. —— Rigorous and approximate theories of electrical transmission along wires. *Bell Syst. Tech. J.* **7**, 11, 1928.

37. —— Asymptotic solution of an operational equation. *Trans. Amer. Math. Soc.* **31**, 782, 1929.

38. —— Notes on Heaviside's operational calculus. *Bell Syst. Tech. J.* **9**, 150, 1930.

39. —— An extension of operational calculus. *Bell Syst. Tech. J.* **15**, 340, 1936.

40. CARTER, F. W. Surges of voltage and current in transmission lines. *Proc. Roy. Soc. A*, **156**, 1, 1936.

368   REFERENCES

41. CASPER, L. Zur formel von Heaviside für Einschaltvorgänge. *Arch. Elektrotech.* **15**, 95, 1925.
42. —— Zum Beweis der Formel von Heaviside. *Arch. Elektrotech.* **15**, 545, 1926.
43. CAUER, W. and BRUNE, O. *J. Math. Physics*, Massachusetts Inst. Tech. p. 131, 1931.
44. CAUER, W. *Bull. Amer. Math. Soc.* p. 713, October 1932.
45. CHU, W. and CHANG, C.-K. Transients in dissipative low-pass electric wave filters with a terminating resistance. *Chinese J. Phys.* **2**, 76, 1936.
46. —— —— Transients in resistance terminated dissipative low-pass and high-pass electric wave filters. *Chinese J. Phys.* **2**, 154, 1936.
47. COHEN, L. Electrical oscillations in lines. *J. Franklin Inst.* **195**, 45, 1923.
48. —— Alternating current cable telegraphy. *J. Franklin Inst.* **195**, 165, 1923.
49. —— Applications of Heaviside's expansion theorem. *J. Franklin Inst.* **195**, 319, 1923.
50. COLOMBO, S. Sur quelques nouvelles correspondances symboliques. *Bull. Sci. Math.* (2), **67**, 1943.
51.† COPSON, E. T. Operational calculus and the evaluation of Kapteyn integrals. *Proc. Lond. Math. Soc.* **33**, 145, 1931.
52. CROSSLEY, A. F. Operational solution of some problems in viscous fluid motion. *Proc. Camb. Phil. Soc.* **24**, 231, 1928.
53. —— On the motion of a rotating circular cylinder filled with viscous fluid. *Proc. Camb. Phil. Soc.* **24**, 480, 1928.
54. DAHR, K. Beitrag zur allgemeinen Theorie der Vierpole und Kettenleiter. *Ann. Phys., Lpz.*, **21**, 182, 1934.
55. —— Asymptotiska formen på insvängningsförloppet vid en rationell homogen kedjeledning. *Tekn. Tidskr. Elektrotek.*, Stockholm, **8**, 1935.
56.† DAHR, S. C. Operational representation of confluent hypergeometric functions and their integrals. *Phil. Mag.* **21**, 1082, 1936.
57.† —— Operational representation of *M*-functions of confluent hypergeometric type. *Phil. Mag.* **25**, 416, 1938.
58. DALZELL, D. P. Heaviside's operational method. *Proc. Phys. Soc. Lond.* **42**, 75, 1930.
59. DOETSCH, G. Die Integrodifferentialgleichungen vom Faltungstypus. *Math. Ann.* **89**, 192, 1923.
60. —— Über das Problem der Wärmeleitung. *Jber. dtsch. MatVer.* **33**, 1925.
61. —— Elektrische Schwingungen in einem anfänglichen strom- und spannunsglosen Kabel unter dem Einfluss einer Randerregung. *Festschr. d. Tech. Hochschule, Stuttgart*, 1929.
62. —— *Math. Z.* **32**, 587, 1930.
63. —— *Z. reine angew. Math.* **167**, 274, 1932.

64. DOETSCH, G. Die Anwendung von Funktionaltransformationen in der Theorie der Differentialgleichungen und die symbolische Methode. *Jber. dtsch. MatVer.* **43**, 238, 1934.

65. EKELÖF, S. Einige einfache Impedanztransformationen von elektrischen Netzwerken und deren Anwendung auf Wellenfilter *Elektr. Nachr. Tech.* **12**, 100, 1935.

66. —— Transients in inductively shunted transmission line, with reference to impulse transmission in selective-calling telephone systems. *Ericsson Technics,* pp. 107–58, 161–205, 1937.

67. ELIAS, G. J. Über den Stand unserer Kenntnisse über die Heavisideschicht. *Elekt. Nachr. Tech.* **2**, 351, 1925.

68. FAN KY. Le calcul symbolique. *Rev. sci., Paris,* 1942.

69. FRY, T. C. Solution of circuit problems. *Phys. Rev.* **14**, 115, 1919.

70. GEMANT, A. Wanderwellen in stetig veränderlichen Kabeln. *Arch. Elektrotech.* **24**, 11, 1932.

71. —— Method of analysing experimental results obtained from elasto-viscous bodies. *Physics,* **7**, 311, 1936.

72. GOLDSTEIN, S. Two dimensional diffusion problems with circular symmetry. *Proc. Lond. Math. Soc.* (2), **34**, 51, 1932.

73.† —— Operational forms of Whittaker's confluent hypergeometric function and Weber's parabolic cylinder function. *Proc. Lond. Math. Soc.* (2), **34**, 103, 1932.

74. —— Operational solution of an integral equation. *J. Lond. Math. Soc.* **6**, 262, 1932.

75. —— Application of Heaviside's operational method to problems in heat conduction. *Z. angew. Math. Mech.* **12**, 234, 1932.

76. —— Calculation of surface temperature of geometrically simple bodies. *Z. angew. Math. Mech.* **14**, 158, 1934.

77. HARTREE, D. R., CALLENDER, A. and PORTER, A. Time lag in a control system. *Philos. Trans.* A, **235**, 415, 1936.

78. HAZEN, H. L. Theory of servo-mechanisms. *J. Franklin Inst.* **218**, 279, 1934.

79. HEAVISIDE, O. Operators in mathematical physics. *Proc. Roy. Soc.* A, **52**, 504, 1893; **54**, 105, 1894.

80. HORTON, G. K. and GRIFFITH, M. V. Transient flow of heat through two layer wall. *Proc. Phys. Soc. Lond.* **58**, 481, 1946.

81.† HOWELL, W. T. Products of Laguerre polynomials. *Phil. Mag.* **24**, 396, 1937.

82.† —— Operational forms of products of parabolic cylinder functions and products of Laguerre functions. *Phil. Mag.* **24**, 1082, 1937.

83. —— Functions reciprocal in the Hankel transform. *Phil. Mag.* **25**, 622, 1938.

84.† HUMBERT, P. Les fonctions hypergéométriques et le calcul symbolique. *Ann. Soc. sci. Brux.,* Sci. math., série A, **105**, 53, 1933.

85.† —— Le calcul symbolique à deux variables. *Ann. Soc. sci. Brux.,* Sci. math., série A, **56**, 26, 1936.

370   REFERENCES

86.† HUMBERT, P. *Proc. Edinb. Math. Soc.* **3**, 276, 1933.

87.† —— Some new operational representations. *Proc. Edinb. Math. Soc.* **4**, 232, 1935.

88.† —— Nouvelles remarques sur les fonctions de Bessel du troisième ordre. *Act. P. Acad. Sci. Nov. Lync.* **87**, 323, 1934.

89.† —— Le calcul symbolique à deux variables. *C.R. Acad. Sci., Paris*, **199**, 657, 1934.

90.† —— Sur les intégrales de Fresnel. *Mathematica*, **10**, 32, 1934.

91.† —— Sur le logarithme intégral. *Soixante-neuvième Congrès des Sociétés savantes*, 1935.

92.† —— Bessel function products. *Phil. Mag.* **24**, 888, 1937.

93.† —— Formules nouvelles pour le calcul symbolique. *Bull. Soc. Math. Fr.* **65**, 119, 1937.

94. HUMBERT, P. and POLI, L. *Bull. Sci. math.* **67**, 104, 1943.

95. IKEDA, Y. Anwendung der Operatorengleichung auf die lineare Differentialgleichung. *Tôhoku Math. J.* **37**, 202, 1933.

96. ITOO, T. Pseudo-Operational method. *Tôhoku Math. J.* **42**, 230 1936.

97. JEFFREYS, H. Compressional waves in two superposed layers. *Proc. Camb. Phil. Soc.* **23**, 472, 1927.

98. —— Wave propagation in strings with continuous and concentrated loads. *Proc. Camb. Phil. Soc.* **23**, 768, 1927.

99. —— The earth's thermal history. *Beitr. Geophys.* **1**, 18, 1927.

100. KAUCKY, T. *Proc. Edinb. Math. Soc.* **43**, 115, 1925.

101. KERMACK, W. O. and McCREA, W. H. *Proc. Edinb. Math. Soc.* **2**, 205, 220, 1931.

102. —— —— *Proc. Roy. Soc. Edinb.* **51**, 176, 1931.

103. KLEMIN, A. and RUFFNER, B. F. Operator solutions in airplane dynamics. *Amer. J. Aero. Sci.* **3**, 252, 1936.

104. KOIZUMI, S. Heaviside's operational solution of a Volterra integral equation. *Phil. Mag.* **11**, 432, 1931.

105. —— A new method of evaluation of the Heaviside operational expression by Fourier series. *Phil. Mag.* **19**, 1061, 1935.

106. —— Asymptotic evaluation of operational expressions. *Phil. Mag.* **21**, 265, 1936.

107. KOPPENFELS, W. Der Faltungssatz und seine Anwendung bei der Integration linearer Differentialgleichungen mit konstanten Koeffizienten. *Math. Ann.* **105**, 694, 1931.

108. KÜPFMÜLLER, K. Über Beziehungen zwischen Frequenzcharakteristiken und Ausgleichsvorgängen in linearen Systemen. *Elekt. Nachr.-Tech.* **5**, 30, 1928.

109. LÉVY, P. *Bull. Sci. Math.* (2), **50**, 174, 1926.

110. LOWAN, A. N. Operational determination of Green's function in theory of heat conduction. *Phil. Mag.* **24**, 62, 1937.

111.† LOWRY, H. V. Operational calculus. *Phil. Mag.* **13**, 1033, 1932; **13**, 1144, 1932.

112. LUIKOV, A. Application of Heaviside-Bromwich method to a heat conduction problem. *Phil. Mag.* **22**, 239, 1936.

113. Mächler, W. Laplacesche Integraltransformation und Integration partieller Differentialgleichungen vom hyperbolischen und parabolischen Typus. *Comment. math. Helvet.* 5, 256, 1933.

114. March, H. W. The Heaviside operational calculus. *Bull. Amer. Math. Soc.* 33, 311, 1927.

115. McCrea, W. H. Operational proofs of some identities. *Math. Gaz.* 17, 43, 1933.

116. McKay, A. T. Diffusion into an infinite plane sheet subject to a surface condition. *Proc. Phys. Soc. Lond.* 42, 547, 1930.

117. —— Diffusion for the infinite plane sheet. *Proc. Phys. Soc. Lond.* 44, 17, 1932.

118. —— Absorption and classical diffusion. *Trans. Faraday Soc.* 28, 721, 1932.

119. —— Die einfache Diffusionsfunktion. *Z. angew. Math. Mech.* 16, 183, 1936. Tabular values given.

120. McLachlan, N. W. and McKay, A. T. Transient oscillations in a loud speaker horn. *Proc. Camb. Phil. Soc.* 32, 265, 1936.

121. —— —— Die Wiedergabe von Ausgleichsvorgängen durch einen Trichterlautsprecher. *Elekt. Nachr.-Tech.* 13, 251, 1936.

122. McLachlan, N. W. Reproduction of transients by a television amplifier. *Phil. Mag.* 22, 481, 1936.

123. —— Behaviour of a thermionic valve output circuit to transients. *Wireless Engr,* 13, 630, 1936.

124. McLachlan, N. W. and Meyers, A. L. Ster and stei functions. *Phil. Mag.* 21, 425, 1936.

125. —— —— Contour integral expressions for Bessel functions. *Phil. Mag.* 23, 762, 1937.

126.† —— —— Operational forms of Bessel and Struve functions. *Phil. Mag.* 23, 918, 1937.

127. McLachlan, N. W. Fourier expansions obtained operationally. *Phil. Mag.* 24, 1055, 1937.

128. —— Operational systems. *Phil. Mag.* 25, 259, 1938.

129. —— Submarine cable problems solved by contour integration. *Math. Gaz.* 22, 37, 1938.

130. —— Historical note on Heaviside's operational method. *Math. Gaz.* 22, 255, 1938.

131.† —— Operational forms and contour integrals for Bessel functions with argument $a\sqrt{(t^2-b^2)}$. *Phil. Mag.* 26, 394, 1938.

132.† —— Operational forms and contour integrals for Struve and other functions. *Phil. Mag.* 26, 457, 1938.

133. —— Operational form of $f(t)$ for a finite interval, with application to impulses. *Phil. Mag.* 26, 695, 1938.

134. —— Application of Mellin inversion theorem to impulses. *Math. Gaz.* 23, 270, 1939.

135. —— General theorem in Laplace transforms. *Math. Gaz.* 30, 85, 1946.

136. MᶜLEOD, A. R. Thermometer lag with spherical and cylindrical bulbs in a medium whose temperature varies at a constant rate. *Phil. Mag.* 37, 134, 1919.

137. MELLIN, H. Abriss einer einheitlichen Theorie der Gamma und der hypergeometrischen Functionen. *Math. Ann.* 68, 305, 1910.

138. MITRA, S. C. Integrals and expansions involving Bessel functions. *Bull. Calcutta Math. Soc.* 25, 81, 1933.

139.† —— Operational representation of $D_n(x)$ and $D^2_{-(n+1)}(ix)$ $-D^2_{-(n+1)}(-ix)$. *Proc. Edinb. Math. Soc.* 4, 33, 1933.

140. —— Integrals involving Lommel's function of two variables. *Bull. Calcutta Math. Soc.* 25, 173, 1933.

141. —— Certain definite integrals. *Bull. Calcutta Math. Soc.* 25, 185, 1933.

142. MOELLER, F. Die Abflachung steiler Wellenstirnen unter Berücksichtigung der Stromverdrängung im Leiter. *Arch. Elektrotech.* 15, 1926.

143. MURNAGHAN, F. D. The Cauchy-Heaviside formula and the Boltzmann-Hopkinson principle of superposition. *Bull. Amer. Math. Soc.* 33, 81, 1927.

144. —— The operational calculus. *Math. Mag.* (U.S.A.), 21, 117, 1947.

145. NEUFELD, J. Extension of Heaviside's calculus to circuits whose parameters vary with time. *Phil. Mag.* 15, 170, 1933.

146. —— Operational solution of linear mixed difference equations. *Proc. Camb. Phil. Soc.* 30, 389, 1934.

147.† NIESSEN, K. F. A contribution to symbolic calculus. *Phil. Mag.* 20, 977, 1935.

148. PARODI, M. *Bull. Sci. math.* 69, 174, 1945.

149. —— *C.R. Acad. Sci., Paris,* 220, 870, 1945.

150. —— Sur des équations intégrales singulières ayant des solutions communes. *C.R. Acad. Sci., Paris,* 225, 35, 1947.

151. —— Sur un type d'équations intégrales résolubles par le calcul symbolique. *C.R. Acad. Sci., Paris,* 226, 43, 1948.

152. —— Remarque sur l'équation intégrale de seconde espèce à noyan singulier de Weyl. *C.R. Acad. Sci., Paris,* 226, 153, 1948.

153. PATERSON, S. Heating or cooling of a sphere in a well-stirred fluid. *Proc. Phys. Soc. Lond.* 59, 50, 1947.

154. PIPES, L. A. Operational and matrix methods in linear variable networks. *Phil. Mag.* 25, 585, 1938.

155. —— Operational theory of solid friction. *Phil. Mag.* 25, 950, 1938.

156. —— Operational solution of the wave equation. *Phil. Mag.* 26, 333, 1938.

157. PLEIJEL, H. and LILJEBLAD, R. Operatorkalkylens Samband med den Symboliska metoden. *Tekn. Tidskr.* p. 25, February 1919.

158. PLEIJEL, A. Über asymptotische Reihenentwicklungen in der Operatorenrechnung. *Z. angew. Math. Mech.* 15, 300, 1935.

159. POL, B. VAN DER. Simple proof and extension of Heaviside's operational calculus for invariable systems. *Phil. Mag.* **7**, 1153, 1929.

160.† —— Operational solution of differential equations. *Phil. Mag.* **8**, 861, 1929.

161. POL, B. VAN DER and NIESSEN, K. F. Simultaneous operational calculus. *Phil. Mag.* **11**, 368, 1931.

162.† —— —— Symbolic calculus. *Phil. Mag.* **13**, 537, 1932.

163. POL, B. VAN DER and WEYERS, TH. J. Tchebycheff polynomials and their relation to circular functions, Bessel functions and Lissajous figures. *Physica*, **1**, 78, 1933.

164. POL, B. VAN DER. Theorem on electrical networks, with application to filters. *Physica*, **1**, 521, 1934.

165. —— Symbolic calculus with applications to radio telegraphy. *Tijdschr. ned. Radiogenoot.* **7**, 18, 1935.

166. —— Theorem on electrical networks. *Physica*, **4**, 585, 1937.

167. —— Discontinuous phenomena in radio communication. *J. Instn Elec. Engrs*, **81**, 381, 1937.

168. POLI, L. Équations intégrales et calcul symbolique. *Ann. Soc. sci. Brux.* série A, **55**, 111.

169. —— Sinus du $n^{\text{ième}}$ ordre et calcul symbolique. *Ann. Soc. sci. Brux.* série **60**, 1, 5.

170. —— Le calcul symbolique à deux variables. *Rev. sci., Paris*, 1947.

171. POLLACZEK, F. Das Einschaltproblem für das homogene Kabel bei beliebiger Endschaltung. *Elect. Nachr-Tech.* **1**, 80, 1924.

172. POMEY, J. B. Heaviside's symbolic calculus. *Rev. gén. Elect.* **13**, 813, 1923; **13**, 859, 1923; **24**, 699, 1928.

173. PRESS, A. Generalised division and Heaviside's operators. *Phil. Mag.* **14**, 78, 1932.

174. SAKURAI, T. An extension of Heaviside's operational method. *Proc. Phys.-Math. Soc. Japan*, **18**, 356, 1936.

175. —— Relation between 'spectra' and poles of operator. *Proc. Phys.-Math. Soc. Japan*, **19**, 108, 1937.

176. —— Complementary Heaviside operator. *Proc. Phys.-Math. Soc. Japan*, **20**, 355, 1938.

177. SALINGER, H. Die Heavisidesche Operatorenrechnung. *Elekt. Nachr.-Tech.* **2**, 365, 1925.

178. SBRANA, F. Characteristic equation for propagation in one dimension. *R.C. Circ. mat. Palermo*, **56**, 58, 1932.

179. SCHOUTEN, J. P. Over de grondslagen van de operatorenrekening volgens Heaviside. Thesis, Delft, 1933.

180. —— Theorem in operational calculus and an application thereof. *Physica*, **2**, 75, 1935.

181. SCHULZ, H. *Telegraphen und Fernsprech. Techn.* **19**, 231, 1930; **19**, 374, 1930.

182. SHARMA, J. L. On Whittaker's confluent hypergeometric function. *Phil. Mag.* **25**, 491, 1938.

183.† SHASTRI, N. A. On Lommel functions. *Phil. Mag.* **25**, 930, 1938.

374 REFERENCES

184. SLADE, J. *Bull. Amer. Math. Soc.* **40**, 339, 1934.
185. SMITH, J. J. Solution of differential equations by a method similar to that of Heaviside. *J. Franklin Inst.* **195**, 815, 1923.
186. —— Analogy between pure mathematics and the operational mathematics of Heaviside. *J. Franklin Inst.* **200**, 519, 1925; **200**, 635, 1925; **200**, 775, 1925.
187. STARR, A. T. Thermometer lag with varying temperature. *Phil. Mag.* **9**, 901, 1930.
188. —— Ballistic and perfect balance in electrical bridges. *Phil. Mag.* **12**, 265, 1931.
189. STEPHENS, E. Bibliography on General Differentiation. *Washington Univ. Studies*, **12**, 137, 1925.
190. SUMPNER, W. E. Heaviside's fractional differentiator. *Proc. Phys. Soc. Lond.* **41**, 404, 1929.
191. —— Impulse functions. *Phil. Mag.* **11**, 345, 1931.
192. —— Index operators. *Phil. Mag.* **12**, 201, 1931.
193. —— The work of Oliver Heaviside. *J. Instn Elect. Engrs*, **71**, 837, 1932.
194. SUTTON, W. The asymptotic expansion of a function whose operational form is known. *J. Lond. Math. Soc.* **9**, 131, 1934.
195. TS'EN, M.-K. Differential pulse generator. *Chinese J. Phys.* **1**, no. 1, 87, 1935; **1**, no. 3, 68, 1935.
196. —— Differential indicial admittances (currents produced by unit differential pulse voltage as in Fig. 50b). *Chinese J. Phys.* **2**, 43, 1936.
197. VAHLEN, K. T. Über den Heaviside-Kalkul. *Z. angew. Math. Mech.* **13**, 283, 1933.
198. VALLARTA, M. A. Heaviside's proof of his expansion theorem. *Proc. Amer. Inst. Elect. Engrs*, **65**, 383, 1926.
199. VARMA, R. S. Self-reciprocal function in the Hankel transform. *Proc. Lond. Math. Soc.* **42**, 9, 1936.
200.† —— Operational representation of the parabolic cylinder function. *Phil. Mag.* **22**, 29, 1936; **23**, 926, 1937.
201. —— Infinite integral involving Bessel and parabolic cylinder functions. *Proc. Camb. Phil. Soc.* **32**, 210, 1937.
202. —— Summation of some infinite series of Weber's parabolic cylinder functions. *J. Lond. Math. Soc.* **12**, 26, 1937.
203. —— Infinite integrals involving Weber's parabolic cylinder functions. *J. Indian Math. Soc.* **3**, 25, 1938.
204. VOGT, H. Sur le calcul symbolique et ses applications à l'intégration des équations différentielles. *Rev. gén. Élect.* **2**, 483, 563, 1917; **5**, 581, 907, 1919.
205. WAGNER, K. W. Über eine Formel von Heaviside zur Berechnung von Einschaltvorgängen. *Arch. Elektrotech.* **4**, 159, 1916.
206. —— Oliver Heaviside. *Elekt. Nachr.-Tech.* **2**, 345, 1925.
207. —— Der Satz von der wechselseitigen Energie. *Elekt. Nachr.-Tech.* **2**, 376, 1925.

208. WARREN, A. G. and FRIEND, R. G. Effect of a magnetomotive force applied for a short time to a steel cylinder. *Phil. Mag.* **21**, 980, 1936.

209. WHITEHEAD, S. Approximate method for calculating heat flow in an infinite medium heated by a cylinder. *Proc. Phys. Soc. Lond.* **56**, 357, 1944.

210. WHITTAKER, E. T. Oliver Heaviside. *Bull. Calcutta Math. Soc.* **20**, 199, 1929.

211. WIDDER, D. V. Inversion of Laplace integral and related moment problem. *Trans. Amer. Math. Soc.* **36**, 107, 1934.

212. WIENER, N. Operational calculus. *Math. Ann.* **95**, 557, 1925.

B. BOOKS

213.† BERG, E. J. *Heaviside's operational calculus.* 2nd edition (1936).

214. BOREL, E. *Leçons sur les séries divergentes.* 2nd edition (1928).

215. BROMWICH, T. J. I'A. *Theory of infinite series.* 2nd edition, revised by T. M. MacRobert (1926).

216.† BUSH, V. *Operational circuit analysis.* (1929.)

217.† CARSLAW, H. S. and JAEGER, J. C. *Operational methods in applied mathematics.* 2nd ed. (1948).

218.† CARSON, J. R. *Electric circuit theory and operational calculus.* (1926.)

219. COHEN, L. *Heaviside's electric circuit theory.* (1928.)

220. COPSON, E. T. *Theory of functions of a complex variable.* (1935.)

221. COURANT, R. and HILBERT, D. *Methoden der Mathematischen Physik.* Vol. 1 (1931).

222.† DAHR, K. *Integrational and operational calculus.* (1935.)

223. DAVIS, H. T. *Theory of linear operators.* (1936.)

224. DOETSCH, G. *Handbuch der Laplace-Transformation.* 3 vols. (1950/56).

225.† GARDNER, M. F. and BARNES, J. L. *Transients in linear systems studied by the Laplace transformation.* (1942.)

226. GOURSAT, E. *Cours d'analyse mathématique.* Vol. 1 (1923).

227. HEAVISIDE, O. *Electromagnetic theory.* Vol. 1 (1893); vol. 2 (1899); vol. 3 (1912); new edition (1922).

228. —— *Electrical papers.* 2 vols. (1892).

229.† HUMBERT, P. *Le calcul symbolique.* (1934.)

230.† HUMBERT, P. and COLOMBO, S. *Le calcul symbolique et ses applications à la physique mathématique.* Mém. des Sci. Mathématiques, Fascicule 105, 1947.

231. JANET, P. *Le calcul symbolique d'Heaviside et ses applications à l'électrotechnique.* (1938.)

232.† JEFFREYS, H. *Operational methods in mathematical physics.* 2nd edition. (1931.)

233. MᶜLACHLAN, N. W. *Loud speakers, theory, performance, testing and design.* (1934.)

234. MᶜLACHLAN, N. W. *Bessel functions for engineers.* 2nd ed. (1955).
235a.† MᶜLACHLAN, N. W. and HUMBERT, P. *Formulaire pour le calcul symbolique.* Mém. des Sci. Math. Fascicule 100 (1941), 2nd edition (1950).
235b.† —— —— and Poli, L. *Supplement au Formulaire pour le calcul symbolique.* Fascicule 113 (1950).
236.† MᶜLACHLAN, N. W. *Modern operational calculus.* (1962.)
237. MACROBERT, T. M. *Functions of a complex variable.* 2nd edition (1933).
238. PEARSON, K. *Tables for statisticians and biometricians.* Part 1 (1924); part 2 (1931).
239. STEPHENS, E. *Elementary theory of operational mathematics.* (1937.)
240. TITCHMARSH, E. *Theory of functions.* (1932.)
241. —— *Introduction to theory of Fourier integrals.* (1937.)
242. WAGNER, K. W. *Elektromagnetische Ausgleichsvorgänge in Freileitungen und Kabeln.* (1908.)
243. WHITTAKER, E. T. and WATSON, G. N. *Modern analysis.* (1927.)

C. ADDITIONAL REFERENCES

244. BOURGIN, D. G. and DUFFIN, R. J. Heaviside operational calculus. *Amer. J. Math.* 59, 489, 1937.
245. —— —— Laplace-Heaviside method for boundary value problems. *Bull. Amer. Math. Soc.* 45, 859, 1939.
246. BRUNETTI, C. Operational solution of electric circuits. *Trans. Amer. Inst. Elect. Engrs,* 55, 158, 1936.
246a. CHERRY, C. *Pulses and Transients in Communication Circuits.* (1950.)
247. COSSAR, J. and ERDÉLYI, A. *Dictionary of Laplace transforms.* (1944–46.)
248. DOETSCH, G. *Einführung in Theorie und Anwendung der Laplace Transformation.* (1958.)
249. ERDÉLYI, A. Generalisation of the Laplace transformation. *Proc. Edinburgh Math. Soc.* (2) 10, 53, 1954.
250. ERDÉLYI, A., MAGNUS, W., OBERHETTINGER, F. and TRICOMI, F. G. *Tables of integral transforms,* vol. 1, 1954.
250a. ESTRIN, T. A. and HIGGINS, T. J. Solution of boundary value problems by multiple Laplace transformations. *Jour. Frank. Inst.* 252, 153, 1951.
250b.† GOLDMAN, S. *Transformation calculus and electrical transients.* (1949.)
251. HEINS, A. E. Solution of linear difference-differential equations. *J. Math. Phys.* 19, 153, 1940.
252. —— Solution of partial difference equations. *Amer. J. Math.* 63, 435, 1941.
253. —— A mixed boundary value problem. *Bull. Amer. Math. Soc.* 49, 130, 1943.

254. HIGGINS, T. J. History of operational calculus as used in electric circuit analysis. *Elect. Engng, New York*, **68**, 12, 1949.

255.† HUMBERT, P. Images des fonctions de Mathieu. *C.R. Acad. Sci.*, *Paris*, **225**, 715, 1947.

256. —— Fonctions de Bessel et Calcul symbolique. *Ann. Soc. Scient. de Bruxelles*, **64**, 55, 1950.

257. KÁRMÁN, T. VON and BIOT, M. A. *Mathematical methods in engineering*. (1940.)

258.† LABIN, E. *Calcul opérationnel*. (1949.)

259. LOWAN, A. N. Problem of heat recuperator. *Phil. Mag.* **17**, 914, 1934.

260. —— Operational treatment of mechanical and electrical problems. *Phil. Mag.* **17**, 1134, 1934.

261. —— Two dimensional problems in heat conduction. *Phil. Mag.* **24**, 410, 1937.

262. —— Problems in diffraction of heat. *Phil. Mag.* **29**, 93, 1940.

263. —— Problem of wave motion for wedge of an angle. *Phil. Mag.* **32**, 373, 1941.

264.† MAGNUS, W. and OBERHETTINGER, F. *Formeln und Sätze für die Speziellen Funktionen der Mathematischen Physik*. (1948.)

265. MᶜLACHLAN, N. W. Solution of cable problem by Laplace transform. *Math. Gaz.* **30**, 291, 1946.

266. —— *Theory of vibrations*. (1951.)

267. MIKUSIŃSKI, JAN G. Sur le fondements du calcul opératoire. *Studia math.* **11**, 41, 1949.

267 a. PARODI, M. *Application physiques de la transformation de Laplace*. (1948.)

268. PIPES, L. A. The operational calculus. *J. Appl. Phys.* **10**, nos. 3, 4 and 5, 1939.

269. —— Application of the operational calculus to the theory of structures. *J. Appl. Phys.* September 1943.

270.† —— *Applied mathematics for engineers and physicists*. (1946.)

271. POL, B. VAN DER and BREMMER, H. *Operational calculus based on the two-sided Laplace integral*. 2nd ed. (1955).

272.† THOMSON, W. T. *Laplace transformation; theory and engineering applications*. (1950.)

273.† WAGNER, K. W. *Operatorenrechnung*. (1940.)

Additional references will be found from time to time in *Mathematical Reviews* and *Zentralblatt für Mathematik*.

An errata list for 235 a, b is given in *Mathematical Tables and Other Aids to Computation*, **6**, 100, 1952; **7**, 13, 1953; **7**, No. 41, 1953.

INDEX